MODERN PROBABILITY AND STATISTICS

ROBUSTNESS IN DATA ANALYSIS: CRITERIA AND METHODS

Also available in
Modern Probability and Statistics:

Asymptotic Theory of Testing Statistical Hypotheses: Efficient Statistics, Optimality, Power Loss and Deficiency
V.E. Bening

Selected Topics in Characteristic Functions
N.G. Ushakov

Chance and Stability. Stable Distributions and their Applcations
V.M. Zolotarev and V.V. Uchaikin

Normal Approximation: New Results, Methods and Problems
V.V. Senatov

Modern Theory of Summation of Random Variables
V.M. Zolotarev

Modern
Probability and Statistics

Robustness in Data Analysis: criteria and methods

Georgy L. Shevlyakov and Nikita O. Vilchevski
St. Petersburg State Technical University

///VSP///
Utrecht • Boston • Köln • Tokyo
2002

VSP BV
P.O. Box 346
3700 AH Zeist
The Netherlands

Tel: +31 30 692 5790
Fax: +31 30 693 2081
vsppub@compuserve.com
www.vsppub.com

First published in 2002

ISBN 90-6764-351-3

Printed in The Netherlands by Ridderprint bv, Ridderkerk.

Robustness in data analysis: criteria and methods

Georgy L. Shevlyakov Nikita O. Vilchevski

St. Petersburg State Technical University

To our wives, Elena and Nelli

Contents

Foreword

This book is the sixth in the series of monographs 'Modern Probability and Statistics' following the books

- V.M. Zolotarev, *Modern Theory of Summation of Random Variables*;
- V.V. Senatov, *Normal Approximation: New Results, Methods and Problems*;
- V.M. Zolotarev and V.V. Uchaikin, *Chance and Stability. Stable Distributions and their Applications*;
- N.G. Ushakov, *Selected Topics in Characteristic Functions.*
- V.E. Bening, *Asymptotic Theory of Testing Statistical Hypotheses: Efficient Statistics, Optimality, Power Loss, and Deficiency.*

The Russian school of probability theory and mathematical statistics made a universally recognised contribution to these sciences. Its potentialities are not only very far from being exhausted, but are still increasing. During last decade there appeared many remarkable results, methods and theories which undoubtedly deserve to be presented in monographic literature in order to make them widely known to specialists in probability theory, mathematical statistics and their applications.

However, due to recent political changes in Russia followed by some economic instability, for the time being, it is rather difficult to organise the publication of a scientific book in Russia now. Therefore, a considerable stock of knowledge accumulated during last years yet remains scattered over various scientific journals. To improve this situation somehow, together with the VSP publishing house and first of all, its director, Dr. Jan Reijer Groesbeek who with readiness took up the idea, we present this series of monographs.

The scope of the series can be seen from both the title of the series and the titles of the published and forthcoming books:

- Yu.S. Khokhlov, *Generalizations of Stable Distributions: Structure and Limit Theorems*;
- A.V. Bulinski and M.A. Vronski, *Limit Theorems for Associated Random Variables*;

- V.E. Bening and V.Yu. Korolev, *Generalized Poisson Models and their Applications in Insurance and Finance*;
- E.V. Morozov, *General Queueing Networks: the Method of Regenerative Decomposition*;
- G.P. Chistyakov, *Analytical Methods in the Problem of Stability of Decompositions of Random Variables*;
- A.N. Chuprunov, *Random Processes Observed at Random Times*;
- D.H. Mushtari, *Probabilities and Topologies on Linear Spaces*;
- V.G. Ushakov, *Priority Queueing Systems*;
- V.Yu. Korolev and V.M. Kruglov, *Random Sequences with Random Indices*;
- Yu.V. Prokhorov and A.P. Ushakova, *Reconstruction of Distribution Types*;
- L. Szeidl and V.M. Zolotarev, *Limit Theorems for Random Polynomials and Related Topics*;
- E.V. Bulinskaya, *Stochastic Inventory Systems: Foundations and Recent Advances*;

as well as many others.

To provide high-qualified international examination of the proposed books, we invited well-known specialists to join the Editorial Board. All of them kindly agreed, so now the Editorial Board of the series is as follows:

We hope that the books of this series will be interesting and useful to both specialists in probability theory, mathematical statistics and those professionals who apply the methods and results of these sciences to solving practical problems. Of course, the choice of authors primarily from Russia is due only to the reasons mentioned above and by no means signifies that we prefer to keep to some national policy. We invite authors from all countries to contribute their books to this series.

V. Yu. Korolev,
V. M. Zolotarev,
Editors-in-Chief

Moscow, December 2000.

Preface

The field of mathematical statistics called robust statistics appeared due to the pioneer works of J. W. Tukey (1960), P. J. Huber (1964), and F. R. Hampel (1968); it has been intensively developed since the sixties and is rather definitely formed by present. The term 'robust' (strong, sturdy) as applied to statistical procedures was proposed by G. E. P. Box (1953).

The basic reasons of research in this field are of a general mathematical character. 'Optimality' and 'stability' are the mutually complementary characteristics of any mathematical procedure. It is a well-known fact that the behaviour of many optimal decisions is rather sensible to 'small deviations' from prior assumptions. In mathematical statistics, the remarkable example of such unstable optimal procedure is given by the least squares method: its performance may become extremely poor under small deviations from normality.

Roughly speaking, robustness means stability of statistical inference under the variations of the accepted distribution models.

Nearly at the same time with robust statistics, there appeared another direction in statistics called exploratory or probability-free data analysis, which also partly originated from J. W. Tukey (1962). By definition, data analysis techniques aim at practical problems of data processing. Although robust statistics involves mathematically highly refined asymptotic tools, robust methods exhibit a satisfactory behaviour in small samples being quite useful in applications.

Our work represents new results related to robustness and data analysis technologies, having definite accents both on theoretical aspects and practical needs of data processing: we have written the book to be accessible to users of statistical methods, as well as for professional statisticians.

In addition, we would like to dwell on the appreciable contribution of Russian authors to robust statistics and data analysis, though most of their original results were published in Russian. Here we acknowledge the following names: S. A. Aivazyan, V. Ya. Katkovnik, Yu. S. Kharin, V. Ya. Kreinovich, V. P. Kuznetsov, A. V. Makshanov, L. D. Meshalkin, B. T. Polyak, V. P. Shulenin, A. M. Shurygin, A. B. Tsybakov, Ya. Z. Tsypkin.

Chapter 1 represents a general description of main approaches in robust statistics. Chapter 2 gives a new probability-free approach to constructing optimisation criteria in data analysis. Chapter 3 contains new results on robust minimax (in the Huber sense) estimation of location over the distribution classes with bounded variances and subranges, as well as for the classes of lattice distributions. Chapter 4 is confined to robust estimation of scale. Chapter 5 deals with robust regression and autoregression problems. Chapter 6 covers the particular case of L_1-norm estimation. Chapter 7 treats robust estimation of correlation. Chapter 8 introduces and discusses data analysis technologies, and Chapter 9 represents applications.

We would like to express our deep appreciation to I. B. Chelpanov, E. P. Guilbo, B. T. Polyak, and Ya. Z. Tsypkin who attracted our attention to robust statistics.

We are grateful to S. A. Aivazyan and L. D. Meshalkin for discussions and comments of our results at their seminars in the Central Institute of Economics and Mathematics of Russian Academy of Sciences (Moscow).

We are very grateful to A. V. Kolchin for his great help in the preparation of this book.

Finally, we highly appreciate V. Yu. Korolev and V. M. Zolotarev for their attention to our work and, in general, to such an important field of mathematical statistics as robustness.

Georgy L. Shevlyakov

Nikita O. Vilchevski

St. Petersburg, December 2000

1

Introduction

1.1. General remarks

Our personal experience in data analysis and applied statistics is relatively wide and long. It refers to the problems of data processing in medicine (cardiology, ophthalmology), economics and finances (financial mathematics), industry (electromechanics and energetics), defense (detection of air targets). Besides and due to those problems, we have been working on the methods of theoretical statistics, mainly in robustness and optimization. Now we briefly outline our vision of data analysis problems with their ideological environment and indicate the place of the problems touched in this book. Let us only keep in mind that any classification is a convention, such are the forthcoming ones.

1.1.1. Data, their forms of representation, characteristics, and related aims

The forms of data representation. We begin with the customary forms of data representation:

(i) as a sample $\{x_1, \dots, x_n\}$ of real numbers $x_i \in \mathbf{R}$ being the easiest form to handle;

(ii) as a sample $\{\mathbf{x}_1, \dots, \mathbf{x}_n\}$ of real-valued vectors $\mathbf{x}_i \in \mathbf{R}^m$;

(iii) as a realization $x(t)$, $t \in [0, T]$ of a real-valued continuous process (function);

(iv) as a sample of 'non-numerical nature' data representing qualitative variables; and finally,

(v) as semantic type data (statements, texts, pictures, etc.).

The first three possibilities widely occur in the physical sciences with the measurement techniques being well developed, clearly defined and largely standardized. In the social sciences, the last forms are relatively common.

In our study we deal with the first three forms; the multivariate case, especially of high dimension (greater than three), has proved to be the most important and difficult in our experience of solving real-life data analysis problems.

The types of data statistical characteristics. The experience of treating various statistical problems shows that nearly all of them are solved with the use of only a few qualitatively different types of data statistical characteristics (DSCs). Now we do not discuss how to use them tackling statistical problems: only note that their solutions result in evaluating some of the DSCs, and final decision making essentially depends on their values (Aivazyan *et al.*, 1989; Tukey, 1962). These DSCs may be classified as follows:

- the measures of location (central tendency, mean values);
- the measures of spread (dispersion, scale, scatter);
- the measures of interdependence (association, correlation);
- the DSC of extreme values;
- the DSC of data distributions or the measures of shape.

In this book we work with all of these DSCs.

The main aims of data analysis. These aims may be formulated as follows:

(A1) compact representation of data,

(A2) estimation of model parameters describing mass phenomena,

(A3) prediction.

A human mind cannot efficiently work with large volumes of information, since there exist natural psychological bounds of perception. Thus it is necessary to provide a compact data output of information: only in this case we may expect a satisfactory final decision. Note that data processing often begins and ends with this first item (A1).

The next step (A2) is to suggest an explanatory underlying model for the observed data and phenomena. It may be a regression model, or a distribution model, or any other, desirably a simple one: an essentially multiparametric model is usually a bad model. Parametric models refer to the first to be considered and examined.

Finally, all previous aims are only the steps to the last (A3): here we have to state that this aim remains a main challenge to statistics and to science as a whole.

In this book we pursue the aims (A1) *and* (A2).

1.1.2. Evaluation of the DSC: an optimization approach and related requirements

An optimization approach. In statistics, many problems can be posed in optimization settings, for example, the problems of point and interval estimation, hypotheses testing, regression, etc. In fact, an optimization approach is natural and, moreover, convenient as it allows to use the elaborated optimization technique.

In this book we use optimization settings of statistical problems whenever it is possible.

A probability-free approach. The crucial point of an optimization approach is the choice of an optimization criterion. In mathematical statistics this choice is completely determined by the choice of the underlying stochastic model. So, in this case the stochastic model is primary and the optimization criterion is secondary.

Another situation occurs when a stochastic model cannot be applied to the data, for example, when only one *unique* or *few* data samples can be observed and thus there are no any grounds to regard the data originating from some population. In other words, the stability of frequencies (this fundamental condition of applicability of a statistical approach (Cramér, 1946)) cannot be verified in this case.

In general, the problems of the medical data processing often refer to the above case: a patient is mainly interested in the 'individual diagnosis' but not in the 'average-statistical over ensemble.'

The methods aimed at the data analysis problems under the conditions of inapplicability of the statistical approach are called *probability-free* and they have been intensively developed since the seventies: the fuzzy approach (Klir and Folger, 1990; Schum, 1994; Zadeh, 1965; Zadeh, 1975); the methods of exploratory data analysis (Mosteller and Tukey, 1977; Tukey, 1977); the logical-algebraic and geometrical methods of multivariate classification and cluster-analysis (Aivazyan *et al.*, 1989; Diday, 1972; Lebart *et al.*, 1984; Papadimitriou and Steiglitz, 1982); projection pursuit (Huber, 1985); the interval probability models (Kuznetsov, 1991; Walley, 1990; Weichselberger, 1995).

In this book we propose a probability-free approach to the choice of optimization criteria in data analysis.

The requirements towards the DSC. As a rule, an optimization criterion is exogenously postulated in the above case, so it may seem that the problem of its choice does not arise. Our experience shows that, as a rule, it is rather senseless to discuss the choice of an optimization criterion with a user, say the choice between the L_1- and L_2-norm criteria. But it is quite reasonable to discuss and specify the requirements towards the solution of the corresponding optimization problem, in other words, on the concrete type of the DSC. Hence these requirements may considerably narrow the possible class of criteria and sometimes completely determine its choice.

Now we dwell on these requirements towards the DSC-algorithms, especially for the medical data. The first specific peculiarity of these problems is a weak metrological support of measurement procedures in medicine as compared to the physical sciences. Thus it is important to take into account some **metrological requirements** on the DSC-algorithms providing equivariancy of the DSC-algorithms under some kinds of data and measurement scale transformations. The second specific feature is the presence of outliers or gross errors in the data. This requires the **stability** of statistical inference under the uncontrolled deviations from the assumed models of signals and noises. We have already mentioned the 'individual' character of the medical data. For example, when filtering electrocardiogram (ECG) noises, the components of this noise appear to be particularly individual, their characteristics are different not only for different patients but for different ECGs of the same patient. Thus filtering of ECG signals should be performed adaptively to the individual characteristics of noises in each ECG. In general, the requirement of **adaptation** of the DSC-algorithms on the individual behavior of the examined object is quite desirable.

Further we use the requirements of a metrological character, stability, and adaptation while choosing optimization criteria in data analysis and distribution classes in robustness studies.

1.1.3. Formalization of uncertainty

The requirement of stability of statistical inferences directly leads to the use of robust statistical methods. It may be said that, with respect to the form of information on underlying distributions, robust statistical methods occupy the intermediate place between classical parametric and nonparametric methods.

In parametric statistics, the shape of an underlying distribution is assumed known up to the values of unknown parameters. In nonparametric statistics, we suppose that the underlying distribution belongs to some sufficiently 'wide' class of distributions (continuous, symmetric, etc.). In robust statistics, at least within the Huber minimax approach, we also consider distribution classes but with more detailed information about the underlying distribution, say, in the form of some neighborhood of the normal distribution. The latter peculiarity allows to raise efficiency of robust procedures as compared to nonparametric

Table 1.1. Forms and levels of uncertainty of information about distributions and related approaches

Uncertainty and Information: Forms and Levels	Approaches and Methods
Given $f(x;\theta)$, random θ	Bayesian Methods
Given $f(x;\theta)$, unknown θ	Parametric Approach: Maximum Likelihood and Related Methods
$f(x;\theta) \in \mathscr{F}$ Class $\mathscr{F}$—a neighborhood of the normal distribution	Robust Methods
$f(x;\theta) \in \mathscr{F}$ General classes $\mathscr{F}$	Nonparametric Methods
Unique Sample Frequency Instability	Probability-Free Methods: Fuzzy, Exploratory, Interval Probability, Logical-Algebraic, Geometrical

methods simultaneously preserving their high stability.

At present, there are two main approaches in robustness:

- the Huber minimax approach—quantitative robustness (Huber, 1964; Huber, 1981);
- the Hampel approach based on influence functions—qualitative robustness (Hampel, 1968; Hampel *et al.*, 1986).

The topics of robustness are treated in many books beside the abovementioned; here we only enlist those comprising extended surveys: (Rey, 1978; Rey, 1983; Tsypkin, 1984; Makshanov *et al.*, 1991; Kharin, 1996a; Shulenin, 1993; Shurygin, 2000).

Table 1.1 classifies the methods of point estimation for the parameter θ of the underlying distribution density $f(x;\theta)$ in their dependence on the form of information about $f(x;\theta)$. Note that the upper and lower levels of this classifications, namely the Bayesian and probability-free methods, are being intensively developed at present.

1.2. Huber minimax approach

1.2.1. Some general remarks on robustness

The first robust estimators based on rejection of outliers are dated by the second half of the eighteenth century, namely they originate from Boscovich (1757) and Daniel Bernoulli (1777). Several outstanding scientists of the late nineteenth and early twentieth century (the astronomer Newcomb (1886), the chemist Mendeleyev (1895), the astrophysicist Eddington (1914), and the geophysicist Jeffreys (1932) among them) understood the weakness of the standard estimators under heavy-tailed error distributions and proposed some robust alternatives to them (for details, see (Stigler, 1973)). It would not be out of place to note the paper (Kolmogorov, 1931), in which he compared the behavior of the sample mean and sample median and recommended to use the latter under heavy-tailed distributions.

The convincing arguments for robust statistics are given in (Tukey, 1960; Huber, 1981; Hampel *et al.*, 1986). Here we only recall that the classical examples of robust and non-robust estimators of location are given by the sample median and sample mean, respectively.

As it was said above, robust statistics deal with the consequences of possible deviations from the assumed statistical model and suggests the methods protecting statistical procedures against such deviations. Thus the statistical models used in robust statistics are chosen so that to account possible violations of the assumptions about the underlying distribution. For description of these violations, the concrete forms of neighborhoods of the underlying model are formed with the use of an appropriately chosen metric, for example, the Kolmogorov, Prokhorov, or Lévy (Hampel *et al.*, 1986; Huber, 1981). Hence the initial model (basic or ideal) is enlarged up to the so-called *supermodel* that describes both the ideal model and the deviations from it.

Defining a robust procedure, it is useful to answer three main questions:

- Robustness of what?
- Robustness against what?
- Robustness in what sense?

The first answer defines the type of a statistical procedure (point or interval estimation, hypotheses testing, etc.); the second specifies the supermodel, and the third introduces the criterion of quality of a statistical procedure and some related requirements towards its behavior. The wide spectrum of the problems observed in robust statistics can be explained by the fact that there exists a variety of answers to each of the above questions.

Now we briefly enlist main supermodels in robust statistics (Bickel, 1976; Hampel *et al.*, 1986; Huber, 1981; Shulenin, 1993). In general, there is a great variety of supermodels but here we are mostly interested in the supermodels describing possible changes of the distribution shape. For supermodels, we may distinguish two types: local and global (Bickel, 1976).

A local type suggests setting an ideal (basic) model, and then the related supermodel is defined as a neighborhood of this ideal model. A global supermodel represents some class F of distributions with given properties that also comprises an ideal model. For example, Hodges and Lehmann (1963) consider the supermodel in the form of all absolutely continuous symmetric distributions. Birnbaum and Laska (1967) propose the supermodel as a finite collection of distribution functions: $\mathsf{F} = \{F_1, F_2, ..., F_k\}$. Andrews *et al.* (1972) examine estimators in the supermodels containing distributions with heavier tails than the normal. In particular, they use the Tukey supermodel based on the quantile function, the inverse to the distribution function. This supermodel comprises rather accurate approximations to the normal, Laplace, logistic, Cauchy, and Student distributions.

Various supermodels are used to study deviations from normality: the family of power-exponential distributions with the normal, Laplace, and uniform distributions as particular cases; the family of the Student t-distributions with the normal and Cauchy distributions; also the influence of non-normality can be studied with the use of the measures of asymmetry and kurtosis, the positive values of the latter indicate gross errors and heavy tails.

For describing gross errors and outliers, the most popular is the Tukey supermodel (Tukey, 1960)

$$\mathsf{F} = \left\{F\colon F(x) = (1-\varepsilon)\Phi(x) + \varepsilon\Phi\left(\frac{x-\theta}{k}\right),\ 0 \le \varepsilon \le 1,\ k \ge 1\right\}. \tag{1.2.1}$$

The generalization of this supermodel

$$\mathsf{F} = \{F\colon F(x) = (1-\varepsilon)F_0(x) + \varepsilon H(x),\ 0 \le \varepsilon \le 1\}, \tag{1.2.2}$$

where F_0 is some given distribution (the ideal model) and $H(x)$ is an arbitrary continuous distribution, is considered in (Huber, 1964). Supermodel (1.2.2) has the following natural interpretation: the parameter ε is the probability of gross errors in the data.

In general, a supermodel can be defined with the use of some suitable metric $d(F_0, F)$ in the space of all distributions: $\mathsf{F} = \{F\colon d(F_0, F) \le \varepsilon)\}$. The Prokhorov metric (Prokhorov, 1956) and its particular case, the Lévy metric, are rather convenient choices, since the supermodels based on them describe simultaneously the effects of gross errors, grouping, and rounding-off in the data (for details, see (Huber, 1981)).

The use of other metrics for constructing supermodels is discussed in (Bickel, 1976). The relations between various metrics can be found in (Huber, 1981; Zolotarev, 1997).

Summarizing the above, we may answer the second question: 'Robustness against what?' as follows: 'Robustness against extension of ideal models to supermodels.'

Now we are in position partly to answer the third question: 'Robustness in what sense?'

1.2.2. M-estimators of location

The first general approach to robust estimation is based on the minimax principle (Huber, 1964; Huber, 1972; Huber, 1981). The minimax approach aims at the least favorable situation for which it suggests the best solution. Thus, in some sense, this approach provides a guaranteed result, perhaps too pessimistic. However, being applied to the problem of estimation of the location parameter, it yields a robust modification of the principle of maximum likelihood.

Let $x_1, \dots, x_n$ be a random sample from a distribution F with density $f(x-\theta)$ in a convex class $\mathcal{F}$, where θ is the location parameter. Assume that F is a symmetric unimodal distribution, hence θ is the center of symmetry to be estimated. Then the M-estimator $\widehat{\theta}_n$ of the location parameter is defined as some solution of the following minimization problem

$$\widehat{\theta}_n = \arg\min_{\theta} \sum_{i=1}^{n} \rho(x_i - \theta), \tag{1.2.3}$$

where $\rho(u)$ is an even non-negative function called the *contrast function* (Pfanzagl, 1969); $\rho(x_i - \theta)$ is the measure of discrepancy between the observation x_i and the estimated center.

Choosing $\rho(u) = u^2$, we have the least squares (LS) method with the sample mean $\overline{x}_n$ as the estimator; for $\rho(u) = |u|$, we have the least absolute values (LAV) method with the sample median $\operatorname{med} x$ as the estimator, and, what is most important, for a given density f, the choice $\rho(u) = -\log f(u)$ yields the maximum likelihood estimator (MLE).

It is convenient to formulate the properties of M-estimators in terms of the derivative of the contrast function $\psi(u) = \rho'(u)$ called the *score function*. In this case, the M-estimator is defined as a solution of the following implicit equation

$$\sum_{i=1}^{n} \psi(x_i - \widehat{\theta}_n) = 0. \tag{1.2.4}$$

Under rather general regularity conditions imposed on the class Ψ of score functions ψ and on the related class $\mathcal{F}$ of densities f (their various forms can be found in (Huber, 1964; Huber, 1967; Deniau *et al.*, 1977a; Deniau *et al.*, 1977c; Huber, 1981; Hampel *et al.*, 1986)), M-estimators are consistent, asymptotically normal with the asymptotic variance

$$\mathrm{Var}\, n^{1/2}\widehat{\theta}_n = V(\psi, f) = \frac{\mathsf{E}_F\psi^2}{(\mathsf{E}_F\psi')^2} = \frac{\int \psi^2(x)\, dF(x)}{\left(\int \psi'(x)\, dF(x)\right)^2}, \tag{1.2.5}$$

and satisfy the minimax property

$$V(\psi^*, f) \le V(\psi^*, f^*) = \sup_{f\in\mathscr{F}} \inf_{\psi\in\Psi} V(\psi, f), \tag{1.2.6}$$

where the least favorable (informative) density f^* minimizes the Fisher information for location over the class $\mathscr{F}$

$$f^* = \arg\min_{f\in\mathscr{F}} I(f), \qquad I(f) = \int \left[\frac{f'(x)}{f(x)}\right]^2 f(x)\, dx, \tag{1.2.7}$$

whereas the optimal contrast function and score function are given by the maximum likelihood method for the least favorable density f^*

$$\rho^* = -\log f^*, \qquad \psi^* = -f^{*\prime}/f^*. \tag{1.2.8}$$

For most of our aims, the following regularity conditions defining the classes $\mathscr{F}$ and Ψ are sufficient (for details, see (Hampel *et al.*, 1986, pp. 125–127)):

($\mathscr{F}$1) f is twice continuously differentiable and satisfies $f(x) > 0$ for all x in $\mathbf{R}$.

($\mathscr{F}$2) the Fisher information for location satisfies $0 < I(f) < \infty$.

(Ψ1) ψ is well-defined and continuous on $\mathbf{R} \setminus C(\psi)$, where $C(\psi)$ is finite. At each point of $C(\psi)$ there exist finite left and right limits of ψ which are different. Moreover, $\psi(-x) = -\psi(x)$ if $(-x, x) \subset \mathbf{R} \setminus C(\psi)$, and $\psi(x) \ge 0$ for $x \ge 0$ not belonging to $C(\psi)$.

(Ψ2) The set $D(\psi)$ of points at which ψ is continuous but in which ψ' is not defined or not continuous is finite.

(Ψ3) $\int \psi^2\, dF < \infty$.

(Ψ4) $0 < \int \psi'(x)\, dF(x) = -\int \psi(x) f'(x)\, dx < \infty$.

The key point of this approach is the solution of variational problem (1.2.7): various classes $\mathscr{F}$ (supermodels) with the corresponding least favorable densities f^* and minimax estimators are given in Section 3.1. Here we only recall the Huber solution for the supermodel of gross errors

$$\mathscr{F} = \{f : f(x) = (1-\varepsilon) f_0(x) + \varepsilon h(x),\ 0 \le \varepsilon < 1\}, \tag{1.2.9}$$

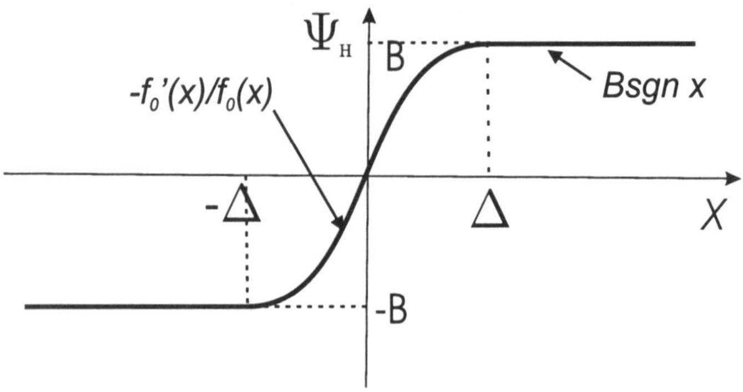

Figure 1.1. Huber ψ-function

where f_0 is a given density, $h(x)$ is an arbitrary density satisfying conditions ($\mathscr{F}1$) and ($\mathscr{F}2$) along with the additional logconvexity condition. Then the least favorable density f^* and the optimal score function are of the forms

$$f^*(x) = f_H(x) = \begin{cases} (1-\varepsilon)f_0(x), & |x| \le \Delta, \\ A\exp(-B|x|), & |x| > \Delta, \end{cases} \tag{1.2.10}$$

$$\psi^*(x) = \psi_H(x) = \begin{cases} -f_0'(x)/f_0(x), & |x| \le \Delta, \\ B\,\mathrm{sgn}\,x, & |x| > \Delta, \end{cases} \tag{1.2.11}$$

where the parameters A, B, and Δ are determined from the conditions of normalization, continuity, and differentiability of the solution at $x = \Delta$

$$\int f^*(x)\,dx = 1, \qquad f^*(\Delta - 0) = f^*(\Delta + 0), \qquad f^{*\prime}(\Delta - 0) = f^{*\prime}(\Delta + 0).$$

Figure 1.1 illustrates the Huber score function yielding a robustified version of the MLE: in the central zone $|x_i - \theta| \le \Delta$, the data are processed by the ML method, and they are trimmed within distribution tails. In the limiting case of a completely unknown density as $\varepsilon \to 1$, the minimax variance M-estimator of location is the sample median.

Within this approach, robustness is measured in terms of efficiency, namely by the supremum of asymptotic variance in the supermodel $\mathscr{F}$: $\sup_{f\in\mathscr{F}} V(\psi, f)$. Obviously, the smaller this characteristic, the more robust M-estimator is. Observe that the asymptotic normality of M-estimators allows to use the asymptotic variance as a characteristic for both efficiency and robustness.

Another measure of robustness is given by the supremum of asymptotic bias $\sup_{f\in\mathscr{F}} |b(\psi, f)|$ under asymmetric distributions (Huber, 1981; Rychlik,

1987; Smolyak and Titarenko, 1980; Zieliński, 1987), where

$$b(\psi, f) = \lim_{n\to\infty} \mathsf{E}(\widehat{\theta}_n - \theta) = \frac{\mathsf{E}_F\psi}{\mathsf{E}_F\psi'} = \frac{\int \psi(x)\,dF(x)}{\int \psi'(x)\,dF(x)}.$$

In particular, for the normal density $f_0(x) = \mathscr{N}(x;\theta,\sigma)$ and asymmetric contaminating density h in the supermodel of gross errors (1.2.9), the minimax bias estimator determined from the condition $\psi^* = \arg\inf_\psi \sup_h |b(\psi, f)|$ is the sample median (Huber, 1981; Smolyak and Titarenko, 1980).

1.2.3. *L*-estimators of location

The linear combinations of order statistics (*L*-estimators) are defined as

$$\widehat{\theta}_n = \sum_{i=1}^{n} C_i x_{(i)}, \qquad \sum_{i=1}^{n} C_i = 1, \tag{1.2.12}$$

where $x_{(i)}$ is the ith order statistic. The normalization condition in (1.2.12) provides equivariancy of *L*-estimators under translation. The trimmed mean

$$\overline{x}_{tr}(k) = \frac{1}{n-2k} \sum_{i=k+1}^{n-k} x_{(i)} \tag{1.2.13}$$

and the Winsorized mean

$$\overline{x}_W(k) = \frac{1}{n}\left[(k+1)x_{(k+1)} + \sum_{i=k+2}^{n-k-1} x_{(i)} + (k+1)x_{(n-k)}\right] \tag{1.2.14}$$

belong to this class. In asymptotics, the fraction α of censored observations is used: $k = [\alpha n]$.

L-estimators were proposed by in (Daniel, 1920) and since then they have been forgotten for thirty years, being revived in robustness studies. The description of *L*-estimators can be formalized with the use of the *weight function*.

Let $h\colon [0,1] \to \mathbf{R}$ be a given function satisfying the following conditions: $h(t) = h(1-t)$ for all $t \in [0,1]$, $\int_0^1 h(t)\,dt = 1$, and h is a function of bounded variation on $[0,1]$. The estimator

$$\widehat{\theta}_n = \frac{1}{n}\sum_{i=1}^{n} h\left(\frac{i}{n+1}\right) x_{(i)} \tag{1.2.15}$$

is called the *L*-estimator with weight function $h(t)$. The above regularity conditions on h along with the conditions ($\mathscr{F}$1) and ($\mathscr{F}$2) on distributions provide consistency and asymptotic normality of *L*-estimators (1.2.12) with asymptotic variance

$$\mathsf{Var}\, n^{1/2}\widehat{\theta}_n = A_L(h, F) = \int_0^1 K^2(t)\,dt, \tag{1.2.16}$$

where

$$K(t) = \int_{1/2}^{1} \frac{h(u)}{f(F^{-1}(u))}\,du, \qquad \int_0^1 K(t)\,dt = 0.$$

1.2.4. R-estimators of location

R-estimators proposed in (Hodges and Lehmann, 1963) are based on rank tests. There are several methods of their construction. Now we briefly describe one of those (Azencott, 1977a; Azencott, 1977b; Huber, 1981).

Let $y_1, \dots, y_n$ and $z_1, \dots, z_n$ be independent samples from the distributions $F(x)$ and $F(x - \theta)$ respectively. For testing the hypothesis $\theta = 0$ against the alternative $\theta > 0$, the following statistic is used:

$$W_n(y_1, \dots, y_n, z_1, \dots, z_n) = \sum_{i=1}^{n} J\left(\frac{s_i}{2n+1}\right),$$

where s_i is the rank of y_i, $i = 1, \dots, n$, in the united sample of size $2n$. Let $J(t)$, $0 \le t \le 1$, satisfy the following conditions: $J(t)$ is increasing; $J(t) + J(1-t) = 0$ for all $t \in [0,1]$; $J'(t)$ is defined on $(0,1)$; the functions J' and $f(F^{-1})$ are of bound variation on $[0,1]$, and $\int_0^1 J'(t) f(F^{-1}(t))\,dt \ne 0$.

Under these conditions, the test with the critical region $W_n > c$ has certain optimal in power properties (Hájek and Šidák, 1967). The R-estimator $\widehat{\theta}_n$ based on this test is defined as a solution of the equation

$$W_n(x_1 - \widehat{\theta}_n, \dots, x_n - \widehat{\theta}_n, -(x_1 - \widehat{\theta}_n), \dots, -(x_n - \widehat{\theta}_n)) = 0.$$

Under the above conditions, $\widehat{\theta}_n$ is consistent and asymptotically normal with asymptotic variance

$$\mathrm{Var}\, n^{1/2}\widehat{\theta}_n = A_R(J,F) = \frac{\int_0^1 J^2(t)\,dt}{\left[\int J'(F(x)) f^2(x)\,dx\right]^2}. \tag{1.2.17}$$

For any fixed function $F(x)$, it is possible to find the function $J(t)$ minimizing asymptotic variance $A_R(J,F)$. The test based on such function $J(t)$ also has optimal properties for given F. In particular, the logistic distribution $F(x) = (1+e^{-x})^{-1}$ leads to the Wilcoxon test. The corresponding estimator of location is the Hodges–Lehmann median

$$\widehat{\theta}_n = \operatorname{med}\left\{\frac{x_{(i)} + x_{(k)}}{2}\right\}, \qquad 1 \le i \le k \le n. \tag{1.2.18}$$

1.2.5. The relations between M-, L- and R-estimators of location

In (Jaeckel, 1971b), the asymptotic equivalence of these estimators was established. Let F and ψ be fixed. Set

$$h(t) = \frac{\psi'(F^{-1}(t))}{\int \psi'(x)f(x)\,dx}, \qquad t \in [0,1],$$
$$J(t) = \psi(F^{-1}(t)), \qquad t \in [0,1].$$

Then $V(\psi, f) = A_L(h,F) = A_R(J,F)$. However, M-estimators are most convenient for analysis and L-estimators are the simplest for computing.

1.3. Hampel approach

The main advantage of robust methods is their lower sensitivity to possible variations of data statistical characteristics. Thus it is necessary to have specific mathematical tools allowing to analyze the sensitivity of estimators to outliers, rounding-off errors, etc. On the other hand, such tools make it possible to solve the inverse problem: to design estimators with the required sensitivity. Now we introduce the above-mentioned apparatus, namely the sensitivity curves and the influence functions.

1.3.1. The sensitivity curve

Let $\{T_n\}$ be some sequence of statistics. Let $T_n(X)$ denote the statistic from $\{T_n\}$ on the sample $X = (x_1,\dots,x_n)$, and let $T_{n+1}(x,X)$ denote the same statistic on the sample $(x_1,\dots,x_n,x)$. Then the function

$$SC_n(x; T_n, X) = (n+1)[T_{n+1}(x,X) - T_n(X)] \qquad (1.3.1)$$

characterizes the sensitivity of T_n to the addition of one observation at x and is called the *sensitivity curve* for this statistic (Tukey, 1977). In particular,

$$SC_n(x; \overline{x}, X) = x - \frac{1}{n}\sum_{i=1}^{n} x_i = x - \overline{x}$$

for the sample mean $\overline{x}$;

$$SC_n(x; \operatorname{med} x, X) = \begin{cases} 0.5(n+1)[x_{(k)} - x_{(k+1)}], & x \le x_{(k)}, \\ 0.5(n+1)[x - x_{(k+1)}], & x_{(k)} \le x \le x_{(k+2)}, \\ 0.5(n+1)[x_{(k)} - x_{(k+1)}], & x \ge x_{(k+2)} \end{cases}$$

for the sample median $\operatorname{med} x$ with $n = 2k+1$; for the trimmed mean (1.2.13) with the two removed extreme order statistics, the main part of the sensitivity

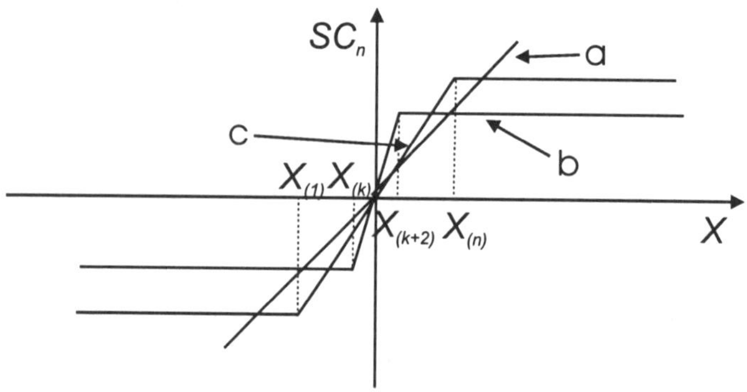

Figure 1.2. Sensitivity curves of the sample mean, median, and trimmed mean

curve is of the form

$$SC_n(x, \overline{x}_{tr}(1), X) = \begin{cases} x_{(1)}, & x \le x_{(1)}, \\ x, & x_{(1)} \le x \le x_{(n)}, \\ x_{(n)}, & x \ge x_{(n)}. \end{cases}$$

The sensitivity curves for the sample mean, median, and trimmed mean are given in Fig. 1.2. We can see that the sensitivity curve of the sample mean is unbounded, hence only one extreme observation can completely destroy the estimator. In addition, the maximal error of the trimmed mean is of order $(x_{(n)} - x_{(1)})/n$.

The derivative characteristics of $SC_n(x; T_n, F)$ such as $\sup_x |SC_n(x; T, X)|$ or the difference $SC_n(x; T, X) - SC_n(y; T, X)$ allow to compare the impact of adding new observations to the data on estimators. In particular, the sample median is sensitive to the occurrence of new sample elements in the interval $(x_{(k)}, x_{(k+2)})$. There exist some other characteristics describing the influence of the data perturbations on estimators. It is desirable to have such a characteristic that does not depend on the specific sample X. The most convenient for asymptotic analysis is the influence function (curve) introduced in (Hampel, 1974).

1.3.2. The influence function and its properties

Let F be a fixed distribution and $T(F)$ be a functional defined on some set F of distributions satisfying conditions ($\mathscr{F}1$) and ($\mathscr{F}2$), and let the estimator $T_n = T(F_n)$ of $T(F)$ be that functional of the sample distribution function F_n. Then the influence function $IF(x; T, F)$ is defined as

$$IF(x; T, F) = \lim_{t \to 0} \frac{T((1-t)F + t\delta_x) - T(F)}{t}, \tag{1.3.2}$$

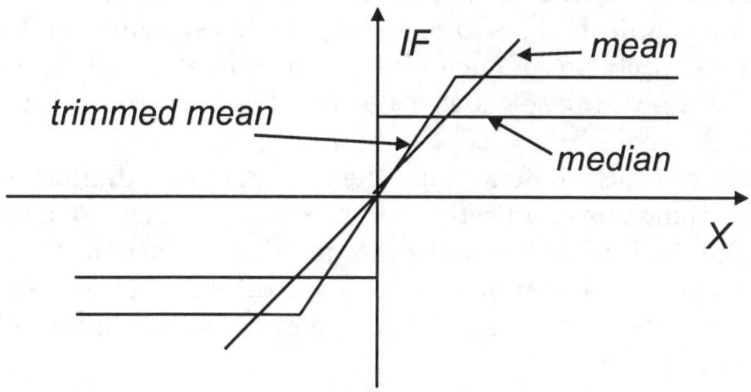

Figure 1.3. Influence functions of the sample mean, median, and trimmed mean

where δ_x is the degenerate distribution at x.

For the sample mean $\overline{x} = T(F_n) = \int x\,dF_n(x)$, the influence function is

$$IF(x;\overline{x},F) = x - T(F) = x - \int x\,dF(x);$$

for the α-trimmed mean, the functional is

$$T(F) = \frac{1}{1-2\alpha}\int_{F^{-1}(\alpha)}^{F^{-1}(1-\alpha)} x\,dF(x),$$

and

$$IF(x;\overline{x}_{tr},F) = \begin{cases} F^{-1}(\alpha)/(1-2\alpha), & x \le F^{-1}(\alpha), \\ x/(1-2\alpha), & F^{-1}(\alpha) \le x \le F^{-1}(1-\alpha), \\ F^{-1}(1-\alpha)/(1-2\alpha), & x \ge F^{-1}(1-\alpha); \end{cases}$$

whereas for the sample median $\operatorname{med} x$, the functional is $T(F) = F^{-1}(1/2)$ and

$$IF(x;\operatorname{med} x,F) = \frac{\operatorname{sgn} x}{2f(0)}.$$

Comparing Fig. 1.2 and Fig. 1.3, we see that the forms of influence and sensitivity curves are similar: as a rule, $SC_n(x;T,F) \to IF(x;T,F)$ as $n \to \infty$.

Under conditions ($\mathscr{F}$1), ($\mathscr{F}$2), (Ψ1)-(Ψ4), the influence function for the M-estimator with the score function ψ is of the form (Hampel *et al.*, 1986; Huber, 1981)

$$IF(x;\psi,F) = \frac{\psi(x)}{\int \psi(x)\,dF(x)}. \tag{1.3.3}$$

For M-estimators, the relation between the influence function and the score function is the simplest. This allows to apply M-estimators to solving some specific extremal problems of maximization of estimators' efficiency over their sensitivity to outliers, the so-called problems of optimal Huberization (Deniau *et al.*, 1977d; Hampel *et al.*, 1986; Huber, 1972).

The influence function measuring the impact of an infinitesimal contamination at x on the value of an estimator, however, is a delicate and useful tool having deep intrinsic relations with other important statistical notions (Hampel, 1968; Hampel, 1974; Hampel *et al.*, 1986; Huber, 1981). For example, with the use of $IF(x;T,F)$, the functional $T(F)$ can be linearized in the neighborhood of the ideal model F_0 as

$$T(F) = T(F_0) + \int IF(x;T,F_0)\,d[F(x) - F_0(x)] + \text{remainder};$$

$\sqrt{n}\,[T_n(F_n) - T(F)]$ tends to $\int IF(x;T,F)\,dF_n(x)$ in probability so that

$$T_n(F_n) = T(F) + \int IF(x;T,F)\,dF_n(x) + \text{remainder}.$$

Further,

$$\sqrt{n}(T_n - T(F)) = \frac{1}{\sqrt{n}}\sum_{i=1}^{n} IF(x_i;T,F) + \text{remainder}.$$

Since in most cases the remainder is negligible as $n \to \infty$, $\sqrt{n}\,T_n$ is asymptotically normal with asymptotic variance

$$V(T,F) = \int IF^2(x;T,F)\,dF(x). \tag{1.3.4}$$

This line of reasoning is accurately verified in (Fernholz, 1983).

1.3.3. The local measures of robustness

From the influence function, the following robustness measures can be defined (Hampel, 1968; Hampel, 1974):

THE SUPREMUM OF THE ABSOLUTE VALUE

$$\gamma^*(T,F) = \sup_x |IF(x;T,F)|, \tag{1.3.5}$$

called the *gross-error sensitivity* of T at F. This general characteristic of sensitivity being an upper bound to the asymptotic bias of the estimator measures the worst influence of an infinitesimal contamination on the value of the estimator. The estimators T having finite $\gamma^*(T,F)$ are called B-robust (Rousseeuw, 1981), and those for which there exists a positive minimum of γ^* are the *most B-robust* estimators (Hampel *et al.*, 1986).

THE LOCAL-SHIFT SENSITIVITY

$$\lambda^*(T,F) = \sup_{x \neq y} |IF(y;T,F) - IF(x;T,F)|/|y-x|$$

accounts the effects of rounding-off and grouping of the observations.

THE REJECTION POINT

$$\rho^*(T,F) = \inf\{r > 0 \colon IF(x;T,F) = 0 \text{ where } |x| > r\}$$

defines the observations to be rejected completely.

At present, the influence function is the main heuristic tool for designing estimators with given robustness properties (Hampel *et al.*, 1986; Huber, 1981; Rey, 1978). For example, slightly changing the maximum likelihood estimator, it is possible to improve considerably its sensitivity to gross errors by lessening γ^* and its sensitivity to local effects of rounding-off and grouping types by bounding the slope of $IF(x;T,F)$ (i.e., λ^*) above. Setting $IF(x;T,F)$ tend to zero as $n \to \infty$ leads to the stabilization of the asymptotic variance while bounding the slope above stabilizes the asymptotic bias.

By analogy with the influence function, the *change-of-variance function* $CVF(x;T,F)$ is defined as

$$CVF(x;T,F) = \lim_{t\to 0}[V(T,(1-t)F + t\delta_x) - V(T,F)]/t,$$

where $V(T,F)$ is the functional of asymptotic variance (Hampel *et al.*, 1986).

Further, the *change-of-variance sensitivity* is defined as

$$\kappa^*(T,F) = \sup_x CVF(x;F,T)/V(T,F),$$

and the estimator $T_n = T(F_n)$ of the functional $T(F)$ is called *V-robust* if $\kappa^*(T,F) < \infty$.

1.3.4. Global robustness: the breakdown point

All the above-introduced measures of robustness based on the influence function and its derivatives are of a local character being evaluated at the model distribution F. Hence it is desirable to have a measure of the *global* robustness of the estimator over the chosen class of distributions, in other words, in the chosen supermodel $\mathscr{F}$. Since the general definition of a supermodel is based on the concept of a distance (Kolmogorov, Lévy, Prokhorov) in the space of all distributions (for details, see (Hampel *et al.*, 1986; Huber, 1981)), the same concept is involved into the construction for a measure of the global robustness. Let d be such a distance. Then the *breakdown point* ε^* of the estimator $T_n = T(F_n)$ for the functional $T(F)$ at F is defined by

$$\varepsilon^*(T,F) = \sup\{\varepsilon \le 1 \colon \sup_{F \colon d(F,F_0)<\varepsilon} |T(F) - T(F_0)| < \infty\}. \tag{1.3.6}$$

The breakdown point characterizes the maximal deviation (in the sense of a metric chosen) from the ideal model F_0 that provides the boundedness of the estimator bias.

For our further aims, the concept of the *gross-error breakdown point* suffices:

$$\varepsilon^*(T,F) = \sup\{\varepsilon\colon \sup_{F\colon F=(1-\varepsilon)F_0+\varepsilon H} |T(F) - T(F_0)| < \infty\}. \tag{1.3.7}$$

This notion defines the largest fraction of gross errors that still keeps the bias bounded.

EXAMPLE 1.3.1. Such a famous estimator of location as the sample median possesses many optimal robustness properties which hold simultaneously: the Huber minimax variance, the Huber minimax bias, B-robust, V-robust, and globally robust with the maximal value of the breakdown point $\varepsilon^* = 1/2$.

REMARK 1.3.1. The basic relations between the Hampel and Huber approaches, are thoroughly analyzed in (Hampel *et al.*, 1986, pp. 172–178), between the concepts of continuity and qualitative robustness, in (Hampel, 1971). In particular, for sufficiently small ε, the Huber minimax solution ψ^* minimizing asymptotic variance $V(\psi, f)$ of M-estimators in the gross-error supermodel turns out to be optimal V-robust minimizing $V(\psi, f)$ under the condition that κ^* is an upper bound of the change-of-variance sensitivity. (Rousseeuw, 1981). The similar assertion holds for the Huber minimax bias solution and the optimal B-robust estimator, namely the sample median.

REMARK 1.3.2. In our further robustness studies, we mainly use **the Huber minimax variance approach in the global type supermodels** allowing to complement the desired **stability** of estimation with the **adaptation** of its **efficiency** to the underlying model. However, the **influence function tools** along with such a global measure of robustness as the **gross-error breakdown point** are also involved.

2

Optimization criteria in data analysis: a probability-free approach

Here we obtain the characteristics of location for the univariate and multivariate data using prior requirements towards the properties of these characteristics under the conditions of uncertainty of optimization criteria. The problem settings connected with replacing of a data collection by a unique characteristic quantity are introduced in this chapter. To a certain extent, this characteristic is equivalent to the initial data and thus it can be interpreted as an estimator for all of the data.

Henceforth, this characteristic is called the *'typical' representative* (the measure of location or mean values in the Cauchy sense) and it is defined as the solution of the problem of minimization of some measure of total discrepancy between observations and their 'typical' representative. It is shown that such a measure of discrepancy must satisfy certain *a priori* postulated natural requirements towards the properties of a 'typical' representative. These requirements mainly follow from metrological restrictions. The latter commonly are the requirements of translation, scale, orthogonal, and affine equivariancy.

In this chapter we show that taking into account such metrological requirements we can narrow the admissible classes of measures of discrepancy and in some cases to reduce them to parametric dependences. In particular, the requirement of scale equivariancy results in the L_p-norm estimates with arbitrary values of p (Bickel and Lehmann, 1975; Gehrig and Hellwig, 1982; Kreinovich, 1986) and the requirement of affine equivariancy leads to the method of least squares.

2.1. Introductory remarks

2.1.1. Prior requirements towards the DSC of location

The problem of data processing is often reduced to the maximal compression of the initial information. As a rule, this compression means the replacement of the initial data $x_1, \ldots, x_n$ by a 'typical' representative m. In this case, a measure of discrepancy between the 'typical' representative and the observation x_i is given by the value of a criterion $\rho(x_i, m)$, i.e., the value of the contrast function $\rho(x, m)$ (see Section 1.2).

Henceforth, we assume that the measure of discrepancy between all the data $x_1, \ldots, x_n$ and its typical representative m is $\sum \rho(x_i, m)$ of individual criteria. Thus we define the value m to be some solution of the following minimization problem

$$m^* = \arg\min_m \sum_{i=1}^n \rho(x_i, m), \tag{2.1.1}$$

or, in other words, m is the M-estimator for the data $x_1, \ldots, x_n$.

REMARK 2.1.1. Henceforth we put $m^* = m$ and this convention will not cause any ambiguity.

If $\rho(x, m)$ satisfies certain regularity conditions (differentiability, convexity, etc.) then the solution of problem (2.1.1) can be determined from the following equation

$$\sum_{i=1}^n \varphi(x_i, m) = 0, \tag{2.1.2}$$

where $\varphi(x, m) = \partial\rho(x, m)/\partial m$ is the score function for the contrast function $\rho(x, m)$.

All the abovesaid can be easily extended to the problems of multivariate data processing, i.e., to the case where observations are vectors, for example, when in each experiment there are several qualitatively different characteristics of the data. Obviously, in this case, a 'typical' representative is also a vector.

Let the results of an experiment be some collection of vectors $\mathbf{x}_1, \ldots, \mathbf{x}_n$ and let a 'typical' representative of this collection be a vector $\mathbf{m}$

$$\mathbf{x}_i = (x_i^1, \ldots, x_i^M)^T, \qquad \mathbf{m} = (m^1, \ldots, m^M).$$

Now we introduce a measure of discrepancy between the 'typical' representative and the observation $\mathbf{x}_i$ as follows:

$$\rho(\mathbf{x}_i, \mathbf{m}) = \rho(x_i^1, \ldots, x_i^M;\ m^1, \ldots, m^M).$$

As above, let a measure of discrepancy between the data $\mathbf{x}_1, \dots, \mathbf{x}_n$ and its 'typical' representative $\mathbf{m}$ be the sum of individual criteria. In this case, we, in a natural way, define $\mathbf{m}$ as the solution of the minimization problem

$$\mathbf{m}^* = \arg\min_{\mathbf{m}} \sum_{i=1}^{n} \rho(\mathbf{x}_i, \mathbf{m}). \tag{2.1.3}$$

As before, set $\mathbf{m}^* = \mathbf{m}$ (see Remark 2.1.1). The vector $\mathbf{m}$ determined from (2.1.3) is the M-estimator for the multivariate data $\mathbf{x}_1, \dots, \mathbf{x}_n$.

In the case where a contrast function $\rho(\mathbf{x}, m)$ satisfies certain regularity conditions, minimization problem (2.1.3) is reduced to the solution of the following simultaneous equations (the *score system*)

$$\sum_{i=1}^{n} \varphi_s(\mathbf{x}_i, \mathbf{m}) = 0, \qquad s = 1, 2, \dots, M, \tag{2.1.4}$$

where $\varphi_s(\mathbf{x}, \mathbf{m}) = \partial\rho(\mathbf{x}, \mathbf{m})/\partial m^s$.

If the contrast function $\rho(\mathbf{x}, m)$ is *a priori* given then the problem of determination of the 'typical' representative is completely defined and its solution may face only computational difficulties. Nevertheless, some natural questions may arise:

- How to ground the choice of the contrast function?
- What assumptions are associated with that choice? etc.

It is important indeed to pose these questions, since the value of the 'typical' representative essentially depends on the contrast function chosen. Some answers to these questions are presented in this chapter.

In Chapter 1 we have already observed that there exist such situations where the use of the probabilistic approach cannot be grounded, for instance, where it is necessary to deal with results of unique experiments and thus the hypothesis of stability of frequencies cannot be verified. In these cases, we must pose some conditions in order to choose the method of data processing. One part of these conditions is connected with purely mathematical requirements such as convexity, differentiability, etc. Other conditions should be formulated due to those or either requirements towards the properties of the parameter m sought for. Requirements of metrological character seem to be the most natural. For example, it is often necessary to provide the adaptation of the value m to the changes of the starting point and scale of a measuring device; while processing the geodesic data, the adaptation to the rotations of coordinate axes is desirable.

In the subsequent sections we show that similar requirements considerably reduce the variety of possible forms for the contrast function and in some situations these requirements determine them uniquely or within a parametric structure (Vilchevski and Shevlyakov, 1987; Shevlyakov, 1991; Vilchevski and Shevlyakov, 1995a).

Now we introduce the notion of equivariancy of a 'typical' representative under some transformation $\mathbf{f}(\mathbf{x})$.

DEFINITION 2.1.1. Let $\mathbf{x}_1, \dots, \mathbf{x}_n$ and $\mathbf{m}_x = \mathbf{m}(\mathbf{x}_1, \dots, \mathbf{x}_n)$ be some collection of the initial data and a 'typical' representative, respectively. We call the rule of determination of a 'typical' representative equivariant under the transformation $\mathbf{f}(\mathbf{x})$ if a 'typical' representative of the transformed data is the transformed 'typical' representative of the initial data, i.e., if the following relation holds for each collection of the data:

$$\mathbf{m}(\mathbf{f}(\mathbf{x}_1), \dots, \mathbf{f}(\mathbf{x}_n)) = \mathbf{f}(\mathbf{m}(\mathbf{x}_1, \dots, \mathbf{x}_n)).$$

Now we describe some general requirements on the contrast function and the score function. These requirements mainly follow from the intuitively obvious assumptions about the dependence between the 'typical' representative and data.

1. Let $x_1 = x_2 = \dots = x_n = x$ be n equal observations. Then it is natural to assume that $m = x$: the 'typical' representative coincides with the observed value. This is equivalent to the assumption

$$n \min_m \rho(x, m) = n\rho(x, x) = 0. \tag{2.1.5}$$

2. From (2.1.5) we obtain

$$\varphi(x, m) = \frac{\partial \rho(x, m)}{\partial m} \begin{cases} < 0, & m < x, \\ > 0, & m > x. \end{cases} \tag{2.1.6}$$

If the score function is continuous, then from (2.1.6) we obtain $\varphi(x, x) = 0$.

3. UNIQUENESS. The requirement of uniqueness of the 'typical' representative can be formally introduced as the requirement of concavity of the contrast function. In the case of a differentiable score function, this condition can be written as

$$\frac{\partial^2 \rho(x, m)}{\partial m^2} = \frac{\partial \varphi(x, m)}{\partial m} > 0.$$

4. SYMMETRY. Now assume that the change of the sign of every observation induces the change of the sign of the 'typical' representative

$$m(-x_1, \dots - x_n) = -m(x_1, \dots x_n).$$

REMARK 2.1.2. Note that it is not expedient to assume the hypothesis of symmetry for all data collections. For instance, if the distances between some objects are measured, or the quantitative characteristics of objects, or the ranks of objects in accordance with a chosen scale, etc., then the requirement of symmetry is not natural.

5. Symmetry also yields the oddness of a score function. Indeed, since we can arbitrarily choose x_i, take them symmetric

$$x_1 = \cdots = x_{n/2} = -x, \quad x_{n/2+1} = \cdots = x_{n/2} = x$$

(the number of observations is set even, if this number is odd then one observation is taken equal to zero).

For this variant of the distribution of observations, we have the following equation for m:

$$\frac{n}{2}(\varphi(-x,m) + \varphi(x,m)) = 0.$$

By symmetry, we have

$$\frac{n}{2}(\varphi(x,-m) + \varphi(-x,-m)) = 0.$$

Obviously, the initial and symmetric collections are equivalent. Hence the solutions of these two equations should coincide, i.e., $m = -m$. Thus, for the symmetric data, we have $m = 0$ and, therefore, the score function is odd:

$$\varphi(x,0) = -\varphi(-x,0). \tag{2.1.7}$$

6. TRANSLATION EQUIVARIANCY. For many problems of data processing, it is desirable to provide the coordination between the changes of a 'typical' representative and the data under the changes of the starting point of a measuring device.

Let $\{x_i\}, i = 1,\dots,n$, and $\{y_i = x_i + \lambda\}, i = 1,\dots,n$ be the initial and transformed data with their 'typical' representatives m_x and m_y, respectively. Then the 'typical' representative m is said to be *translation equivariant* if $m_y = m_x + \lambda$.

Naturally, the requirement of translation equivariancy impose certain restrictions on the structure of the contrast function, namely the contrast function should be a decreasing function of the absolute value of the difference between an observation and the 'typical' representative, which is stated in the following theorem.

THEOREM 2.1.1. *Let a solution to equation* (2.1.2) *be translation equivariant for any* λ *and* $\{x_i\}$

$$\sum_{i=1}^{n} \varphi(x_i - \lambda, m - \lambda) = 0, \tag{2.1.8}$$

where $\varphi(x,m)$ *is differentiable with respect to* x *and* m. *Then minimization problem* (2.1.1) *is equivalent to the minimization problem with the contrast function* $\rho(x,m) = A(|x-m|)$, *where* $A(u)$ *is a decreasing function of* u *for* $u > 0$ *and* $A(0) = 0$.

REMARK 2.1.3. It is easy to see that with the additional requirement of uniqueness of the 'typical' representative, the assertion of Theorem 2.1.1 can be reformulated as the condition of translation equivariancy for the contrast function $\rho(x,m) = A(|x-m|)$ so that $A(0) = 0$, $A(u) > 0$, $A'(0) = 0$, and $A''(u) > 0$ for $u > 0$.

These conditions can be generalized in an obvious way for the case of vector observations.

Let the data and 'typical' representative be

$$\mathbf{x}_i = (x_i^1, \dots, x_i^M)^T, \qquad i = 1, \dots, n; \qquad \mathbf{m} = (m^1, \dots, m^M)^T.$$

Then the analogs of the above conditions are of the form

1. $$\min_{m^1,\dots,m^M} \rho(x^1, \dots, x^M;\, m^1, \dots, m^M) = \rho(x^1, \dots, x^M; x^1, \dots, x^M) = 0.$$

2. $$\varphi_i(x^1, \dots, x^M; m^1, \dots, m^M) = \frac{\partial \rho(\dots)}{\partial m^i} \begin{cases} < 0, & \text{if } m^i < x^i, \quad m^j = x^j, \quad j \neq i, \\ > 0, & \text{if } m^i > x^i, \quad m^j = x^j, \quad j \neq i. \end{cases} \tag{2.1.9}$$

3. From (2.1.9) we have for continuous score functions:

$$\varphi_i(x^1, \dots, x^M; m^1, \dots, m^{i-1}, x^i, m^{i+1}, \dots, m^M) = \frac{\partial \rho(x^1, \dots, x^M; m^1, \dots, m^{i-1}, x^i, m^{i+1}, \dots, m^M)}{\partial m^i} = 0. \tag{2.1.10}$$

4. $$\varphi_i(x^1, \dots, x^{i-1}, -x^i, x^{i+1}, \dots, x^M; m^1, \dots, m^{i-1}, 0, m^{i+1}, \dots, m^M) = -\varphi_i(x^1, \dots, x^{i-1}, x^i, x^{i+1}, \dots, x^M; m^1, \dots, m^{i-1}, 0, m^{i+1}, \dots, m^M). \tag{2.1.11}$$

5. TRANSLATION EQUIVARIANCY is established by the following result.

THEOREM 2.1.2. *Let a solution of score system* (2.1.4) *satisfy the property of translation equivariancy*

$$\mathbf{m}(\mathbf{x}_1 + \boldsymbol{\lambda}, \dots, \mathbf{x}_n + \boldsymbol{\lambda}) = \boldsymbol{\lambda} + \mathbf{m}(\mathbf{x}_1, \dots, \mathbf{x}_n),$$

where $\mathbf{m}$, $\mathbf{x}_i$, *and* $\boldsymbol{\lambda}$ *are M-dimensional vectors and the contrast function is twice differentiable.*

Then the contrast function is of the form

$$\rho(\mathbf{x}, \mathbf{m}) = \rho(\mathbf{x} - \mathbf{m}) = \rho(|x^1 - m^1|, \dots, |x^M - m^M|).$$

Furthermore, $\rho(\mathbf{u})$ *satisfies the following relations:* $\rho(\mathbf{0}) = 0$ *and* $\partial\rho(\mathbf{u})/\partial u^s > 0$ *for* $\mathbf{u} > \mathbf{0}$.

REMARK 2.1.4. To guarantee uniqueness, we may add the requirement of concavity of $\rho(\mathbf{u})$ to the above conditions.

2.1.2. Proofs

PROOF OF THEOREM 2.1.1. By differentiating (2.1.4) with respect to λ, we obtain

$$\sum_{i=1}^{n} \left(\frac{\partial \varphi(x_i - \lambda, m - \lambda)}{\partial (x_i - \lambda)} \left(\frac{dx_i}{d\lambda} - 1 \right) + \frac{\partial \varphi(x_i - \lambda, m - \lambda)}{\partial (m - \lambda)} \left(\frac{dm}{d\lambda} - 1 \right) \right) = 0,$$

and taking into account the independence of x_i and m of λ, we have

$$\sum_{i=1}^{n} \left(\frac{\partial \varphi(x_i - \lambda, m - \lambda)}{\partial (x_i - \lambda)} + \frac{\partial \varphi(x_i - \lambda, m - \lambda)}{\partial (m - \lambda)} \right) = 0. \tag{2.1.12}$$

Equation (2.1.12) should hold for each $\{x_i\}$, so we choose the data in such a way that $x_i = x$, $i = 1, \dots, n$. Hence (2.1.12) takes the form

$$n \left(\frac{\partial \varphi(u, v)}{\partial u} + \frac{\partial \varphi(u, v)}{\partial v} \right) = 0,$$

where $u = x_i - \lambda$ and $v = m - \lambda$. The solution of this partial differential equation is given by $\varphi(u, v) = F(u - v)$, where $F(u)$ is an arbitrary function. Taking into account the condition of oddness for the score function $\varphi(-u, 0) = -\varphi(u, 0)$, we obtain that $F(u)$ can be written as $F(u) = \Psi(|u|) \operatorname{sgn} u$. As a result, the score function takes the form

$$\varphi(u, v) = \Psi(|u - v|) \operatorname{sgn}(u - v),$$

or for the variables x and m,

$$\varphi(x, m) = \Psi(|x - m|) \operatorname{sgn}(x - m). \tag{2.1.13}$$

The subsequent restrictions on the properties of the function $\Psi(u)$ are connected with the condition of minimum of the contrast function at $x = m$. This condition can be formulated in the form of requirement (2.1.6) and hence the function $\Psi(u)$ should be negative, or in the form of the condition $\Psi'(u) < 0$.

By $\varphi = \partial \rho(x, m)/\partial m$, we obtain after integrating from (2.1.13) that

$$\rho(x, m) = \int_0^m \Psi(|x - z|) \operatorname{sgn}(x - z)\, dz + C(x).$$

The form of $C(x)$ is determined from the condition $\rho(x, x) = 0$, hence we have

$$\rho(x, m) = \int_x^m \Psi(|x - z|) \operatorname{sgn}(x - z)\, dz.$$

Changing the variables $u = (x - z)/(x - m)$, we rewrite the integral as

$$\int_x^m \Psi(|x - z|)\,\mathrm{sgn}(x - z)\,dz = -\frac{1}{x - m}\int_0^1 \Psi(|x - m|u)\,\mathrm{sgn}((x - z)u)\,dz$$
$$= -\frac{\mathrm{sgn}(x - m)}{x - m}\int_0^1 \Psi(|x - m|u)\,du$$
$$= -\int_0^{|x-m|} \Psi(z)\,dz.$$

Therefore,

$$\rho(x, m) = -\int_0^{|x-m|} \Psi(z)\,dz, \tag{2.1.14}$$

where $\Psi(z)$ is an arbitrary negative function, which completes the proof. □

PROOF OF THEOREM 2.1.2. The proof practically coincides with the proof of the above theorem for the univariate case. In fact, the score system is of the form

$$\sum_1^n \frac{\partial \rho(\mathbf{x}_i - \boldsymbol{\lambda}, \mathbf{m} - \boldsymbol{\lambda})}{\partial m^s} = 0, \qquad s = 1, \dots, M,$$

where $\mathbf{x}$ and $\mathbf{m}$ do not depend on $\boldsymbol{\lambda}$.

By differentiating the above equation with respect to λ^l, we obtain

$$-\sum_1^n \left[\frac{\partial^2 \rho(\mathbf{x}_i - \boldsymbol{\lambda}, \mathbf{m} - \boldsymbol{\lambda})}{\partial m^s \partial x^l} + \frac{\partial^2 \rho(\mathbf{x}_i - \boldsymbol{\lambda}, \mathbf{m} - \boldsymbol{\lambda})}{\partial m^s \partial m^l}\right] = 0.$$

This equation should hold for each $\mathbf{x}_i$. Choose $\mathbf{x}_i = \mathbf{x}, i = 1, \dots, n$. Denote $\mathbf{x} - \boldsymbol{\lambda}$ as $\mathbf{u}$ and $\mathbf{m} - \boldsymbol{\lambda}$ as $\mathbf{v}$, hence,

$$n\left[\partial^2 \rho(\mathbf{u}, \mathbf{v})/\partial u^l \partial v^s + \partial^2 \rho(\mathbf{u}, \mathbf{v})/\partial v^l \partial v^s\right] = 0.$$

The solutions of these partial differential equations are

$$\frac{\partial \rho(\mathbf{u}, \mathbf{v})}{\partial v^l} = \Phi_s^l(u^1, \dots, u^{s-1}, u^s - v^s, u^{s+1}, \dots, u^M; v^1, \dots, v^{s-1}, v^{s+1}, \dots, v^M),$$
$$l = 1, \dots, M \qquad s = 1, \dots, M.$$

From these relations we easily obtain

$$\frac{\partial \rho(\mathbf{u}, \mathbf{v})}{\partial v^l} = \Phi^l(|\mathbf{u} - \mathbf{v}|), \qquad l = 1, \dots, M.$$

By symmetry condition (2.1.11),

$$\frac{\partial \rho(\mathbf{u},\mathbf{v})}{\partial v^l} = \Phi^l(|\mathbf{u}-\mathbf{v}|)\,\mathrm{sgn}(u^l - v^l), \qquad l = 1,\dots,M.$$

Since $\partial^2\rho(\mathbf{u},\mathbf{v})/\partial v^l \partial v^s = \partial^2\rho(\mathbf{u},\mathbf{v})/\partial v^s \partial v^l$, for $\mathbf{y} > 0$, there exists the potential function $\Phi(\mathbf{y})$ with the partial derivatives equal to $\Phi^l(\mathbf{y})$. Therefore, $\rho(\mathbf{x},\mathbf{m}) = \Phi(|\mathbf{x}-\mathbf{m}|)$. Finally, by $\Phi(0) = 0$ and the condition of minimum at $\mathbf{x} = \mathbf{m}$, we see that $\partial\Phi(\mathbf{y})/\partial y^l > 0$ for $\mathbf{y} > 0$.

Obviously, uniqueness of the solution of the score system holds in the case where $\Phi(\mathbf{y})$ is concave. This remark completes the proof. □

2.2. Translation and scale equivariant contrast functions

For the problems of data processing, it is often necessary to provide the automatic adaptation of the DSC-algorithms to the changes of starting points and scales of measuring devices. A typical example of such problems is given by the problem of processing of temperature measurements. Temperature is measured by thermometers with different scales: absolute, Celsius or Fahrenheit. It is natural to require that the result of processing of temperature measurements, the 'typical' temperature, should be determined by the same scale as the results of initial measurements. Obviously, this condition imposes some restrictions on the admissible structures of contrast functions.

These restrictions are connected with two typical situations appearing in multivariate data processing, for example, in processing of temperature measurements taken from different places or at different time, or with measuring different coordinates (length, width and height) of an object. The following cases are possible.

1. The measurements in every coordinate are made by measuring devices of the same type but the type of a device has not been fixed beforehand. For instance, it is *a priori* known that temperature is measured in the absolute scale, or in the Celsius scale, or in the Fahrenheit scale; distances are measured only in centimeters or in meters, etc. In this case, it is necessary to provide the adaptation of the results of data processing to *equal changes* of scale of measuring devices.

2. The measurements in every coordinate can be made with the use of different measuring devices, and the type of a measuring device is not known beforehand. For instance, the measurements in one coordinate are made by a ruler scaled in centimeters, and in other coordinate in meters, etc., and which coordinate is measured by which ruler is not known beforehand. In this case, it is necessary to provide the adaptation of the results of data processing to *independent changes* of scale.

It is obvious that the second variant of data receiving should impose stronger restrictions on the structure of admissible contrast functions and the corresponding score systems.

2.2.1. Translation and scale equivariant contrast functions: the equal changes of scale

Let a 'typical' representative of the experimental data possess such a property that the changes of starting points of measuring scales and simultaneously equal for all coordinates changes of the graduations of these measuring scales imply the analogous changes of the value of the 'typical' representative. The structure of the corresponding contrast function is given by the following result.

THEOREM 2.2.1. *Let a solution of the score system satisfy the following conditions:*

- *the translation equivariancy* $\mathbf{m}(\mathbf{x} + \boldsymbol{\lambda}) = \boldsymbol{\lambda} + \mathbf{m}(\mathbf{x})$;
- *the scale equivariancy under simultaneous and equal for all coordinates changes of scale* $\mathbf{m}(\mu\mathbf{x}) = \mu\mathbf{m}(x)$, *where* $\mu > 0$ *is a scalar parameter;*
- *the uniqueness of the solution for all collections of the data* $\mathbf{x}_1, \dots, \mathbf{x}_n$.

If there exist the second derivatives $\partial^2\rho/\partial m^s \partial x^l$ *and* $\partial^2\rho/\partial m^s \partial m^l$, $s, l = 1, \dots, M$, *then the contrast function* $\rho(\mathbf{x}, \mathbf{m})$ *is the sum of homogeneous functions of* $|x^s - m^s|$.

More precisely, the contrast function has the following structure. Let the variables $\mathbf{u}$ *be separated into disjoint subgroups* I_k, $r = 1, \dots, L$, *and the vectors* $\mathbf{u}(k)$ *contain only those variables that belong to the subgroup* I_k. *Then*

$$\rho(\mathbf{u}) = \sum_{k=1}^{L} \Phi_k(\mathbf{u}(k)),$$

where $\Phi_k(\mathbf{u}^{(k)})$ *are homogeneous functions of order* α_k *of the variables belonging to the subgroup* I_k, *i.e., this function has the following property*

$$\Phi_k(tv^1, tv^2, \dots, tv^{M_k}) = t^{\alpha_k}\Phi(v^1, v^2, \dots, v^{M_k}) \quad and \quad u^s = |x^s - m^s|.$$

Moreover,

$$\Phi_k(0, 0, \dots, 0) = 0, \qquad \frac{\partial \Phi_k(v^1, v^2, \dots, v_k^M)}{\partial v^s} > 0, \quad for \quad v_s > 0,$$

$s = 1, \dots M_k$, *and* $\Phi(v^1, v^2, \dots, v^{M_k})$ *are convex functions of their arguments.*

REMARK 2.2.1. A rather wide class of homogeneous functions, which can be used for description of contrast functions, is defined as follows:

$$\Phi(\mathbf{x}, m) = \sum_{\sum k_i = \alpha} C_{k_1,k_2,\dots,k_M} \prod_{s=1}^{M} |x^s - m_s|^{k_s}$$

for integer α;

$$\Phi(\mathbf{x}, \mathbf{m}) = \sum_{\sum k_i = L} C_{k_1,k_2,\dots,k_M} \prod_{s=1}^{M} |x^s - m_s|^{k_s/K}$$

for rational α equal to L/K.

For irrational α, the contrast function $\Phi(\mathbf{x}, \mathbf{m})$ is defined as the corresponding limit transition in the latter relation.

2.2.2. Translation and scale equivariant contrast functions: the independent changes of scale

In order to guarantee the translation and scale equivariancy of a 'typical' representative under arbitrary independent scale changes, the class of admissible functions of contrast must become more narrow.

THEOREM 2.2.2. *Let a solution of a score system satisfy the following conditions:*

- *the translation equivariancy* $\mathbf{m}(\mathbf{x} + \boldsymbol{\lambda}) = \boldsymbol{\lambda} + \mathbf{m}(\mathbf{x})$;
- *the scale equivariancy under independent changes of scale for all coordinates* $\mathbf{m}(\mathcal{M}\mathbf{x}) = \mathcal{M}\mathbf{m}(\mathbf{x})$, *where* $\mathcal{M} = \{\mu_i\}$ *is a diagonal* $M \times M$ *matrix;*
- *the uniqueness of the solution for all collections of the data* $\mathbf{x}_1, \dots, \mathbf{x}_n$.

If there exist the second derivatives $\partial^2\rho/\partial m^s \partial x^l$ *and* $\partial^2\rho/\partial m^s \partial m^l$, $s, l = 1, \dots, M$, *then the contrast function is of the form*

$$\rho(\mathbf{x}, \mathbf{m}) = \sum_{s=1}^{M} A_s |x^s - m^s|^{\gamma_s}, \tag{2.2.1}$$

with $\gamma_s \geq 1$ *and* $A_s \geq 0$.

2.2.3. Proofs

PROOF OF THEOREM 2.2.1. From Theorem 2.1.1 it follows that translation equivariancy means $\rho(\mathbf{u}, \mathbf{v}) = \rho(|\mathbf{u} - \mathbf{v}|)$. Choose $\mathbf{x}_i = \mathbf{x}$, $i = 1, \dots, k$, and $\mathbf{x}_i = \mathbf{y}$, $i = k + 1, k + 2, \dots, k + l$. Hence the problem of determination of the 'typical' representative is reduced to

$$\min_{\mathbf{m}} [k\rho(|\mathbf{x} - \mathbf{m}|) + l\rho(|\mathbf{y} - \mathbf{m}|)].$$

Changing the scale of all the variables by the factor $\mu > 0$ and taking into account the requirement of scale equivariancy, we see that the vector $\mathbf{m}$ must be changed by exactly the same factor. Therefore, we can rewrite the above minimization problem as

$$\min_{\mathbf{m}}[k\rho(\mu|\mathbf{x}-\mathbf{m}|)+l\rho(\mu|\mathbf{y}-\mathbf{m}|)].$$

Set $\mu(|\mathbf{x}-\mathbf{m}|) = \mathbf{u}_1$ and $\mu(|\mathbf{y}-\mathbf{m}|) = \mathbf{u}_2$. Then the necessary condition of minimum is given by

$$k\frac{\partial\rho(\mathbf{u}_1)}{\partial u_1^s}\frac{\partial u_1^s}{\partial m^s}+l\frac{\partial\rho(\mathbf{u}_2)}{\partial u_2^s}\frac{\partial u_2^s}{\partial m^s}=0, \qquad s=1,\dots,M,$$

or

$$-k\frac{\partial\rho(\mathbf{u}_1)}{\partial u_1^s}=l\frac{\partial\rho(\mathbf{u}_2)}{\partial u_2^s}, \qquad s=1,\dots,M. \tag{2.2.2}$$

As $\mathbf{m}$ is independent of μ, by differentiating the above terms with respect to μ we obtain

$$-k\sum_l\frac{\partial^2\rho(\mathbf{u}_1)}{\partial u_1^l\partial u_1^s}u_1^l=l\frac{\partial^2\rho(\mathbf{u}_2)}{\partial u_2^l\partial u_2^s}u_2^l, \qquad s=1,\dots,M. \tag{2.2.3}$$

Dividing (2.2.3) by (2.2.2), we obtain

$$\sum_l\frac{\partial^2\rho(\mathbf{u}_1)}{\partial u_1^l\partial u_1^s}u_1^l\Bigg/\frac{\partial\rho(\mathbf{u}_1)}{\partial u_1^s}=\frac{\partial^2\rho(\mathbf{u}_2)}{\partial u_2^l\partial u_2^s}u_2^l\Bigg/\frac{\partial\rho(\mathbf{u}_2)}{\partial u_2^s}=\gamma_s, \qquad s=1,\dots,M,$$

or

$$\sum_l\frac{\partial^2\rho(\mathbf{u})}{\partial u^l\partial u^s}u^l=\gamma_s\frac{\partial\rho(\mathbf{u})}{\partial u^s}, \qquad s=1,\dots,M. \tag{2.2.4}$$

Set

$$\frac{\partial\rho(\mathbf{u})}{\partial u^s}=\varphi_s, \qquad s=1,\dots,M. \tag{2.2.5}$$

Then equations (2.2.4) take the form

$$\sum_l u^l\frac{\partial\varphi_s(\mathbf{u})}{\partial u^l}=\gamma_s\varphi_s, \qquad s=1,\dots,M. \tag{2.2.6}$$

The solution of this partial differential equation is the homogeneous function of order γ_s

$$\varphi_s=(u^s)^{\gamma_s}A_s\left(\mathbf{u}/u^s\right), \tag{2.2.7}$$

where $A_s(\mathbf{u}/u^s)$ is an arbitrary function of argument u^i/u^s.

With the use of the equality of the mixed partial derivatives

$$\frac{\partial^2\rho(\mathbf{u})}{\partial u^l \partial u^s} = \frac{\partial^2\rho(\mathbf{u})}{\partial u^s \partial u^l}, \qquad l,s = 1,\dots,M,$$

we arrive at

$$\frac{\partial \varphi_s(\mathbf{u})}{\partial u^l} = \frac{\partial \varphi_l(\mathbf{u})}{\partial u^s}$$

or by (2.2.6),

$$(u^s)^{\gamma_s-1}\frac{\partial A_s(\mathbf{u}/u^s)}{\partial u_l} = (u^l)^{\gamma_l-1}\frac{\partial A_l(\mathbf{u}/u^l)}{\partial u_s}, \qquad l,s = 1,\dots,M. \tag{2.2.8}$$

Two cases are possible when these equations hold:

$$\frac{\partial A_s(\mathbf{u}/u^s)}{\partial u_l} \neq 0, \quad \text{or} \quad \frac{\partial A_s(\mathbf{u}/u^s)}{\partial u_l} = 0, \quad l,s = 1,\dots,M.$$

Let us separate all variables into subgroups in such a way that, for the variables belonging to each subgroup, their mixed partial derivatives are non-zero, and for the variables belonging to different groups, the mixed partial derivatives are zero. We denote these subgroups as $I_1, I_2, \dots, I_L$.

First, consider the particular case where all variables belong to one and the same group. Hence, as the partial derivatives in (2.2.8) are homogeneous functions of order zero, the relation $\gamma_s = \gamma_l = \gamma$ must hold.

Now we show that the solution of system (2.2.4) with $\gamma_i = \gamma$ is a homogeneous function of order $\gamma + 1$. Indeed, we integrate the sth equation of system (2.2.4) from 0 to u^s with respect to u^s. Integrating the sth term by parts, we obtain

$$\sum_1^M u^l \frac{\partial\rho(u^1,\dots,u^M)}{\partial u^l} - (\gamma+1)\rho(u^1,\dots,u^M)$$
$$= \sum_{l\neq s} u^l \frac{\partial\rho(u^1,\dots,u^{s-1},0,u^{s+1},\dots,u^M)}{\partial u^l}$$
$$- (\gamma+1)\rho(u^1,\dots,u^{s-1},0,u^{s+1},\dots,u^M), \quad s = 1,\dots,M.$$

As the left-hand sides of these equations are the same and the right-hand sides do not depend on u^s, we obtain

$$\sum_{l\neq s} u^l \frac{\partial\rho(u^1,\dots,u^{s-1},0,u^{s+1},\dots,u^M)}{\partial u^l}$$
$$- (\gamma+1)\rho(u^1,\dots,u^{s-1},0,u^{s+1},\dots,u^M) = d = \text{const}$$

and

$$\sum_{1}^{M} u^l \frac{\partial \rho(u^1, \dots, u^M)}{\partial u^l} - (\gamma + 1)\rho(u^1, \dots, u^M) = d.$$

From the latter equation by $\rho(0, 0, \dots, 0) = 0$, we have $d = 0$ and, hence,

$$\sum_{l=1}^{M} u^l \frac{\partial \rho(u^1, \dots, u^M)}{\partial u^l} - (\gamma + 1)\rho(u^1, \dots, u^M) = 0.$$

This partial differential equation defines the contrast function as a homogeneous function of order $\gamma + 1$.

Thus, if the condition $\partial A_s(\mathbf{u}/u^s)/\partial u_l \neq 0$ holds for all s and l, then the contrast function $\rho(u^1, \dots, u^M)$ is a homogeneous function of order $(\gamma + 1)$.

To provide uniqueness, we require the concavity of this function for $\mathbf{u} > \mathbf{0}$.

In the general case, we separate all variables into the subgroups in such a way that, for the variables belonging to each subgroup, their mixed partial derivatives are non-zero, and for the variables belonging to different groups, the mixed partial derivatives are zero. Let these subgroups be $I_1, I_2, \dots, I_L$.

From (2.2.8) we obtain

$$\gamma_s = \gamma_l = \alpha_k, \quad \text{for} \quad s, l \in I_k.$$

Therefore, the order of homogeneity of the functions $\varphi_s = \partial\rho(\mathbf{u})/\partial u^s$ for the variables belonging to one subgroup is one and the same, and it is equal to α_k.

Taking this into account, repeating the above reasoning word for word, we easily obtain that, in the general case, the contrast function is

$$\rho(\mathbf{u}) = \sum_{k=1}^{L} \Phi_k(\mathbf{u}^{(k)}),$$

where $\Phi_k(\mathbf{u}^{(k)})$ are homogeneous functions of order $(\alpha_k + 1)$ of the variables belonging to the subgroup I_k. Here $\Phi_k(\mathbf{u}^{(k)})$ are concave functions for $\mathbf{u}^{(k)} > \mathbf{0}$ and $\Phi_k(\mathbf{0}) = 0$.

Returning to the variables $\mathbf{x}$ and $\mathbf{m}$ and replacing $(\alpha_k + 1)$ by α_k, we complete the proof. □

PROOF OF THEOREM 2.2.2. By Theorem 2.1.2, translation equivariancy means that the contrast function is

$$\rho(\mathbf{u}, \mathbf{v}) = \rho(|\mathbf{u} - \mathbf{v}|).$$

Using the arbitrariness of the data choice, we take

$$\mathbf{x}_i = \mathbf{x}, \quad i = 1, \dots, k, \quad \text{and} \quad \mathbf{x}_i = \mathbf{y}, \quad i = k + 1, \dots, k + l.$$

In this case, the problem of determination of the 'typical' representative is reduced to the minimization problem

$$\min_{\mathbf{m}}[k\rho(|\mathbf{x}-\mathbf{m}|)+l\rho(|\mathbf{y}-\mathbf{m}|)].$$

Changing the scale of the ith coordinate by the factor μ_i and using the condition of scale equivariancy according to which the ith coordinate of the vector $\mathbf{m}$ should be also changed by the factor μ_i, we can rewrite the above minimization problem as follows:

$$\min_{\mathbf{m}}[k\rho(|\boldsymbol{\mu}(\mathbf{x}-\mathbf{m})|)+l\rho(|\boldsymbol{\mu}(\mathbf{y}-\mathbf{m})|)].$$

Set $|\boldsymbol{\mu}(\mathbf{x}-\mathbf{m})| = \mathbf{u}_1$ and $|\boldsymbol{\mu}(\mathbf{y}-\mathbf{m})| = \mathbf{u}_2$. Then the necessary condition of minimum is given by

$$k\frac{\partial\rho(\mathbf{u}_1)}{\partial u_1^s}\frac{\partial u_1^s}{\partial m^s}+l\frac{\partial\rho(\mathbf{u}_2)}{\partial u_2^s}\frac{\partial u_2^s}{\partial m^s}=0, \qquad s=1,\dots,M,$$

or

$$-k\frac{\partial\rho(\mathbf{u}_1)}{\partial u_1^s}=l\frac{\partial\rho(\mathbf{u}_2)}{\partial u_2^s}, \qquad s=1,\dots,M. \tag{2.2.9}$$

Since $\mathbf{m}$ is independent of μ_l, after differentiating (2.2.9) with respect to μ_l we arrive at

$$-k\frac{\partial^2\rho(\mathbf{u}_1)}{\partial u_1^l\partial u_1^s}u_1^l=l\frac{\partial^2\rho(\mathbf{u}_2)}{\partial u_2^l\partial u_2^s}u_2^l \qquad s,l=1,\dots,M. \tag{2.2.10}$$

Dividing (2.2.10) by the sth equation of (2.2.9), we obtain

$$\frac{\partial^2\rho(\mathbf{u}_1)}{\partial u_1^l\partial u_1^s}u_1^l\bigg/\frac{\partial\rho(\mathbf{u}_1)}{\partial u_1^s}=\frac{\partial^2\rho(\mathbf{u}_2)}{\partial u_2^l\partial u_2^s}u_2^l\bigg/\frac{\partial\rho(\mathbf{u}_2)}{\partial u_2^s}=\gamma_{sl}, \qquad s,l=1,\dots,M,$$

or

$$\frac{\partial^2\rho(\mathbf{u})}{\partial u^l\partial u^s}u^l=\gamma_{sl}\frac{\partial\rho(\mathbf{u})}{\partial u^s}, \qquad s,l=1,\dots,M.$$

Setting

$$\frac{\partial\rho(\mathbf{u})}{\partial u^s}=\varphi_s, \tag{2.2.11}$$

we obtain

$$\frac{\partial\varphi_s(\mathbf{u})}{\partial u^l}=\gamma_{sl}\varphi_s(\mathbf{u}) \qquad s,l=1,\dots,M. \tag{2.2.12}$$

The solution of equation (2.2.12) for $s = 1, l = 1$

$$\frac{\partial \varphi_1(\mathbf{u})}{\partial u^l} = \gamma_{1l}\varphi_1(\mathbf{u})$$

is determined by

$$\varphi_1(\mathbf{u}) = C_1(u^2, \dots, u^M)(u^1)^{\gamma_{11}},$$

where $C_1(u^2, \dots, u^M)$ are arbitrary functions.

Substituting this expression into (2.2.12) with $l = 1$, we obtain

$$\frac{\partial C_1(u^2, \dots, u^M)}{\partial u^s}\left(u^1\right)^{\gamma_{11}+1} = \gamma_{s1}C_1(u^2, \dots, u^M)\left(u^1\right)^{\gamma_{11}}, \qquad s = 2, \dots, M.$$

Further,

$$\frac{\partial C_1(u^2, \dots, u^M)}{\partial u^s} = 0, \quad \gamma_{s1} = 0, \quad s = 2, \dots, M.$$

Therefore, $C_1(u^2, \dots, u^M) = C_1 = \text{const}$, and thus, $\varphi_1 = C_1\left(u^1\right)^{\gamma_{11}}$. Integrating with respect to u_1, we arrive at

$$\rho(\mathbf{u}) = A_1\left(u^1\right)^{\gamma_1} + \rho(0, u^2, \dots, u^M), \tag{2.2.13}$$

where $A_1 = C_1/(\gamma 11 + 1), \gamma_1 = \gamma_{11} + 1$. By substituting (2.2.13) into (2.2.12) and repeating the above procedure, we obtain

$$\rho(\mathbf{u}) = \sum_{s=1}^{M} A_s\left(u^s\right)^{\gamma_s} + \rho(0, 0, \dots, 0).$$

By $\rho(0, 0, \dots, 0) = 0$, $u^s = |x^s - m^s|$, and by concavity, we get $\gamma_s \geq 1$, and the required assertion

$$\rho(\mathbf{x}, m) = \sum_{s=1}^{M} A_s|x^s - m^s|^{\gamma_s}, \qquad \gamma_s \geq 1.$$

□

2.3. Orthogonally equivariant contrast functions

There are sufficiently many problems of data processing, in particular, with the geodesic data, when it is necessary to provide the equivariancy of the 'typical' representative under rotation of coordinate axes in which the measurements are made. Indeed, let the location of some object be determined by measuring its coordinates relative to two orthogonal axes defined by their directions at given bench-marks. It is natural to require that, with the rotation of axes by some angle (the choice of other field bench-marks), the coordinates of a 'typical' representative in new axes must be rotated by the same angle. Obviously, such a requirement imposes sufficiently strict restrictions on the structure of a contrast function. The results below show that

- if the measurements are made in a plane, then the contrast function depends only on two measurement characteristics:
  - the Euclidean distance between the measured object and the 'typical' representative;
  - the angle determining the direction at the 'typical' representative from the measured object;
- if the measured object is of dimension greater than two, then the contrast function depends only on the Euclidean distance between the measured object and the 'typical' representative.

2.3.1. Translation and orthogonally equivariant contrast functions: the bivariate data

THEOREM 2.3.1. *Let the data* $X_i = (x_i, y_i)^T$, $i = 1, \dots, n$, *and the solution* $(m(x,y), n(x,y))^T$ *of the score system satisfy the following conditions:*

- *the condition of translation equivariancy*

$$\begin{pmatrix} m(x+\lambda_1, y+\lambda_2) \\ n(x+\lambda_1, y+\lambda_2) \end{pmatrix} = \begin{pmatrix} \lambda_1 \\ \lambda_2 \end{pmatrix} + \begin{pmatrix} m(x,y) \\ n(x,y) \end{pmatrix};$$

- *the condition of equivariancy under rotation of coordinate axes*

$$\begin{pmatrix} \cos\phi & \sin\phi \\ -\sin\phi & \cos\phi \end{pmatrix} \begin{pmatrix} m(x,y) \\ n(x,y) \end{pmatrix} = \begin{pmatrix} m(x\cos\phi + y\sin\phi, -x\sin\phi + y\cos\phi) \\ n(x\cos\phi + y\sin\phi, -x\sin\phi + y\cos\phi) \end{pmatrix}.$$

Then the contrast function is of the form

$$\rho(x,y;m,n) = F(\sqrt{(x-m)^2 + (y-n)^2}) \exp\left(\alpha \arctan \frac{x-m}{y-n}\right),$$

where $F(u)$ *is a twice differentiable function,* $F(0) = 0$, *and* $\partial^2 F(u)/\partial u^2 > 0$.

PROOF. The sufficiency of the assertion is obvious. We prove the necessity. By Theorem 2.1.1, translation equivariancy means that the contrast function is $\rho(x,y;m,n) = \rho(x-m, y-n)$. Choosing the data, we take

$$(x_i, y_i)^T = (x_1, y_1)^T, \quad i = 1, \dots, k; \quad (x_i, y_i)^T = (x_2, y_2)^T, \quad i = k+1, \dots, k+l.$$

In this case, the problem of determination of the 'typical' representative takes the form of the minimization problem

$$\min_{m,n} [k\rho(x_1 - m, y_1 - n) + l\rho(x_2 - m, y_2 - n)].$$

After rotating the coordinate axes by the angle ϕ, the coordinates of the 'typical' representative must be rotated by the same angle. Taking this into account, we rewrite the minimization problem as

$$\min_{m,n} \big[k\rho((x_1 - m)\cos\phi + (y_1 - n)\sin\phi, -(x_1 - m)\cos\phi + (y_1 - n)\sin\phi)$$
$$+ l\rho((x_2 - m)\cos\phi + (y_2 - n)\sin\phi, -(x_2 - m)\cos\phi + (y_2 - n)\sin\phi)\big].$$

For the bivariate case $i = 1, 2$, set

$$(x_i - m)\cos\phi + (y_i - n)\sin\phi = u_i, \qquad -(x_i - m)\sin\phi + (y_i - m)\cos\phi = v_i.$$

Equating the derivatives of the minimized function with respect to m and n to zero, we obtain

$$k\frac{\partial\rho(u_1,v_1)}{\partial u_1}\frac{\partial u_1}{\partial m} + k\frac{\partial\rho(u_1,v_1)}{\partial v_1}\frac{\partial v_1}{\partial m} + l\frac{\partial\rho(u_2,v_2)}{\partial u_2}\frac{\partial u_2}{\partial m} + l\frac{\partial\rho(u_2,v_2)}{\partial v_2}\frac{\partial v_2}{\partial m} = 0,$$
$$k\frac{\partial\rho(u_1,v_1)}{\partial u_1}\frac{\partial u_1}{\partial n} + k\frac{\partial\rho(u_1,v_1)}{\partial v_1}\frac{\partial v_1}{\partial n} + l\frac{\partial\rho(u_2,v_2)}{\partial u_2}\frac{\partial u_2}{\partial n} + l\frac{\partial\rho(u_2,v_2)}{\partial v_2}\frac{\partial v_2}{\partial n} = 0.$$

or

$$-k\frac{\partial\rho(u_1,v_1)}{\partial u_1}\cos\phi + k\frac{\partial\rho(u_1,v_1)}{\partial v_1}\sin\phi = l\frac{\partial\rho(u_2,v_2)}{\partial u_2}\cos\phi - l\frac{\partial\rho(u_2,v_2)}{\partial v_2}\sin\phi, \tag{2.3.1}$$

$$-k\frac{\partial\rho(u_1,v_1)}{\partial u_1}\sin\phi - k\frac{\partial\rho(u_1,v_1)}{\partial v_1}\cos\phi = l\frac{\partial\rho(u_2,v_2)}{\partial u_2}\sin\phi + l\frac{\partial\rho(u_2,v_2)}{\partial v_2}\cos\phi. \tag{2.3.2}$$

As the variables x, y, m, n do not depend on ϕ, we differentiate this system of equations in ϕ

$$-k\left[A(u_1,v_1)\cos\phi - B(u_1,v_1)\sin\phi + \frac{\partial\rho(u_1,v_1)}{\partial u_1}\sin\phi + \frac{\partial\rho(u_1,v_1)}{\partial v_1}\cos\phi\right]$$
$$= l\left[A(u_2,v_2)\cos\phi - B(u_2,v_2)\sin\phi + \frac{\partial\rho(u_2,v_2)}{\partial u_2}\sin\phi + \frac{\partial\rho(u_2,v_2)}{\partial v_2}\cos\phi\right], \tag{2.3.3}$$

$$-k\left[A(u_1,v_1)\sin\phi + B(u_1,v_1)\cos\phi + \frac{\partial\rho(u_1,v_1)}{\partial u_1}\cos\phi - \frac{\partial\rho(u_1,v_1)}{\partial v_1}\sin\phi\right]$$
$$= l\left[A(u_2,v_2)\sin\phi + B(u_2,v_2)\cos\phi + \frac{\partial\rho(u_2,v_2)}{\partial u_2}\cos\phi - \frac{\partial\rho(u_2,v_2)}{\partial v_2}\sin\phi\right]. \tag{2.3.4}$$

Here we have set

$$A(u,v) = \frac{\partial^2\rho}{\partial u\partial u}(-u\sin\phi + v\cos\phi) + \frac{\partial^2\rho}{\partial v\partial u}(-u\cos\phi - v\sin\phi)$$
$$\equiv \frac{\partial^2\rho}{\partial u\partial u}v - \frac{\partial^2\rho}{\partial v\partial u}u, \tag{2.3.5}$$
$$B(u,v) = \frac{\partial^2\rho}{\partial u\partial v}(-u\sin\phi + v\cos\phi) + \frac{\partial^2\rho}{\partial v\partial v}(-u\cos\phi - v\sin\phi)$$
$$\equiv \frac{\partial^2\rho}{\partial u\partial v}v - \frac{\partial^2\rho}{\partial v\partial v}u. \tag{2.3.6}$$

Dividing (2.3.3) by (2.3.1), (2.3.4) by (2.3.2), and taking into account that the right-hand side of this relation is a function of u_1, v_1 only, and the left-hand side is a function of u_2, v_2, we obtain

$$\begin{pmatrix}\cos\phi & -\sin\phi\\ \sin\phi & \cos\phi\end{pmatrix}\begin{pmatrix}A\\ B\end{pmatrix} = \begin{pmatrix}\sin\phi & \cos\phi\\ \cos\phi & -\sin\phi\end{pmatrix}\begin{pmatrix}\partial\rho/\partial u\\ -\partial\rho/\partial v\end{pmatrix}$$
$$+\begin{pmatrix}\alpha & 0\\ 0 & \beta\end{pmatrix}\begin{pmatrix}\cos\phi & -\sin\phi\\ \sin\phi & \cos\phi\end{pmatrix}\begin{pmatrix}\partial\rho/\partial u\\ \partial\rho/\partial v\end{pmatrix}.$$

Solving this, for A and B we obtain

$$\begin{pmatrix}A\\ B\end{pmatrix} = \begin{pmatrix}-\partial\rho/\partial v\\ \partial\rho/\partial u\end{pmatrix} + \begin{pmatrix}\alpha\cos^2\phi + \beta\sin^2\phi & (-\alpha+\beta)\sin\phi\cos\phi\\ (-\alpha+\beta)\sin\phi\cos\phi & \alpha\cos^2\phi + \beta\sin^2\phi\end{pmatrix}\begin{pmatrix}\partial\rho/\partial u\\ \partial\rho/\partial v\end{pmatrix}.$$

Obviously, the solution of this set does not depend on ϕ only if $\alpha = \beta$, and in this case, by (2.3.5) and (2.3.6), it takes the form

$$\frac{\partial^2\rho}{\partial u\partial u}v - \frac{\partial^2\rho}{\partial v\partial u}u = -\frac{\partial\rho}{\partial v} + \alpha\frac{\partial\rho}{\partial u},$$
$$\frac{\partial^2\rho}{\partial u\partial v}v - \frac{\partial^2\rho}{\partial v\partial v}u = \frac{\partial\rho}{\partial u} + \alpha\frac{\partial\rho}{\partial v}.$$

Set $\partial\rho/\partial u = \varphi_1$, $\partial\rho/\partial v = \varphi_2$. Hence the above system is transformed to

$$\frac{\partial\varphi_1}{\partial u}v - \frac{\partial\varphi_1}{\partial v}u = -\varphi_2 + \alpha\varphi_1,$$
$$\frac{\partial\varphi_2}{\partial u}v - \frac{\partial\varphi_2}{\partial v}u = +\varphi_1 + \alpha\varphi_2.$$

Using the polar coordinates $u = r\cos t$, $v = r\sin t$, we obtain

$$\frac{\partial\varphi_1}{\partial t} = -\varphi_2 + \alpha\varphi_1,$$
$$\frac{\partial\varphi_2}{\partial t} = +\varphi_1 + \alpha\varphi_2.$$

The solution of these simultaneous differential equations is of the form

$$\varphi_1 \equiv \frac{\partial \rho}{\partial u} = C_1(r)\exp(\alpha t)\cos t + C_2(r)\exp(\alpha t)\sin t,$$

$$\varphi_2 \equiv \frac{\partial \rho}{\partial v} = C_1(r)\exp(\alpha t)\sin t - C_2(r)\exp(\alpha t)\cos t.$$

By the relations

$$\frac{\partial \rho}{\partial r} = \frac{\partial \rho}{\partial u}\cos t + \frac{\partial \rho}{\partial v}\sin t,$$

$$\frac{\partial \rho}{\partial t} = -r\frac{\partial \rho}{\partial u}\sin t + r\frac{\partial \rho}{\partial v}\cos t,$$

we arrive at

$$\frac{\partial \rho}{\partial r} = C_1(r)\exp(\alpha t), \qquad \frac{\partial \rho}{\partial t} = -C_2(r)\exp(\alpha t).$$

Solving these simple equations and reverting to the variables u, v along with the condition $\rho(0) = 0$, we obtain the required assertion, namely

$$\rho(u, v) = F(\sqrt{u^2 + v^2})\exp\left(\alpha \arctan \frac{u}{v}\right)$$

or

$$\rho(x, y; m, n) = F(\sqrt{(x-m)^2 + (y-n)^2})\exp\left(\alpha \arctan \frac{x-m}{y-n}\right).$$

□

2.3.2. Translation and orthogonally equivariant contrast functions: the multivariate data

THEOREM 2.3.2. *Let the data* $\mathbf{x} = (x_1, \dots, x_M)$ *and and the solution of the score system* $\mathbf{m}(\mathbf{x}) = (m_1(\mathbf{x}), \dots, m_M(\mathbf{x}))$ *satisfy the following conditions:*

- *the condition of translation equivariancy* $\mathbf{m}(\mathbf{x} + \boldsymbol{\Lambda}) = \mathbf{m}(\mathbf{x}) + \boldsymbol{\Lambda}$, *where* $\boldsymbol{\Lambda} = (\lambda_1, \dots, \lambda_M)$*;*

- *the condition of equivariancy under rotation of coordinate axes* $\mathbf{m}(\mathbf{T}\mathbf{x}) = \mathbf{T}\mathbf{m}(\mathbf{x})$, *where* $\mathbf{T}$ *is an orthogonal* $M \times M$ *matrix.*

Then the contrast function is of the following structure:

$$\rho(\mathbf{x}; \mathbf{m}) = \Psi\left(\sqrt{\sum_s (x_s - m_s)^2}\right),$$

where $\Psi(u)$ *is a twice differentiable function satisfying*

$$\Psi(0) = 0, \qquad \frac{\partial \Psi(u)/\partial u}{u} > 0, \qquad \frac{\partial^2 \Psi(u)}{\partial u^2} > 0.$$

PROOF. The condition of translation equivariancy determines the contrast function in the form $\rho(\mathbf{x},\mathbf{m}) = \rho(\mathbf{x}-\mathbf{m})$. Observe that an orthogonal matrix can be represented as the product of matrices of rotation around coordinate axes.

Consider the case $M = 3$. Setting obvious notation, we can write for the contrast function $\rho(x,y,z) \equiv \rho(x_1 - m_1, x_2 - m_2, x_3 - m_3)$. Make a rotation around the axis z. By Theorem 2.3.1, the equivariancy under rotation means that the contrast function is

$$\rho(x,y,z) = F\left(\sqrt{x^2+y^2}, z\right) \exp\left(\alpha \arctan \frac{y}{x}\right),$$

where $F(u,v)$ and α are an arbitrary function and an arbitrary parameter satisfying only the requirement of concavity for the contrast function.

Making a rotation around the axis x, we have the contrast function in the form

$$\rho(x,y,z) = G\left(x, \sqrt{y^2+z^2}\right) \exp\left(\beta \arctan \frac{z}{y}\right).$$

Obviously, these expressions must coincide. Hence let us consider the functional equation

$$F\left(\sqrt{x^2+y^2}, z\right) \exp\left(\alpha \arctan \frac{y}{x}\right) = G\left(x, \sqrt{y^2+z^2}\right) \exp\left(\beta \arctan \frac{z}{y}\right). \tag{2.3.7}$$

For $y = 0$, we have (for simplicity of calculations, set $x, z \geq 0$)

$$F(x,z) = G(x,z) \exp\left(\beta \frac{\pi}{2}\right). \tag{2.3.8}$$

From (2.3.8) it follows that (2.3.7) takes the form

$$F\left(\sqrt{x^2+y^2}, z\right) \exp\left(\alpha \arctan \frac{y}{x}\right) = F\left(x, \sqrt{y^2+z^2}\right) \exp\left(\beta \left(\arctan \frac{z}{y} - \frac{\pi}{2}\right)\right). \tag{2.3.9}$$

Setting $z = 0$ in (2.3.9), we obtain

$$F\left(\sqrt{x^2+y^2}, 0\right) \exp\left(\alpha \arctan \frac{y}{x}\right) = F(x,y) \exp\left(-\beta \frac{\pi}{2}\right)$$

Furthermore,

$$F\left(\sqrt{x^2+y^2}, z\right) = F\left(\sqrt{x^2+y^2+z^2}, 0\right) \exp\left(\alpha \arctan \frac{z}{\sqrt{x^2+y^2}} + \beta \frac{\pi}{2}\right), \tag{2.3.10}$$

$$F\left(x, \sqrt{y^2+z^2}\right) = F\left(\sqrt{(x^2+y^2+z^2}, 0\right) \exp\left(\alpha \arctan \frac{\sqrt{y^2+z^2}}{x} + \beta \frac{\pi}{2}\right).$$

By substituting the latter relations into (2.3.9), excluding the factor $F(\sqrt{x^2+y^2+z^2},0)$, and taking the logarithm, we arrive at

$$\alpha \arctan \frac{z}{\sqrt{x^2+y^2}} + \beta\frac{\pi}{2} + \alpha \arctan \frac{y}{x} = \alpha \arctan \frac{\sqrt{y^2+z^2}}{x} + \beta \arctan \frac{z}{y}.$$

By substituting $x = 0$ into this relation we have $\beta = 0$ and $\alpha = 0$. Hence (2.3.10) takes the form

$$F\left(\sqrt{x^2+y^2}, z\right) = F\left(\sqrt{x^2+y^2+z^2}, 0\right) \equiv \Psi\left(\sqrt{x^2+y^2+z^2}\right).$$

Therefore,

$$\rho(x,y,z) = \Psi\left(\sqrt{x^2+y^2+z^2}\right),$$

i.e., for $M = 3$, the required assertion is true.

In the general case, the proof can be easily made by induction.

Now we show that if $F(u)$ is concave, twice differentiable, and its minimum is attained at $u = 0$, then

$$G(\mathbf{x}) = F\left(\sqrt{x_1^2+x_2^2+\ldots+x_m^2}\right)$$

is also concave.

Indeed,

$$\frac{\partial^2 G}{\partial {x_i}^2} = F''(v)\frac{x_i^2}{v^2} + F'(v)\frac{v^2-x_i^2}{v^{3/2}}, \qquad i = 1,\ldots,M,$$

$$\frac{\partial^2 G}{\partial x_i \partial x_j} = F''(v)\frac{x_i x_j}{v^2} - F'(v)\frac{x_i x_j}{v^{3/2}}, \qquad i \neq j.$$

Here we put $v^2 = \sum_s x_s^2$.

Consider the quadratic form

$$I = \sum_i \sum_j \frac{\partial^2 G}{\partial x_i \partial x_j} y_i y_j.$$

By substituting the partial derivatives into the above formula we have

$$I = F'(v)v^{-1}\sum_i y_i^2 + (F''(v)v^{-2} - F'(v)v^{-3})\left(\sum_i x_i y_i\right)^2.$$

As the minimal eigenvalue of a quadratic form is the solution of the minimization problem

$$\lambda_{\min} = \min_{y_1,\ldots,y_m} I \quad \text{over} \quad \sum_i y_i^2 = 1,$$

after obvious calculations we obtain

$$\lambda_{\min} = \min\left[F''(v), \frac{F'(v)}{v}\right].$$

The condition of concavity of the function $G(\mathbf{x})$ is the condition of positive definiteness of the matrix of second partial derivatives, i.e., the requirement $\lambda_{\min} \geq 0$. By the latter relation,

$$F''(v) > 0; \qquad \frac{F'(v)}{v} > 0. \tag{2.3.11}$$

It is easy to see that these conditions define a concave function decreasing for $v < 0$ and increasing for $v > 0$, i.e., attaining the minimal value at $v = 0$. □

2.4. Monotonically equivariant contrast functions

The strongest requirement on the class of admissible contrast functions is given by the requirement of equivariancy of the 'typical' representative under an arbitrary monotone transformation of the initial data.

Formally this means that if the 'typical' representative $m(x_1, \dots, x_n)$ corresponds to the initial data $x_1, \dots, x_n$, then the 'typical' representative transformed by a monotone function $f(x)$ corresponds to the transformed data $\{y_i = f(x_i)\}$, $i = 1, \dots, n$:

$$m(y_1, \dots, y_n) = m(f(x_1), \dots, f(x_n)) = f(m(x_1, \dots, x_n)). \tag{2.4.1}$$

Here the solution is obvious: any order statistic $x_{(i)}$ $i = 1, \dots, n$ satisfies condition (2.4.1), in particular, the sample median for odd sample sizes. For the even sample sizes $n = 2k$, the equivariant 'typical' representative is given by the kth or $(k+1)$th order statistics.

Taking these considerations into account, it is easy to see that the contrast function providing the equivariancy under arbitrary monotone transformations is given by the sum of absolute deviations

$$\rho(\mathbf{x}, \mathbf{m}) = \sum_{i=1}^{M} |x_i - m_i| \tag{2.4.2}$$

with the additional condition that, for even $n = 2k$, the solution of the minimization problem with the contrast function

$$\min_{\mathbf{m}} \sum_{j=1}^{n} \rho(\mathbf{x}_j, \mathbf{m}) = \min_{\mathbf{m}} \sum_{j=1}^{n} \sum_{i=1}^{M} |x_{ij} - m_i| \tag{2.4.3}$$

is the vector $\mathbf{m}_* = (x_{1(s)}, \dots, x_{M(s)})$, where $s = k$ or $k+1$.

2.5. Minimal sensitivity to small perturbations in the data

One of the natural requirements on the choice of the 'typical' representative is the requirement of the minimal sensitivity of this parameter to small perturbations in the initial data. Such perturbations may be caused, for instance, either by rounding off the observations in accordance with the scale of a measuring device, or by small measurement errors.

Now we characterize the sensitivity of an estimator $m = m(x_1, \dots, x_n)$ based on the data $x_1, \dots, x_n$ to the perturbations of these data by the quadratic criterion

$$I = \sum_{i=1}^{n} \left(\frac{\partial m_n}{\partial x_i} \right)^2 \tag{2.5.1}$$

and choose the structure of the contrast function $\rho(x, m)$ and of the corresponding score function $\phi(x, m) = \partial \rho(x, m)/\partial m$ in such a way that (2.5.1) is minimal.

The solution of this problem is based on the following simple result.

LEMMA 2.5.1. *The minimum of the function*

$$I(a_1, \dots, a_n) = \sum_{i=1}^{n} a_i^2 \Bigg/ \left(\sum_{i=1}^{n} a_i \right)^2 \tag{2.5.2}$$

is attained at $a_1 = \dots = a_n = C = \text{const}$.

PROOF. Setting $y_i = a_i / \sum_1^n a_k$, we arrive at the minimization problem

$$\text{minimize} \quad \sum_{i=1}^{n} y_i^2 \quad \text{under the condition} \quad \sum_{i=1}^{n} y_i = 1.$$

The solution of this simplest problem of conditional minimization is given by $y_1 = \dots = y_n = 1/n$, and this is equivalent to the assertion of the lemma: $a_1 = \dots = a_n = C$, where C is an arbitrary constant. □

Now we prove the theorem that determines the form of the contrast function providing the minimal sensitivity of a 'typical' representative of the data to their small perturbations.

THEOREM 2.5.1. *Let the requirement of translation equivariancy of a 'typical' representative hold.*

Then the minimum of criterion (2.5.1) *is attained at the score functions* $\phi(u) = Cu$ *with the corresponding contrast functions* $\rho(u) = Cu^2$, *i.e., for the estimators of the LS method.*

PROOF. From the requirement of equivariancy (Theorem 2.1.1) it follows that the value of a 'typical' representative is determined by

$$\sum_{i=1}^{n} \varphi(x_i - m) = 0.$$

Differentiating this equation with respect to x_k, $k = 1, \dots, n$,, we obtain

$$-\sum_{i=1}^{n} \varphi'(x_i - m)\frac{\partial m}{\partial x_k} + \varphi'(x_k - m) = 0, \qquad k = 1, \dots, n.$$

Furthermore,

$$\frac{\partial m}{\partial x_k} = \frac{\varphi'(x_k - m)}{\sum_{i=1}^{n} \varphi(x_i - m)}.$$

Hence the criterion of minimal sensitivity takes the form

$$I(\varphi) = \sum_i (\varphi'(x_i - m))^2 \bigg/ \left(\sum_i \varphi(x_i - m)\right)^2.$$

Applying now the assertion of Lemma 2.5.1 to the problem of minimization of this criterion, we have $\phi'(u) = C$ and, by the condition $\phi(0) = 0$, we obtain $\phi(u) = Cu$, which completes the proof. □

Thus, if the influence of measurement errors is characterized by the most often used quadratic criterion (2.5.1), then this influence is minimal when using the least squares method for data processing, i.e., the 'typical' representative is the sample mean. This conclusion also holds true for a more general form of the criterion for estimating the measurement errors.

Now we characterize the influence of errors by the criterion

$$J = G\left(\frac{\partial m}{\partial x_1}, \dots, \frac{\partial m}{\partial x_n}\right), \tag{2.5.3}$$

where $G(u_1, \dots, u_n)$ is a symmetric function such that that $G(0, \dots, 0) = 0$.

Let also the natural requirement of the coordinated augmentation of the criterion value and the augmentation of the error value hold:

$$\frac{\partial G}{\partial u_i} \begin{cases} > 0, & u_i > 0, \\ < 0, & u_i < 0. \end{cases} \tag{2.5.4}$$

In this case, the following is true.

LEMMA 2.5.2. *Let $G(u_1, ..., u_n)$ be a symmetric function and*

$$\frac{\partial^2 G(1/n, ..., 1/n)}{\partial^2 u_1} - \frac{\partial^2 G(1/n, ..., 1/n)}{\partial u_1 \partial u_2} > 0. \tag{2.5.5}$$

Then the minimum of the function

$$J(a_1, ..., a_n) = G\left(\frac{a_1}{\sum_1^n a_i}, ..., \frac{a_n}{\sum_1^n a_i}\right) \tag{2.5.6}$$

is attained at $a_1 = \cdots = a_n = C = \text{const.}$

PROOF. Setting $y_i = a_i/\sum_1^n a_k$, we arrive at the problem of conditional minimization

$$\text{minimize} \quad G(y_1, ..., y_n) \quad \text{under the condition} \quad \sum_{i=1}^n y_i = 1.$$

Excluding the variable y_n, we rewrite the latter problem as

$$\text{minimize} \quad G\left(y_1, ..., y_{n-1}, 1 - \sum_{i=1}^{n-1} y_i\right).$$

Hence the simultaneous equations for determination of the variables sought for are of the form

$$\frac{\partial G(y_1, ..., y_{n-1}, 1 - \sum_1^{n-1} y_i)}{\partial u_k} - \frac{\partial G(y_1, ..., y_{n-1}, 1 - \sum_1^{n-1} y_i)}{\partial u_n} = 0, \ k = 1, ..., n-1.$$

It is easy to see that from (2.5.4) it follows that these equations have the unique solution

$$y_k = 1 - \sum_{i=1}^{n-1} y_i, \qquad k = 1, ..., n-1,$$

and hence we obtain $y_k = 1/n$, $k = 1, ..., n$.

Consider now the conditions of concavity for the function

$$G\left(y_1, ..., y_{n-1}, 1 - \sum_{i=1}^{n-1} y_i\right)$$

at the unique stationary point $y_k = 1/n$, $k = 1, ..., n-1$.

We regard this condition as the requirement of positive definiteness of the quadratic form

$$T(v_1, ..., v_{n-1}) = \sum_{i=1}^{n-1}\sum_{k=1}^{n-1} \frac{\partial^2 G(y_1, ..., y_{n-1}, 1 - \sum_{i=1}^{n-1} y_i)}{\partial y_i \partial y_k} v_i v_k.$$

Evaluating the second derivatives at $y_k = 1/n, k = 1, \dots, n-1$ and taking into account that the function $G(u_1, \dots, u_n)$ is symmetric, we have

$$\left.\frac{\partial^2 G(y_1, \dots, y_{n-1}, 1 - \sum_{i=1}^{n-1} y_i)}{\partial^2 y_i}\right|_{y_1=\dots=y_{n-1}=1/n} = 2\left(\frac{\partial^2 G(1/n, \dots, 1/n)}{\partial^2 u_1} - \frac{\partial^2 G(1/n, \dots, 1/n)}{\partial u_1 \partial u_2}\right).$$

Hence,

$$\left.\frac{\partial^2 G(y_1, \dots, y_{n-1}, 1 - \sum_{i=1}^{n-1} y_i)}{\partial y_i \partial y_k}\right|_{y_1=\dots=y_{n-1}=1/n} = \frac{\partial^2 G(1/n, \dots, 1/n)}{\partial^2 u_1} - \frac{\partial^2 G(1/n, \dots, 1/n)}{\partial u_1 \partial u_2}, \qquad i \neq k.$$

Further, some tedious manipulation yields the following expression for the quadratic form:

$$T(v_1, \dots, v_{n-1}) = \left(\frac{\partial^2 G(1/n, \dots, 1/n)}{\partial^2 u_1} - \frac{\partial^2 G(1/n, \dots, 1/n)}{\partial u_1 \partial u_2}\right) \times \left(\sum_{i=1}^{n-1} v_i^2 + \left(\sum_{i=1}^{n-1} v_i\right)^2\right)$$

and it is positive definite since

$$\frac{\partial^2 G(1/n, \dots, 1/n)}{\partial^2 u_1} - \frac{\partial^2 G(1/n, \dots, 1/n)}{\partial u_1 \partial u_2} > 0.$$

Returning to the variables $\{a_i\}, i = 1, \dots, n$, we obtain the required result: the minimal value of function (2.5.6) is attained at $a_1 = \dots = a_n = C = \text{const}$. □

This lemma makes it possible to generalize Theorem 2.5.1.

THEOREM 2.5.2. *Let the condition of translation equivariancy of a 'typical' representative hold.*

Then the minimum of criterion (2.5.3) *satisfying conditions* (2.5.4) *and* (2.5.5) *is attained at the score functions* $\phi(u) = Cu$, *which correspond to the contrast functions* $\rho(u) = Cu^2$ *of the LS estimators.*

We omit the proof of this theorem because it completely coincides with that of Theorem 2.5.1.

REMARK 2.5.1. The meaning of this subsection is obvious: any reasonable criterion of minimization of the influence of small perturbations in the initial data stipulates the procedure of data processing by the LS method.

Nevertheless, we have to note that in this statement the essential role is played by the supposition of exactly small perturbations. Breaking this assumption, we must use other methods of data processing, that could, generally, be determined from the requirements on the sensitivity curves.

2.6. Affine equivariant contrast functions

In the preceding sections, we considered the restrictions on the admissible structure of the contrast functions connected with the requirement of equivariancy of the 'typical representative' under elementary transformations (translation, scale, orthogonal) of the initial data. Naturally, the requirement of equivariancy under some superposition of elementary transformations makes it necessary to choose the contrast functions under the conditions of the joint structure restrictions specific for separate elementary transformations. Therefore the results presented in this section are corollaries to the results of the preceding sections.

The requirement of affine equivariancy of a 'typical representative' can be written in the form

$$\mathbf{m}(\mathbf{Y}_1, \dots, \mathbf{Y}_M) = \mathbf{C}\mathbf{m}(\mathbf{x}_1, \dots, \mathbf{x}_M), \qquad \mathbf{y}_i = \mathbf{C}\mathbf{x}_i, \quad i = 1, \dots, n, \qquad (2.6.1)$$

where $\mathbf{C}$ is an arbitrary non-degenerate $M \times M$ matrix, $\mathbf{m}$, $\mathbf{x}_i$, $\mathbf{y}_i$ are M-dimensional vectors, and n is a number of observations.

As the following theorem shows, the requirement of affine equivariancy (2.6.1) combined with the requirement of translation equivariancy implies the procedure of data processing by the LS method.

THEOREM 2.6.1. *Let the contrast function be concave and the score function be differentiable. Then the condition of affine equivariancy of a 'typical' representative holds if and only if the contrast function is*

$$\rho(x_1, \dots, x_M; m_1, \dots, m_M) = A \sum_{s=1}^{M} |x_s - m_s|^2, \qquad (2.6.2)$$

where A is an arbitrary positive constant.

PROOF. The sufficiency is obvious. In order to prove the necessity, we observe that the matrix $\mathbf{C}$ of an affine transformation can be represented in the form $\mathbf{C} = \boldsymbol{\mu}\mathbf{T}$, where $\boldsymbol{\mu}$ is a diagonal matrix with positive elements, and $\mathbf{T}$ is an orthogonal matrix, i.e., the matrix $\mathbf{C}$ is the superposition of two transformations: the rotation by some angle and the coordinate-wise independent scale transformation.

From Theorem 2.2.1, Theorem 2.3.1, and Theorem 2.3.2 it follows that the contrast functions providing the equivariancy under the combination of the transformations of translation, scale, and rotation must simultaneously satisfy the conditions

$$\rho(x_1,\dots,x_M;m_1,\dots,m_M) = \sum_{s=1}^{M} A_s|x_s - m_s|_s^{\gamma}, \quad \gamma_s \geq 1, \quad A_s > 0,$$

and

$$\rho(x_1,\dots,x_M;m_1,\dots,m_M) = F\left(\sum_{s=1}^{M} |x_s - m_s|^2\right).$$

The direct comparison of the above expressions shows that the only one possible form of the contrast function is given by the quadratic criterion

$$\rho(x_1,\dots,x_M;m_1,\dots,m_M) = \sum_{s=1}^{M} |x_s - m_s|^2,$$

which completes the proof. □

Now we consider a requirement of equivariancy under the superposition of translation, orthogonality, and a component-wise identical scale transformations, which is less severe than (2.6.1). The restrictions on the structure of the contrast function imposed by this requirement are given by the following theorem.

THEOREM 2.6.2. *Let the contrast function be convex and the score function be differentiable. Then the combined conditions of translation, orthogonal, and component-wise identical scale equivariancy of a 'typical' representative hold if and only if the contrast function is*

$$\rho(x_1,\dots,x_M;m_1,\dots,m_M) = Ar^p, \qquad r = \left(\sum_{i=1}^{M} |x_i - m_i|^2\right)^{1/2}. \tag{2.6.3}$$

PROOF. The sufficiency is obvious. In order to prove the necessity, we observe that from Theorem 2.2.1 and Theorem 2.3.2 it follows that the function sought for should satisfy the relation

$$\rho(x_1,\dots,x_M;m_1,\dots,m_M) = \sum_{k=1}^{L} \Phi_k(|\mathbf{x}_k - \mathbf{m}_k|), \tag{2.6.4}$$

where $\Phi_k(\mathbf{u}_k)$ are homogeneous functions of order α_k of the variables belonging to the disjoint subsets $I_k, k = 1,\dots,L$, and

$$\rho(x_1,\dots,x_M;m_1,\dots,m_M) = F\left(\sum_{s=1}^{M} |x_s - m_s|^2\right). \tag{2.6.5}$$

From the latter relation it follows that such a separation of variables does not exist, and therefore (2.6.4) takes the form

$$\rho(x_1, \dots, x_M; m_1, \dots, m_M) = \Phi(|x_1 - m_1|, \dots, |x_M - m_M|), \qquad (2.6.6)$$

where $\Phi(u_1, \dots, u_M)$ is a homogeneous function of order α, i.e., a function satisfying the condition

$$\Phi(u_1, \dots, u_M) = u_1^\alpha \Phi\left(1, \frac{u_2}{u_1}, \dots, \frac{u_M}{u_1}\right).$$

Setting $|x_1 - m_1| = \dots = |x_M - m_M| = |u|$ in (2.6.5)–(2.6.7), we obtain

$$|u|^\alpha \Phi(1, 1, \dots, 1) = F(Mu^2),$$

and therefore,

$$F(v) = A|v|^{\alpha/2},$$

which completes the proof. □

In the real-life problems of multivariate data processing, certain metrological requirements can be imposed only on a part of the components of the initial data vectors. However, there often appear such situations that different metrological requirements should hold for different groups of components.

For instance, the measurements of some object may be characterized by its space coordinates, by the velocities of its displacement, by the state of environment (temperature, pressure, wind velocity), etc. It may become necessary to require the equivariancy of the estimators of these characteristics under an arbitrary affine transformation of space coordinates and velocities, under changes of scale of measuring devices which register temperature, pressure, etc.

Thus the statements of the above theorems in the cases where the metrological requirements are imposed on a part of the components of the vector data are of some interest.

Now we represent the results which give the structure of the contrast function in the extreme and complementary cases where one part of the components of the initial data vectors is homogeneous—for them it is natural to apply the requirements of orthogonal and identical component-wise scale equivariancy, and for the remaining components, due to their heterogeneity, it is reasonable to assume the requirement of independent component-wise equivariancy.

THEOREM 2.6.3. *The conditions of translation equivariancy for all components and of component-wise independent scale equivariancy for the components with indices* $1, \dots, l$ *hold if and only if the contrast function is*

$$\rho(x_1, \dots, x_M; m_1, \dots, m_M) = \sum_{i=1}^{l} A_i |x_i - m_i|^{p_i} + F(|x_{l+1} - m_{l+1}|, \dots, |x_M - m_M|),$$

where $A_i > 0$, $p_i \geq 1$, *and* $F(u_1, \dots, u_{M-l})$ *is a concave function.*

To prove this, we observe that in this case we obtain the contrast function

$$\rho(x_1, x_2, \dots, x_M; m_1, m_2, \dots, m_M) = \left(\sum_{i=1}^{l} A_i |x_i - m_i|^{p_i}\right)$$
$$\times G(|x_{l+1} - m_{l+1}|, \dots, |x_M - m_M|) + F(|x_{l+1} - m_{l+1}|, \dots, |x_M - m_M|)$$

and, because the product of functions depending on different arguments is concave only if one of the functions is concave and another is constant, we obtain $G(|x_{l+1} - m_{l+1}|, \dots, |x_M - m_M|) = const$, which completes the proof.

THEOREM 2.6.4. *The conditions of translation equivariancy for all components and of combined orthogonal and component-wise identical scale equivariancy for the components with indices* $1, \dots, l$ *hold if and only if the contrast function is*

$$\rho(x_1, \dots, x_M; m_1, \dots, m_M) = \left(\sum_{i=1}^{l} |x_i - m_i|^2\right)^{p/2} + F(|x_{l+1} - m_{l+1}|, \dots, |x_M - m_M|),$$

where $F(u_1, u_2, \dots, u_{M-l})$ *is a concave function.*

To prove this theorem, it suffices to repeat the proof of Theorem 2.6.2.
Combining Theorems 2.6.3 and 2.6.4, we obtain the following

COROLLARY 2.6.1. *In order that the 'typical' representative could possess*

- *translation equivariancy for all components,*
- *independent scale equivariancy for the components with indices* $s = 1, \dots, l$,
- *orthogonal and identical scale equivariancy for the components with indices* $s = l+1, \dots, M$,

it is necessary and sufficient to use the contrast functions

$$\rho(x_1, \dots, x_M; m_1, \dots, m_M) = \sum_{i=1}^{l} A_i |x_i - m_i|^{p_i} + \left(\sum_{i=l+1}^{M} |x_i - m_i|^2\right)^{p/2}, \quad (2.6.7)$$

where $A_i > 0$, $p_i \geq 1$, $p \geq 1$.

REMARK 2.6.1. Using the above approach, it is not difficult to construct as many combinations of particular requirements of equivariancy for various collections of the components of the initial data vectors as desired.

3

Robust minimax estimation of location

3.1. Introductory remarks

The basic stage of the minimax approach is the determination of a least informative distribution minimizing the Fisher information over a given class of distributions. In this section we describe the solutions of this variational problem in some important for theory and applications classes of distribution densities.

3.1.1. A priori information and the classes of data distributions

The ε-neighborhoods of normal distribution, in particular the model of gross errors, are not the only models of interest.

First, we may consider the ε-neighborhoods of other distributions, for example, the uniform, Laplace, or Cauchy. Certainly, the reasons to introduce such classes as supermodels are obviously weaker as compared to that based on normal distribution, but nevertheless, they can be.

Second, in applications rather often there exist a priori information about the dispersion of a distribution, about its central part and/or its tails, about the moments and/or subranges of a distribution. The empirical distribution function and relative estimators of a distribution shape (quantile functions and their approximations, histograms, kernel estimators) along with their confidence boundaries give other examples.

It seems reasonable to use such information in the minimax setting by introducing the corresponding classes $\mathscr{F}$ of distribution densities $f(x)$ in order to increase the efficiency of robust minimax estimation procedures.

In what follows, we mainly deal with symmetric unimodal distribution densities

$$f(-x) = f(x) \tag{3.1.1}$$

satisfying the regularity conditions ($\mathscr{F}1$) and ($\mathscr{F}2$) of Section 1.2. Obviously, distribution densities also satisfy the non-negativeness and normalization conditions

$$f(x) \geq 0, \qquad \int_{-\infty}^{\infty} f(x)\,dx = 1. \tag{3.1.2}$$

For the sake of brevity, we will not write out conditions (3.1.1) and (3.1.2) any time we define a distribution class.

Now we list some typical examples of distribution classes which seem most natural and convenient for the description of a priori knowledge about data distributions (Polyak and Tsypkin, 1978; Polyak and Tsypkin, 1980; Tsypkin, 1984).

$\mathscr{F}_1$, THE CLASS OF NONDEGENERATE DISTRIBUTIONS:

$$\mathscr{F}_1 = \left\{ f\colon\, f(0) \geq \frac{1}{2a} > 0 \right\}. \tag{3.1.3}$$

This class is proposed in (Polyak and Tsypkin, 1978). It is one of the most wide classes: any distribution density with a nonzero value at the center of symmetry belongs to it. The parameter a of this class characterizes the dispersion of the central part of the data distribution, in other words, a is the upper bound for that dispersion. The condition of belonging to this class is very close to the complete lack of information about a data distribution.

$\mathscr{F}_2$, THE CLASS OF DISTRIBUTIONS WITH A BOUNDED VARIANCE:

$$\mathscr{F}_2 = \left\{ f\colon\, \sigma^2(f) = \int_{-\infty}^{\infty} x^2 f(x)\,dx \leq \overline{\sigma}^2 \right\}. \tag{3.1.4}$$

This class is considered in (Kagan *et al.*, 1973). All distributions with variances bounded above are members of this class. Obviously, the Cauchy-type distributions do not belong to it.

$\mathscr{F}_3$, THE CLASS OF APPROXIMATELY NORMAL DISTRIBUTIONS or the gross error model, or the class of ε-contaminated normal distributions, or the Huber supermodel (Huber, 1964):

$$\mathscr{F}_3 = \{ f\colon\, f(x) = (1-\varepsilon)\mathscr{N}(x; 0, \sigma_N) + \varepsilon h(x),\ 0 \leq \varepsilon < 1 \}, \tag{3.1.5}$$

where $h(x)$ is an arbitrary density. The restriction of the mixture form (3.1.5) can be rewritten in the inequality form

$$\mathscr{F}_3 = \{ f\colon\, f(x) \geq (1-\varepsilon)\mathscr{N}(x; 0, \sigma_N),\ 0 \leq \varepsilon < 1 \}, \tag{3.1.6}$$

which is more convenient for solving variational problems.

$\mathcal{F}_4$, THE CLASS OF FINITE DISTRIBUTIONS:

$$\mathcal{F}_4 = \left\{ f: \int_{-l}^{l} f(x)\,dx = 1 \right\}. \tag{3.1.7}$$

The restriction on this class defines the boundaries of the data (i.e., $|x| \le l$ holds with probability one), and there is no more information about the distribution.

$\mathcal{F}_5$, THE CLASS OF APPROXIMATELY FINITE DISTRIBUTIONS:

$$\mathcal{F}_5 = \left\{ f: \int_{-l}^{l} f(x)\,dx = 1 - \beta \right\}. \tag{3.1.8}$$

The parameters l and β, $0 \le \beta < 1$, are given; the latter characterizes the degree of closeness of $f(x)$ to a finite distribution density. The restriction on this class means that the inequality $|x| \le l$ holds with probability $1 - \beta$.

Obviously, the class of finite distributions $\mathcal{F}_4$ is a particular case of the class $\mathcal{F}_5$.

The classes $\mathcal{F}_4$ and $\mathcal{F}_5$ are considered in (Huber, 1981; Sacks and Ylvisaker, 1972).

In what follows, we deal with more narrow classes with the additional restrictions, mainly those which are the intersections of the above:

$$\mathcal{F}_{12} = \mathcal{F}_1 \cap \mathcal{F}_2, \quad \mathcal{F}_{23} = \mathcal{F}_2 \cap \mathcal{F}_3, \quad \mathcal{F}_{25} = \mathcal{F}_2 \cap \mathcal{F}_5.$$

3.1.2. Finding the least informative distribution

We now consider the restrictions defining the classes of densities $\mathcal{F}$. From the above-said it follows that, in general, these restrictions are of the following forms:

$$\int_{-\infty}^{\infty} s_k(x) f(x)\,dx \le \alpha_k, \qquad k = 1, \dots, m, \tag{3.1.9}$$

$$f(x) \ge \varphi(x). \tag{3.1.10}$$

In particular, the normalization condition $\int f(x)\,dx = 1$ $(s(x) = 1)$ and the restriction on the variance $\int x^2 f(x)\,dx \le \overline{\sigma}^2$ $(s(x) = x^2)$ are referred to (3.1.9); the conditions of non-negativeness $f(x) \ge 0$ and of the approximate normality $f(x) \ge (1 - \varepsilon)\mathcal{N}(x; 0, \sigma_N)$ are described by (3.1.10), etc.

The variational problem of minimization of the Fisher information under conditions (3.1.9) and (3.1.10)

$$\text{minimize} \quad I(f) \quad \text{under the condition} \quad f \in \mathcal{F},$$

$$\mathcal{F} = \left\{ f: \int_{-\infty}^{\infty} s_k(x) f(x)\,dx \le \alpha_k,\ k = 1, 2, \dots, m,\ f(x) \ge \varphi(x) \right\} \tag{3.1.11}$$

is non-standard, and by present, there are no general methods of its solution.

Nevertheless, using heuristic and plausible considerations (in the Polya sense), it is possible to find a candidate for the optimal solution of (3.1.11), and then to check its validity. Certainly, such a reasoning must ground on the classical results of the calculus of variations. In general, it may be described as follows: first, use the restrictions of form (3.1.9); solve the Euler equation and determine the family of extremals; second, try to satisfy the restrictions of form (3.1.10) by gluing the pieces of free extremals with the constraints $h(x)$; and finally, verify the obtained solution.

Now we describe a procedure of searching for an eventual candidate for the solution of problem (3.1.11) and final checking proposed in (Tsypkin, 1984).

Consider the classes only with the restrictions of form (3.1.9). In this case, the Lagrange functional is composed as

$$L(f, \lambda_1, \lambda_2, \dots, \lambda_m) = I(f) + \sum_{k=1}^{m} \lambda_k \left(\int_{-\infty}^{\infty} s_k(x) f(x)\, dx - \alpha_k \right), \qquad (3.1.12)$$

where $\lambda_1, \lambda_2, \dots, \lambda_m$ are the Lagrange multipliers. Taking the variation of this functional and equating it to zero, we obtain the Euler equation in the form

$$-2\frac{f''(x)}{f(x)} + \left(\frac{f'(x)}{f(x)}\right)^2 + \sum_{k=1}^{m} \lambda_k s_k(x) = 0. \qquad (3.1.13)$$

Equation (3.1.13), as a rule, cannot be solved in a closed form. Hence one should use numerical methods. But there is a serious obstacle in satisfying the restrictions of the form $f(x) \geq \varphi(x)$.

In what follows, in Section 3.2, we consider some classes $\mathcal{F}$ with analytical solutions for the least informative density.

Another approach is based on direct applying of numerical methods to variational problem (3.1.11). These are associated with some approximation to the distribution density $f(x)$ followed by the subsequent solution of the problem of mathematical programming.

Thus, if to approximate $f(x)$ by a piecewise linear finite function, integrals—by sums and derivatives—by differences, we arrive at the problem of nonlinear programming with linear restrictions

$$\text{minimize} \quad \sum_{i=1}^{N} \frac{(f_{i+1} - f_i)^2}{f_i}$$

$$\sum_{i=1}^{N} s_k(ih) f_i h \leq \alpha_k, \quad k = 1, 2, \dots, m, \quad f_i = f(ih) \geq \varphi(ih), \quad i = 1, 2, \dots, N. \qquad (3.1.14)$$

For this problem of nonlinear programming, there exist quite good methods of solution.

Checking optimality by the Cauchy–Bunyakovskii inequality. Now we assume that there exists an analytical solution for the least informative distribution density $f^*(x)$. In this case, it is possible to use an approach based on applying the Cauchy–Bunyakovskii inequality[1]

$$\left(\int_{-\infty}^{\infty} \psi(x)\phi(x)f(x)\,dx\right)^2 \le \int_{-\infty}^{\infty} \psi^2(x)f(x)\,dx \cdot \int_{-\infty}^{\infty} \phi^2(x)f(x)\,dx. \quad (3.1.15)$$

The equality in (3.1.15) is attained with the proportional functions $\psi(x)$ and $\phi(x)$, i.e., under the condition

$$\psi(x) = -\lambda\phi(x), \quad (3.1.16)$$

where λ is some scalar factor.

Choose now the *informant* as

$$\psi(x) = \frac{f'(x)}{f(x)}. \quad (3.1.17)$$

By the definition of the Fisher information for location (see Section 1.2), we can rewrite inequality (3.1.15) as

$$I(f) \ge \frac{\left(\int_{-\infty}^{\infty} \phi(x)f'(x)\,dx\right)^2}{\int_{-\infty}^{\infty} \phi^2(x)f(x)\,dx}. \quad (3.1.18)$$

The right-hand side of (3.1.18) defines the lower bound for the Fisher information equal to

$$I^* = \min_{f\in\mathcal{F}} \frac{\left(\int_{-\infty}^{\infty} \phi(x)f'(x)\,dx\right)^2}{\int_{-\infty}^{\infty} \phi^2(x)f(x)\,dx}. \quad (3.1.19)$$

Therefore,

$$\min_{f\in\mathcal{F}} I(f) \ge I^*. \quad (3.1.20)$$

If for some distribution density $\widetilde{f} \in \mathcal{F}$ the condition analogous to (3.1.18) holds, i.e.,

$$\frac{\widetilde{f}'(x)}{\widetilde{f}(x)} = -\lambda\phi(x), \quad (3.1.21)$$

---

[1]The algebraic version of this inequality belongs to Cauchy (1821); Bunyakovskii (1856) was the pioneer to use its integral form; Schwartz published it after 1884.

then with $f(x) = \widetilde{f}(x)$ inequality (3.1.18) becomes the equality

$$I(\widetilde{f}) = \frac{\left(\int_{-\infty}^{\infty} \phi(x)\widetilde{f}'(x)\,dx\right)^2}{\int_{-\infty}^{\infty} \phi^2(x)\widetilde{f}(x)\,dx}. \tag{3.1.22}$$

The density $\widetilde{f}(x)$ (3.1.21) depends on the parameter λ; hence $\widetilde{f}(x) = \widetilde{f}(x,\lambda)$. If for some $\lambda = \lambda^*$ the Fisher information $I(\widetilde{f})$ in (3.1.22) equals I^* and $\widetilde{f}(x,\lambda^*) \in \mathscr{F}$, then it follows from (3.1.20) that $f^*(x) = \widetilde{f}(x,\lambda^*)$ is the least informative density in the class $\mathscr{F}$.

The way of searching for the least informative distribution density $f^*(x)$ just described remains valid if to rewrite the right-hand side of (3.1.18) in another form.

If $\lim_{x\to\pm\infty} \phi(x)f(x) = 0$ then integration by parts gives

$$\int_{-\infty}^{\infty} \phi(x)f'(x)\,dx = -\int_{-\infty}^{\infty} \phi'(x)f(x)\,dx;$$

hence it follows from (3.1.18) that

$$I(f) \geq \frac{\left(\int_{-\infty}^{\infty} \phi'(x)f(x)\,dx\right)^2}{\int_{-\infty}^{\infty} \phi^2(x)f(x)\,dx}. \tag{3.1.23}$$

Sometimes this inequality is more convenient for searching for the least informative distribution density $f^*(x)$ than (3.1.18).

Condition (3.1.21) is a differential equation. Integrating it, we obtain

$$\widetilde{f}(x) = \widetilde{f}(0)\exp\left(-\lambda\int_0^x \phi(x)\,dx\right), \tag{3.1.24}$$

where $\widetilde{f}(0)$ is the value of $\widetilde{f}(x)$ at $x = 0$. This value can be determined from the normalization condition (3.1.2), which takes the following form for symmetric densities:

$$2\int_0^{\infty} \widetilde{f}(x)\,dx = 1. \tag{3.1.25}$$

By substituting (3.1.24) into (3.1.25), we obtain

$$\widetilde{f}(0) = \left[2\int_0^{\infty} \exp\left\{\left(-\lambda\int_0^x \phi(x)\,dx\right)\,dx\right\}\right]^{-1}, \tag{3.1.26}$$

and therefore, from (3.1.24) we obtain

$$\widetilde{f}(x) = \frac{\exp\left(-\lambda\int_0^x \phi(x)\,dx\right)}{\left[2\int_0^{\infty} \exp\left\{\left(-\lambda\int_0^x \phi(x)\,dx\right)\,dx\right\}\right]}. \tag{3.1.27}$$

If the minimum of the functional in the right-hand side of inequality (3.1.18) or (3.1.23) is attained at the density $f^* \in \mathscr{F}$ coinciding with density (3.1.27) at some λ, then $f^*(x)$ is the solution of variational problem (3.1.11), and therefore it is the least informative distribution density in the class $\mathscr{F}$.

Thus, in order to determine the least informative density, one can use the following procedure:

(1) choose the function $\phi(x)$ and determine the minimum I^* of the right-hand side of inequality (3.1.18) (or (3.1.23)) over the densities $f(x)$ from the class $\mathscr{F}$;

(2) by formula (3.1.27), determine the density $f(x) = \widetilde{f}(x, \lambda)$ depending on an arbitrary parameter λ and find such $\lambda = \lambda^*$ that minimizes the right-hand side of inequality (3.1.18) (or (3.1.23)) over the densities $f(x) = \widetilde{f}(x, \lambda)$ belonging to the given class $\mathscr{F}$;

(3) verify the equality $I(\widetilde{f}(x, \lambda^*)) = I^*$.

If this equality holds, then the obtained density $\widetilde{f}(x, \lambda^*)$ is the least informative in the class $\mathscr{F}$.

REMARK 3.1.1. The success of this approach completely depends on the lucky choice of the function $\phi(x)$, i.e., on its adequacy to the given class $\mathscr{F}$. Observe that the optimal score function $\psi^* = -f^{*\prime}/f^*$ for M-estimators of location is proportional to the function $\phi(x)$; in other words, one should guess the form of the optimal score function.

Nevertheless, the above approach can be successfully used both for analytical and numerical determination of least informative densities.

Checking optimality by variational methods. Huber (1981) proposed a direct method for final checking the eventual candidate for the least informative density.

Assume that $\mathscr{F}$ is convex, $0 < I(f) < \infty$, and the set where the density f^* is strictly positive is convex. Set also the variation of f^* in the form of the mixture of densities

$$f_t = (1-t)f^* + tf_1, \qquad 0 \le t \le 1,$$

where $f_1 \in \mathscr{F}$. Under these assumptions, Huber (1981, p. 82) shows that f^* minimizes the Fisher information if and only if the inequality

$$\left.\frac{d}{dt} I(f_t)\right|_{t=0} \ge 0 \qquad (3.1.28)$$

holds for any distribution density $f_1 \in \mathscr{F}$.

Condition (3.1.28) can be rewritten in the convenient form

$$\int_{-\infty}^{\infty} (2\psi^{*\prime} - \psi^{*2})(f_1 - f^*)\,dx \ge 0, \tag{3.1.29}$$

where $\psi^*(x) = f^{*\prime}(x)/f^*(x)$ is the optimal score function, or also as

$$-4\int_{-\infty}^{\infty} \frac{(\sqrt{f^*})''}{\sqrt{f^*}}(f_1 - f^*)\,dx \ge 0, \tag{3.1.30}$$

for any $f_1 \in \mathcal{F}$.

Comparing these two approaches, we say that the former has a certain heuristic potential useful for the determination of an optimal solution. The latter gives a direct and explicit rule for verifying the earlier obtained optimal solution.

The extremals of the basic variational problem. In order to maintain any of these approaches, one needs to have an idea about the possible structure of an optimal solution. Now we consider the family of extremals whose constituents would have the minimized Fisher information for location with the only side normalization condition

$$\text{minimize} \quad I(f) = \int_{-\infty}^{\infty} \left(\frac{f'(x)}{f(x)}\right)^2 f(x)\,dx \quad \text{under the condition} \quad \int_{-\infty}^{\infty} f(x)\,dx = 1. \tag{3.1.31}$$

We set $\sqrt{f(x)} = g(x) \ge 0$ and rewrite minimization problem (3.1.31) as

$$\text{minimize} \quad I(f) = 4\int_{-\infty}^{\infty} (g'(x))^2\,dx \quad \text{under the condition} \quad \int_{-\infty}^{\infty} g^2(x)\,dx = 1. \tag{3.1.32}$$

Using the Lagrange multiplier λ together with the normalization condition, we obtain the differential equation

$$4g''(x) + \lambda g(x) = 0. \tag{3.1.33}$$

The general solutions of (3.1.33) are of the following possible forms depending on the sign of λ:

- the exponential form

$$g(x) = C_1 e^{kx} + C_2 e^{-kx}; \tag{3.1.34}$$

- the cosine form

$$g(x) = C_1 \sin kx + C_2 \cos kx; \tag{3.1.35}$$

- the linear form

$$g(x) = C_1 + C_2 x, \tag{3.1.36}$$

where $k = \sqrt{\pm\lambda}/2$.

In what follows, all these forms and their combinations are involved into the structures of optimal solutions for different classes of distribution densities.

3.1.3. The least informative distribution densities

Now we derive the least informative densities over the classes introduced in Subsection 3.1.1 using the approach based on the Cauchy–Bunyakovskii inequality.

The class $\mathscr{F}_1$ of nondegenerate distributions. It is defined by restriction (3.1.3)

$$f(0) \ge \frac{1}{2a} > 0. \tag{3.1.37}$$

Choose

$$\phi(x) = \operatorname{sgn} x. \tag{3.1.38}$$

Then from (3.1.27) we obtain

$$\widetilde{f}(x) = \frac{\lambda}{2}\exp(-\lambda|x|). \tag{3.1.39}$$

By substituting $\phi(x)$ into (3.1.23) we obtain

$$I(f) \ge 4f^2(0) \ge \frac{1}{a^2} \tag{3.1.40}$$

for any distribution density $f(x) \in \mathscr{F}_1$. If $\lambda = 1/a$ then density (3.1.39) belongs to the class $\mathscr{F}_1$, and the Fisher information $I(\widetilde{f})$ attains its minimum and becomes equal to $1/a^2$.

Thus the least informative density in the class $\mathscr{F}_1$ is given by the double-exponential or Laplace density

$$f_1^*(x) = \mathscr{L}(x; 0, a) = \frac{1}{2a}\exp\left(-\frac{|x|}{a}\right). \tag{3.1.41}$$

The optimal score function is $\psi_1^*(x) = |x|$; the minimum of Fisher information is

$$I(f_1^*) = 1/a^2;$$

and the minimax estimator is the sample median.

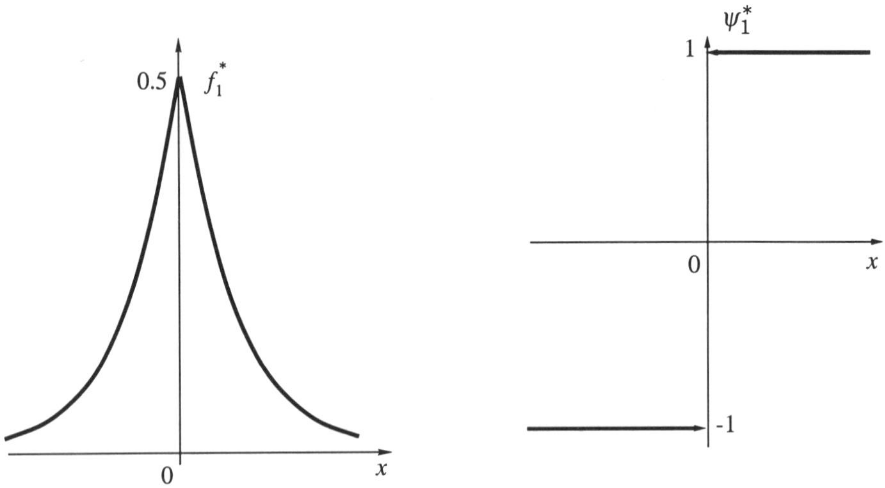

Figure 3.1. The least informative density and optimal score function in the class $\mathscr{F}_1$

The class $\mathscr{F}_2$ of distributions with bounded variance. It is defined by (3.1.4)

$$\sigma^2(f) = \int_{-\infty}^{\infty} x^2 f(x)\,dx \le \overline{\sigma}^2. \tag{3.1.42}$$

We set

$$\phi(x) = x. \tag{3.1.43}$$

Then from (3.1.27) it follows that

$$\widetilde{f}(x) = \frac{\sqrt{\lambda}}{\sqrt{2\pi}} \exp\left(-\frac{\lambda x^2}{2}\right). \tag{3.1.44}$$

By substituting $\phi(x)$ into (3.1.23) we obtain

$$I(f) \ge \frac{1}{\sigma^2(f)} \ge \frac{1}{\overline{\sigma}^2}$$

for any distribution density $f(x) \in \mathscr{F}_2$. For $\lambda = 1/\overline{\sigma}^2$, the distribution density (3.1.44) belongs to the class $\mathscr{F}_2$ and has minimal Fisher information equal to $1/\overline{\sigma}^2$.

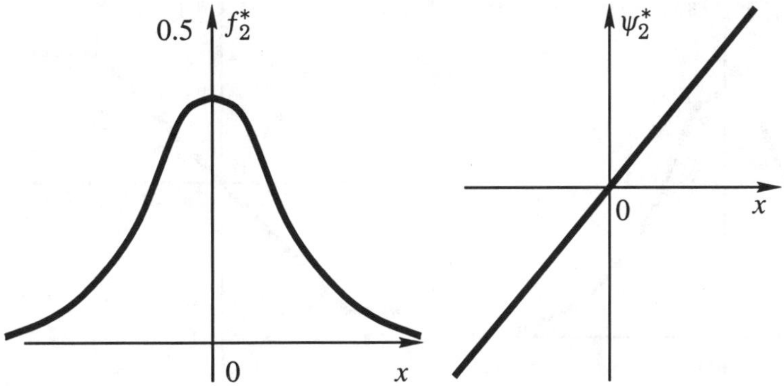

Figure 3.2. The least informative density and optimal score function in the class $\mathscr{F}_2$

Thus the least informative density in the class $\mathscr{F}_2$ is normal

$$f_2^*(x) = \mathcal{N}(x; 0, \overline{\sigma}) = \frac{1}{\sqrt{2\pi}} \exp\left(-\frac{x^2}{2\overline{\sigma}^2}\right). \tag{3.1.45}$$

The optimal score function is $\psi_2^*(x) = x$; the minimum of Fisher information is

$$I(f_2^*) = 1/\overline{\sigma}^2;$$

and the minimax estimator is the sample mean.

Observe that in this case the minimax estimator is extremely non-robust, since its score and influence functions are unbounded.

The class $\mathscr{F}_3$ of approximately normal distributions. The optimal solution in the similar class of ε-contaminated distributions $F = (1-\varepsilon)G + \varepsilon H$ is described earlier in Section 1.2. For the sake of its importance, we write out its particular case where $G = \Phi$ is the standard normal distribution. In this case, the Fisher information is minimized by

$$f_3^*(x) = \begin{cases} \dfrac{1-\varepsilon}{\sqrt{2\pi}} \exp\left(-\dfrac{x^2}{2}\right), & |x| \le k, \\ \dfrac{1-\varepsilon}{\sqrt{2\pi}} \exp\left(-k|x| + \dfrac{k^2}{2}\right), & |x| > k, \end{cases} \tag{3.1.46}$$

where k and ε are related by

$$\frac{2\phi(k)}{k} - 2\Phi(-k) = \frac{\varepsilon}{1-\varepsilon} \tag{3.1.47}$$

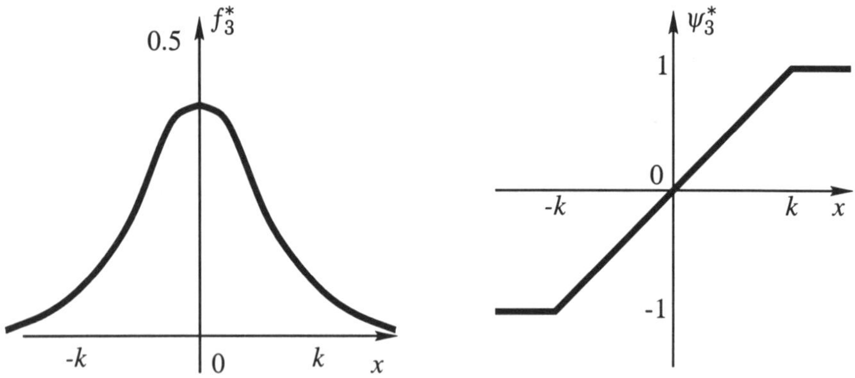

Figure 3.3. The least informative density and optimal score function in the class $\mathscr{F}_3$

where $\phi = \Phi'$ is the standard normal density.

The optimal score function is

$$\psi_3^*(x) = \max\,[-k, \min\,(x, k)]. \tag{3.1.48}$$

The class $\mathscr{F}_4$ of finite distributions. It is defined by restriction (3.1.8)

$$\int_{-l}^{l} f(x)\,dx = 1. \tag{3.1.49}$$

We set

$$\phi(x) = \tan\frac{\pi x}{2l}, \qquad |x| \le l. \tag{3.1.50}$$

For finite distribution densities with finite Fisher information, the following boundary conditions must be satisfied:

$$f(\pm l) = 0, \qquad f'(\pm(l-0)) = 0. \tag{3.1.51}$$

Then inequality (3.1.23) takes the form

$$I(f) \ge \frac{\left(\int_{-l}^{l} \phi(x) f'(x)\,dx\right)^2}{\int_{-l}^{l} \phi^2(x) f(x)\,dx}, \tag{3.1.52}$$

and, after some transformations,

$$I(f) \geq \frac{\left(\frac{\pi}{2l}\right)^2 \left(1 + \int_{-l}^{l} \phi^2(x) f(x)\,dx\right)^2}{\int_{-l}^{l} \phi^2(x) f(x)\,dx} \geq \min_{v>0} \left(\frac{\pi}{2l}\right)^2 \frac{(1+v)^2}{v} = \frac{\pi^2}{l^2}, \quad (3.1.53)$$

where the minimum is attained at $v = 1$.

The equality in (3.1.52) is attained at the densities $\widetilde{f}(x)$ satisfying equation (3.1.21) for the interval $|x| < l$ and boundary conditions (3.1.51), i.e.,

$$\widetilde{f}(x) = \frac{\cos^{\nu} \frac{\pi x}{2l}}{\int_{-l}^{l} \cos^{\nu} \frac{\pi x}{2l}\,dx}, \quad (3.1.54)$$

where $\nu = 2l\lambda/\pi$. Hence it follows that the equality $I(f) = \pi^2/l^2$ can hold true only at densities (3.1.54) satisfying the condition

$$\int_{-l}^{l} \phi^2(x) f(x)\,dx = 1, \quad (3.1.55)$$

which holds only for $\nu = 2$.

Thus the least informative density in the class $\mathscr{F}_4$ is of the form

$$f_4^*(x) = \begin{cases} \frac{1}{l} \cos^2 \frac{\pi x}{2l}, & |x| \leq l, \\ 0, & |x| > l, \end{cases} \quad (3.1.56)$$

The optimal score function is unbounded: $\psi_4^*(x) = \tan \frac{\pi x}{2l}$ for $|x| \leq l$, and the minimum of Fisher information is $I(f_4^*) = \pi^2/l^2$.

REMARK 3.1.2. Optimal solution (3.1.56) was known long ago in the calculus of variations (see (Gelfand and Fomin, 1963)).

The class $\mathscr{F}_5$ of approximately finite distributions. It is characterized by the restriction

$$\int_{-l}^{l} f(x)\,dx = 1 - \beta, \qquad 0 \leq \beta < 1. \quad (3.1.57)$$

Since the determination of the least informative density for this class is associated with cumbersome calculations, we simply formulate the final result.

We set

$$\phi(x) = \begin{cases} \tan B_1 x, & |x| \leq l, \\ \tan B_1 l \operatorname{sgn} x, & |x| > l. \end{cases} \quad (3.1.58)$$

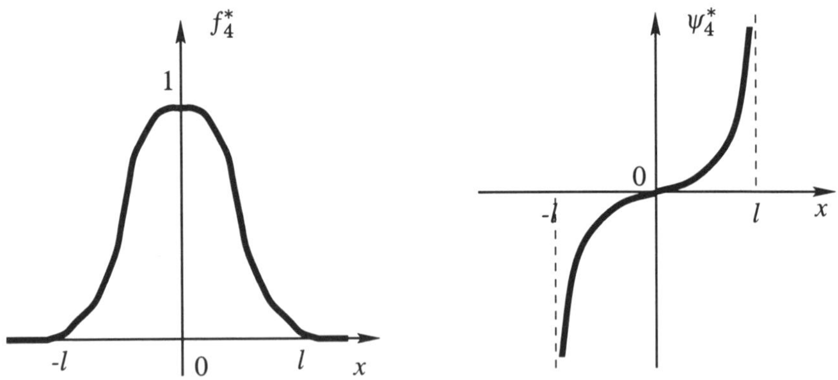

Figure 3.4. The least informative distribution density and optimal score function in the class $\mathscr{F}_4$

Then, following the above method, we can write the least informative density over the class $\mathscr{F}_5$ in the form

$$f_5^*(x) = \begin{cases} A_1 \cos^2 B_1 x, & |x| \le l, \\ A_2 \exp(-B_2|x|), & |x| > l, \end{cases} \tag{3.1.59}$$

where the constants A_1, A_2, B_1, and B_2 are determined from the simultaneous equations characterizing the restrictions of the class $\mathscr{F}_5$, namely the conditions of normalization and approximate finiteness, and the conditions of smoothness at $x = l$:

$$\int_{-\infty}^{\infty} f_5^*(x)\,dx = 1, \qquad \int_{-l}^{l} f_5^*(x)\,dx = 1 - \beta,$$
$$f_5^*(l-0) = f_5^*(l+0), \qquad f_5^{*\prime}(l-0) = f_5^{*\prime}(l+0). \tag{3.1.60}$$

The solution of system (3.1.60) is given by

$$A_1 = \frac{(1-\beta)\omega}{l(\omega + \sin\omega)}, \qquad B_1 = \frac{\omega}{2l},$$
$$A_2 = \frac{\beta\lambda}{2l} e^{\lambda} \qquad B_2 = \frac{\lambda}{l}, \tag{3.1.61}$$

where the parameters ω and β are related by

$$\frac{2\cos^2(\omega/2)}{\omega \tan(\omega/2) + 2\sin^2(\omega/2)} = \frac{\beta}{1-\beta}, \qquad 0 < \omega < \pi,$$

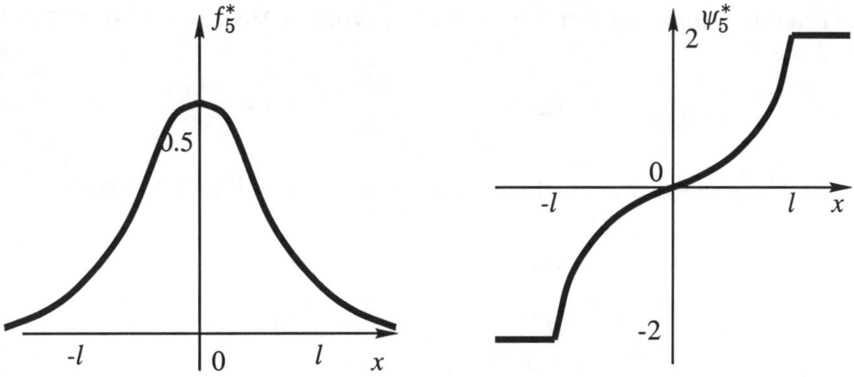

Figure 3.5. The least informative density and optimal score function in the class $\mathscr{F}_5$

and $\lambda = \omega \tan(\omega/2)$.

The optimal score function $\psi_5^*(x)$ is bounded; it has the same shape as the $\phi(x)$ (3.1.58); and the minimum of Fisher information is

$$I(f_5^*) = (1-\beta)\frac{\omega - \sin\omega}{\omega + \sin\omega}\frac{\omega^2}{l^2} + \beta\frac{\lambda^2}{l^2}. \tag{3.1.62}$$

REMARK 3.1.3. The least informative density f_5^* also minimizes the Fisher information in the class with the restriction of inequality form

$$\int_{-l}^{l} f(x)\,dx \ge 1-\beta, \qquad 0 \le \beta < 1. \tag{3.1.63}$$

REMARK 3.1.4. The least informative density over the class $\mathscr{F}_1$ of nondegenerate distributions is the special case of the optimal solution over the class of approximately finite distributions as

$$l \to 0, \quad 1-\beta \to 0, \quad \frac{1-\beta}{2l} \to \frac{1}{2a}.$$

3.2. Robust estimation of location in models with bounded variances

In this section, analytical solutions of the variational problem to minimize the Fisher information are obtained for some new classes of distributions with the restrictions on the variance. These solutions are basic for designing minimax methods and their future study.

3.2.1. The least informative density in the class $\overline{\mathscr{F}}_2$

Let us consider the variational problem to minimize the Fisher information

$$\bar{f}_2^* = \arg\min_{f\in\mathscr{F}} \int_{-\infty}^{\infty} \left(\frac{f'(x)}{f(x)}\right)^2 f(x)\,dx \tag{3.2.1}$$

in the class $\overline{\mathscr{F}}_2$ of symmetric distribution densities with given variance

$$f(x) \ge 0, \qquad f(-x) = f(x),$$
$$\int_{-\infty}^{\infty} f(x)\,dx = 1, \qquad \int_{-\infty}^{\infty} x^2 f(x) dx = d^2. \tag{3.2.2}$$

It follows from (3.1.45) that the solution of problem (3.2.1) under conditions (3.2.2) is given by the normal density (Kagan *et al.*, 1973)

$$\bar{f}_2^*(x) = \mathscr{N}(x; 0, d) = \frac{1}{\sqrt{2\pi}d} \exp\left(-\frac{x^2}{2d^2}\right). \tag{3.2.3}$$

Here we are mainly interested not in the optimal solution (3.2.3) itself but in the structure of the family of extremals of variational problem (3.2.1).

The following statement gives the form of this family of extremals.

LEMMA 3.2.1. *Let $h(x)$ be continuously differentiable on $(0,\infty)$. Then under the conditions*

$$h(x) \ge 0,$$
$$\int_0^{\infty} h(x)\,dx = \frac{1}{2},$$
$$\int_0^{\infty} x^2 h(x)\,dx = \frac{d^2}{2},$$

the extremals of the variational problem

$$h^* = \arg\min_h \int_0^{\infty} \left(\frac{h'(x)}{h(x)}\right)^2 h(x)\,dx \tag{3.2.4}$$

are of the form

$$h^*(x) = \frac{\Gamma(-\nu)\sqrt{2\nu+1+1/S(\nu)}}{\sqrt{2\pi}\,d\,S(\nu)} \mathscr{D}_\nu^2\left(\frac{x}{d}\sqrt{2\nu+1+1/S(\nu)}\right), \tag{3.2.5}$$

where

- *the parameter ν takes its values in $(-\infty, 0]$;*

- $\mathscr{D}_\nu(\cdot)$ *are the Weber–Hermite functions or the functions of the parabolic cylinder* (Abramowitz and Stegun, 1972);

- $S(\nu) = [\psi(1/2 - \nu/2) - \psi(-\nu/2)]/2$,

- $\psi(x) = \dfrac{d \ln \Gamma(x)}{dx}$ *is the digamma function.*

The conditions of symmetry in the setting of problem (3.2.4) are taken into account by the special form of writing out the restrictions.

Thus, we arrive at the following result (Vilchevski and Shevlyakov, 1984; Vilchevski and Shevlyakov, 1990b; Vilchevski and Shevlyakov, 1994).

THEOREM 3.2.1. *The extremals of variational problem* (3.2.1) *are of the form*

$$f(x; \nu, d) = \frac{\Gamma(-\nu)\sqrt{2\nu + 1 + 1/S(\nu)}}{\sqrt{2\pi} d S(\nu)} \mathscr{D}_\nu^2 \left(\frac{|x|}{d} \sqrt{2\nu + 1 + 1/S(\nu)} \right). \quad (3.2.6)$$

Optimal densities (3.2.6) satisfy the characterization conditions of normalization and on a variance of the class $\overline{\mathscr{F}}_{12}$.

This family of the Weber–Hermite distribution densities includes:

- the normal distribution density with $\nu = 0$

$$f(x; 0, d) = \frac{1}{\sqrt{2\pi} d} \exp\left(-\frac{x^2}{2d^2} \right);$$

- the family of the $(k+1)$-modal Hermite distribution densities with $\nu = k$, $k = 0, 1, \dots,$

$$f(x; k, d) = \frac{\sqrt{2k+1}}{\sqrt{2\pi}\, d\, k!\, 2^k} H_k^2 \left(\frac{|x|}{d} \sqrt{k + 1/2} \right) \exp\left(-\frac{(2k+1)x^2}{2d^2} \right),$$

where $H_k(x) = (-1)^k e^{x^2} \dfrac{d^k (e^{-x^2})}{dx^k}$ are the Hermite polynomials;

- the Laplace distribution density as $\nu \to -\infty$

$$f(x; -\infty, d) = L(x; 0, \sqrt{2} d) = \frac{1}{\sqrt{2} d} \exp\left(-\frac{|x|\sqrt{2}}{d} \right);$$

- the unimodal Weber–Hermite densities with $-\infty < \nu < 0$ that are intermediate between the normal and Laplace densities.

Table 3.1. Fisher information for the Weber–Hermite distribution densities with given variance

ν	$-\infty$	-2	-1	-0.5	**0**	0.5	1	2	3
$I(\nu,d)d^2$	2	1.70	1.53	1.28	**1**	2.01	9	25	49

In this case, the Fisher information is

$$I(\nu,d) = \frac{1}{d^2}\left[(2\nu+1)^2 + 4(2\nu+1)S(\nu) + 3/S^2(\nu)\right]. \tag{3.2.7}$$

REMARK 3.2.1. From Table 3.1 it can be seen that the minimum of Fisher information is attained at the normal density with $\nu = 0$.

REMARK 3.2.2. The Weber–Hermite densities (3.2.6) have two free parameters d and ν, thus they can appear in the solutions of the variational problems with two restrictions (one of them should be imposed on a variance).

In Subsection 3.2.2 we show that the Weber–Hermite densities describe the extremals of the variational problem of minimizing Fisher information in the intersection of the distribution classes $\mathcal{F}_1$ and $\mathcal{F}_2$.

3.2.2. The least informative density and the robust minimax estimator in the class $\mathcal{F}_{12}$

We consider the structure of the minimax robust estimator of the location parameter in the class with the restrictions of inequality form on the value of a distribution density at the center of symmetry and on the value of a variance

$$\mathcal{F}_{12} = \left\{f:\quad f(0) \ge \frac{1}{2a} > 0,\ \sigma^2(f) = \int_{-\infty}^{\infty} x^2 f(x)\,dx \le \overline{\sigma}^2\right\}. \tag{3.2.8}$$

The following assertion is the key for further considerations (Vilchevski and Shevlyakov, 1984; Vilchevski and Shevlyakov, 1994).

THEOREM 3.2.2. *In the class $\mathcal{F}_{12}$, the least informative density is of the form*

$$f_{12}^*(x) = \begin{cases} f_2^*(x), & \overline{\sigma}^2/a^2 \le 2/\pi, \\ f(x;\nu,\overline{\sigma}), & 2/\pi < \overline{\sigma}^2/a^2 \le 2, \\ f_1^*(x), & \overline{\sigma}^2/a^2 > 2, \end{cases} \tag{3.2.9}$$

where $f(x;\nu,\overline{\sigma})$ are Weber–Hermite densities (3.2.6)) *with $\nu \in (-\infty; 0]$ determined from the equation*

$$\frac{\overline{\sigma}}{a} = \frac{\sqrt{2\nu+1+1/S(\nu)}\,\Gamma^2(-\nu/2)}{\sqrt{2\pi}\,2^{\nu+1}\,S(\nu)\,\Gamma(-\nu)}. \tag{3.2.10}$$

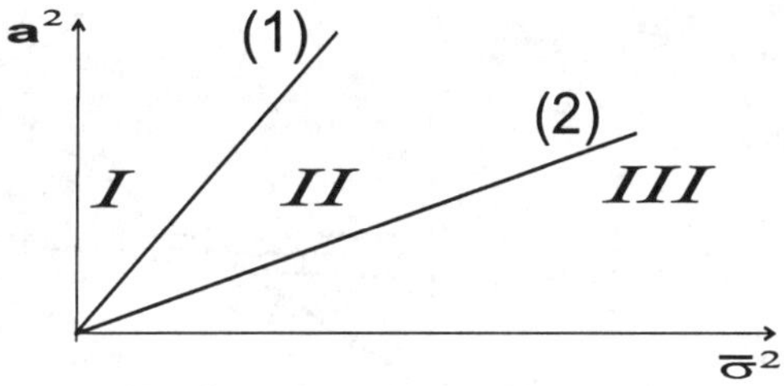

Figure 3.6. The domains of the optimal solution in the class $\mathscr{F}_{12}$

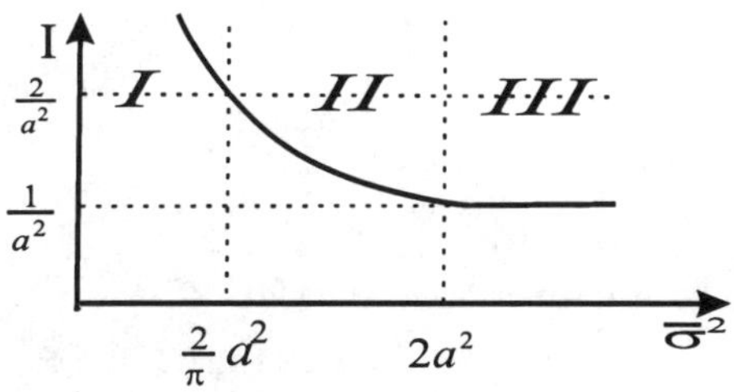

Figure 3.7. The dependence of Fisher information on the parameters of the class $\mathscr{F}_{12}$

The behavior of solution (3.2.9) and of the functional $I(f_{12}^*)$ depends on the parameters $\overline{\sigma}^2$ and a^2, and it is shown in Fig. 3.6 and Fig. 3.7:

- zone I corresponds to the normal density;
- zone II, to the Weber–Hermite densities;
- zone III, to the Laplace density.

The branches of solution (3.2.9) appear due to the degree in which the constraints are taken into account:

- in zone I, $\overline{\sigma}^2 \le 2a^2/\pi$, only the restriction on a variance does matter (the equality $\sigma^2(f_{12}^*) = \overline{\sigma}^2$), and the restriction on the value of a density at the center of symmetry is of the form of the strict inequality ($f_{12}^*(0) > 1/2a$);
- in zone III, $\overline{\sigma}^2 > 2a^2$, only the restriction on the density value is essential: $f_{12}^*(0) = 1/2a$, $\sigma^2(f_{12}^*) < \overline{\sigma}^2$);
- in zone II, both restrictions are of the form of equalities ($f_{12}^*(0) = 1/2a$, $\sigma^2(f_{12}^*) = \overline{\sigma}^2$).

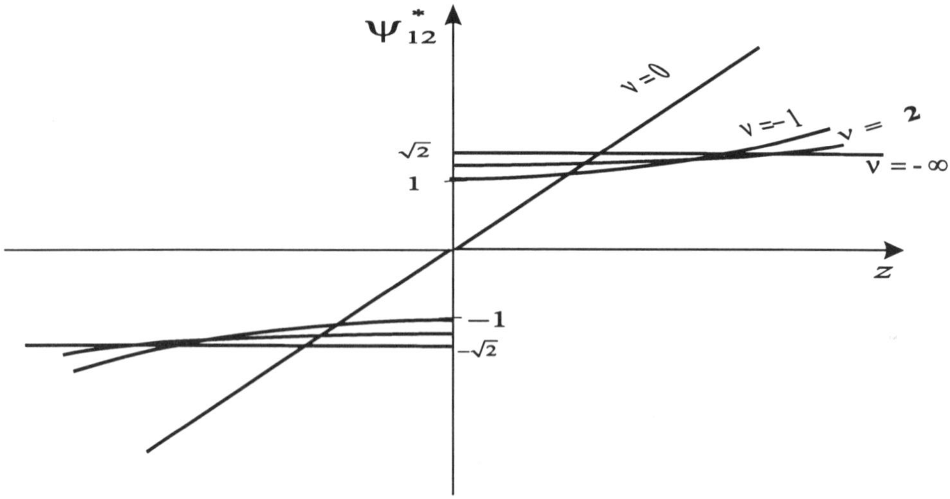

Figure 3.8. The optimal score function in the class $\mathscr{F}_{12}$

The optimal minimax algorithm of data processing is defined by the ML principle (1.2.8)

$$\psi^*_{12}(z) = \begin{cases} z/\overline{\sigma}^2, & \overline{\sigma}^2/a^2 \le 2/\pi, \\ -f'(z;\nu,\overline{\sigma})/f(z;\nu,\overline{\sigma}), & 2/\pi < \overline{\sigma}^2/a^2 \le 2, \\ a^{-1}\operatorname{sgn} z, & \overline{\sigma}^2/a^2 > 2. \end{cases} \tag{3.2.11}$$

From (3.2.11) and Fig. 3.7 we find that:

- in zone I with relatively small variances, the normal density and the corresponding least squares method, $\rho(z) = z^2$, are optimal;
- in zone III with relatively large variances, the Laplace density and the corresponding least absolute values method, $\rho(z) = |z|$, are optimal;
- in zone II with moderate variances, a compromise between the LS and LAV algorithms with the score function $\psi^*_{12}(z)$ is the best: its behavior is displayed in Fig. 3.8.

From Fig. 3.8 we can see that these algorithms of data processing are intermediate between the LS method with $\psi(z) = z$ and the LAV method with $\psi(z) = \operatorname{sgn} z$. The asymptotes of the curves $\psi^* = \psi^*_{12}(z)$ go through the origin of coordinates. The slope of these curves is described by the following: with $\nu = -1$ we have $\psi^{*\prime}_{12}(0) = 0.1$ and $\psi^{*\prime}_{12}(\infty) = 0.44$; with $\nu = -2$ we have $\psi^{*\prime}_{12}(0) = 0.04$ and $\psi^{*\prime}_{12}(\infty) = 0.25$.

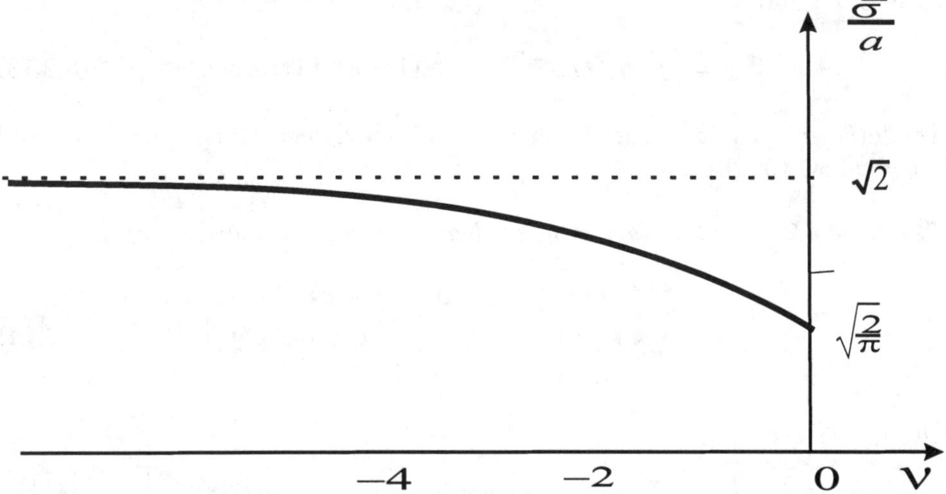

Figure 3.9. The dependence of the parameter ν on the characteristics of the class $\mathscr{F}_{12}$

The proposed algorithm qualitatively differs from the Huber algorithm that is optimal in the class $\mathscr{F}_3$, though both have the LS and the LAV procedures as the limiting cases.

The dependence between the values of the parameter ν and the characteristics $\overline{\sigma}$ and a of the class $\mathscr{F}_{12}$ given by (5.4.10) is shown in Fig. 3.9.

Concluding this section, we note that using the minimax algorithm with the score function $\psi^*_{12}(z)$ provides the guaranteed accuracy of an estimator (in the sense of the supremum of its asymptotic variance) for each distribution in the class $\mathscr{F}_{12}$:

$$\operatorname{Var}\widehat{\theta}_n(\psi^*_{12}, f) \le \sup_{f\in\mathscr{F}_{12}} \operatorname{Var}\widehat{\theta}_n(\psi^*_{12}, f) = \operatorname{Var}\widehat{\theta}_n(\psi^*_{12}, f^*_{12}),$$

$$\operatorname{Var}\widehat{\theta}_n(\psi^*_{12}, f^*_{12}) = \frac{1}{nI(f^*_{12})} = \begin{cases} \overline{\sigma}^2/n, & \overline{\sigma}^2/a^2 \le 2/\pi, \\ 1/[nI(\nu,\overline{\sigma})], & 2/\pi < \overline{\sigma}^2/a^2 \le 2, \\ a^2/n, & \overline{\sigma}^2/a^2 > 2, \end{cases} \tag{3.2.12}$$

where $I(\nu, \overline{\sigma})$ is given by (3.2.7).

3.2.3. The least informative density in the class $\mathscr{F}_{23}$

The family of extremals (3.2.6) is used in the solutions of all problems of minimizing Fisher information in the classes of distributions with the restrictions on a variance.

Now we introduce the class of distributions that is the intersection of the classes $\mathscr{F}_2$ and $\mathscr{F}_3$

$$\mathscr{F}_{23} = \left\{f\colon \sigma^2(f) \le \overline{\sigma}^2,\ f(x) \ge (1-\varepsilon)\mathscr{N}(x; 0, \sigma_N)\right\}. \tag{3.2.13}$$

The following result is true (Vilchevski and Shevlyakov, 1984; Vilchevski and Shevlyakov, 1994).

THEOREM 3.2.3. *In the class $\mathscr{F}_{23}$, the least informative density is of the form*

$$f^*_{23}(x) = \begin{cases} \mathscr{N}(x; 0, \sigma_N), & \sigma_N < \overline{\sigma} < \sigma_N/(1-\varepsilon), \\ \overline{f}^*_{23}(x), & \overline{\sigma}/(1-\varepsilon) \le \overline{\sigma} \le \sigma(f^*_3), \\ f^*_3(x), & \overline{\sigma} > \sigma(f^*_3), \end{cases} \tag{3.2.14}$$

where

$$\overline{f}^*_{23}(x) = \begin{cases} (1-\varepsilon)\mathscr{N}(x; 0, \sigma_N), & |x| \le \Delta, \\ A\mathscr{D}^2_\nu(B|x|), & |x| > \Delta, \end{cases}$$

$$\sigma^2(f^*_3) = \int_{-\infty}^{\infty} x^2 f^*_3(x)\, dx.$$

The values of the parameters A, B, Δ, and ν are determined from the simultaneous equations

- *the condition of normalization*

$$\int_{-\infty}^{\infty} \overline{f}^*_3(x)\, dx = 1;$$

- *the restriction on the variance*

$$\int_{-\infty}^{\infty} x^2 \overline{f}^*_3(x)\, dx = \overline{\sigma}^2; \tag{3.2.15}$$

- *the conditions of smoothness of the optimal solution at $x = \Delta$*

$$A\mathscr{D}^2_\nu(B\Delta) = (1-\varepsilon)\mathscr{N}(\Delta; 0, \sigma_N),$$
$$2AB\mathscr{D}_\nu(B\Delta)\mathscr{D}'_\nu(B\Delta) = (1-\varepsilon)\mathscr{N}'(\Delta; 0, \sigma_N).$$

It seems impossible to obtain a solution of this system in a closed form because of the complexity of the analytical description of the functions $\mathscr{D}_\nu$. Nevertheless, with sufficiently large values of the constraint on the variance

$$\overline{\sigma}^2 > \sigma^2(f^*_3) = \int_{-\infty}^{\infty} x^2 f^*_3(x)\, dx,$$

optimal solution (3.2.14) coincides with the Laplace density: $f^*_{23}(x) = f^*_3(x)$. In this case, the restriction on the variance holds as strict inequality.

If $\overline{\sigma}^2 \le \sigma^2(f^*_3)$, then this restriction becomes quite severe, and the tails of the least informative distribution (3.2.14) cease to be exponential and, hence, they are defined by (3.2.15).

Observe also that if the restriction on the density holds as a strict inequality (with $\sigma_N < \overline{\sigma} < \sigma_N/(1-\varepsilon)$), then the optimal solution coincides with the normal density $f^*_{23}(x) = \mathcal{N}(x; 0, \overline{\sigma})$, and the restriction on the variance holds as equality.

3.2.4. The least informative density in the class $\mathcal{F}_{25}$

Now we consider the structure of the robust minimax estimator in the intersection of the classes $\mathcal{F}_2$ and $\mathcal{F}_5$ with the constraints on the variance and on the mass of the central part of a distribution

$$\mathcal{F}_{25} = \left\{ f\colon \sigma^2(f) \le \overline{\sigma}^2,\ \int_{-l}^{l} f(x)\,dx \ge 1-\beta \right\}. \tag{3.2.16}$$

A lower bound for the mass of the central zone of a distribution is equivalent to an upper bound for its dispersion, or more precisely, for the subrange of a symmetric distribution. In this case, the following result is true (Shevlyakov, 1991; Vilchevski and Shevlyakov, 1994).

THEOREM 3.2.4. *In the class of distributions $\mathcal{F}_{25}$, the least informative density is of the form*

$$f^*_{25}(x) = \begin{cases} f^*_2(x), & \overline{\sigma}^2 \le k_1 l^2, \\ \overline{f}^*_{25}(x), & k_1 l^2 < \overline{\sigma}^2 \le k_2 l^2, \\ f^*_5(x), & \overline{\sigma}^2 > k_2 l^2, \end{cases} \tag{3.2.17}$$

where

- $f^*_2(x), f^*_5(x)$, *and* $\overline{f}^*_{25}(x)$ *are the least informative distribution densities in the classes* $\mathcal{F}_2$, $\mathcal{F}_5$, *and* $\overline{\mathcal{F}}_{25}$;

- *the switching parameters* k_1 *and* k_2 *depend on the parameters of the class* $\mathcal{F}_{25}$

$$\sigma^2(f^*_5) = \int_{-\infty}^{\infty} x^2 f^*_5(x)\,dx = k_2 l^2,$$

$$\frac{1}{\sqrt{2\pi}\sqrt{k_1} l} \int_{-\infty}^{\infty} \exp\left(-\frac{x^2}{2k_1 l^2}\right) dx = 1-\beta;$$

- *the density $\overline{f}^*_{25}(x)$ is defined via the Weber–Hermite functions*

$$\overline{f}^*_{25}(x) = \begin{cases} A_1[\mathscr{D}_{\nu_1}(B_1x) + \mathscr{D}_{\nu_1}(-B_1x)]^2, & |x| \le l, \\ A_2\mathscr{D}^2_{\nu_2}(B_2|x|), & |x| > l; \end{cases} \tag{3.2.18}$$

The values of the parameters A_1, A_2, B_1, B_2, ν_1, and ν_2 in (3.2.18) *are determined by the simultaneous equations*

- *the normalization condition*

$$\int_{-\infty}^{\infty} \overline{f}^*_{25}(x)\,dx = 1, \tag{3.2.19}$$

- *the characterization restrictions of the class $\overline{\mathscr{F}}_{25}$*

$$\int_{-l}^{l} \overline{f}^*_{25}(x)\,dx = 1 - \beta, \tag{3.2.20}$$

$$\int_{-\infty}^{\infty} x^2\overline{f}^*_{25}(x)\,dx = \overline{\sigma}^2; \tag{3.2.21}$$

- *the conditions of smoothness of the optimal solution at $x = l$*

$$\overline{f}^*_{25}(l-0) = \overline{f}^*_{25}(l+0), \quad \overline{f}^{*\prime}_{25}(l-0) = \overline{f}^{*\prime}_{25}(l+0); \tag{3.2.22}$$

- *the additional condition of optimality connecting the solutions in the zones $|x| \le l$ and $|x| > l$*

$$\int_{-l}^{l} \overline{f}^*_{25}(x)\,dx = d_1^{*2} = \arg\min_{d_1^{*2} \le \overline{\sigma}^2} I(f). \tag{3.2.23}$$

As with the solutions of similar problems in Subsections 3.2.2 and 3.2.3, the three branches of solution (3.2.17) appear according to the degree in which the restrictions of the class $\mathscr{F}_{25}$ are taken into account:

- for the first branch $f^*_2(x)$, only the restriction on the variance (3.2.16) does matter taking the form of the equality: $\sigma^2(f^*_2) = \overline{\sigma}^2$;

- for the third branch, the restriction on the central part of a distribution (3.2.16) is essential, and the restriction on the variance has the form of the strict inequality:

$$\int_{-l}^{l} f^*_5(x)\,dx = 1 - \beta, \quad \sigma^2(f^*_5) < \overline{\sigma}^2;$$

- for the second, both restrictions have the form of equalities.

From (3.2.19)–(3.2.23) we have the following particular cases of solution (3.2.18):

- for $\overline{\sigma}^2 = k_1 l^2$,

$$\overline{f}^*_{25}(x) = f^*_2(x) = \frac{1}{\sqrt{2\pi}\overline{\sigma}} \exp\left(-\frac{x^2}{2\overline{\sigma}^2}\right),$$

$$\nu_1 = \nu_2 = 0, \qquad B_1 = B_2 = 1/\sigma;$$

- for $\overline{\sigma}^2 = k_2 l^2$,

$$\overline{f}^*_{25}(x) = f^*_2(x) = \begin{cases} A_1 \cos^2(B_1 x), & |x| \le l, \\ A_2 \exp(-B_2|x|), & |x| > l. \end{cases}$$

The values of the parameters A_1, A_2, B_1, and B_2 can be derived from the simultaneous equations (3.2.19)–(3.2.23) as $\nu_1, \nu_2 \to -\infty$.

We now turn directly to the restrictions of the class $\mathscr{F}_{25}$:

- as $\overline{\sigma} \to \infty$, the first restriction is inessential, and, in this case, we have the optimal solution in the class $\mathscr{F}_5$: $f^*_{25} = f^*_5$;
- as $l \to 0, 1-\beta \to 0$, and $(1-\beta)/(2l) \to 1/(2a)$, we have the restriction of the class $\mathscr{F}_{12}$ ($f(0) \ge 1/(2a) > 0$) and the optimal solution $f^*_{25} = f^*_{12}$ respectively.

We now consider the important particular case of the class $\mathscr{F}_{25}$ where the restriction on the central part of the distribution has the form of an upper bound for the value of the interquartile range of the distribution

$$F^{-1}(3/4) - F^{-1}(1/4) \le \overline{b}. \tag{3.2.24}$$

Restrictions (3.2.24) and (3.2.16) are equivalent for symmetric distributions with $\beta = 1/2$ and $l = \overline{b}/2$. Then from Theorem 3.2.4 we obtain the following.

THEOREM 3.2.5. *In the class*

$$\widetilde{\mathscr{F}}_{25} = \left\{f\colon F^{-1}(3/4) - F^{-1}(1/4) \le \overline{b},\ \sigma^2(f) \le \overline{\sigma}^2\right\},$$

the least informative density is

$$\widetilde{f}^*_{25}(x) = \begin{cases} f^*_2(x), & \overline{\sigma}^2 \le 0.548\overline{b}^2, \\ \overline{f}^*_{25}(x), & 0.548\overline{b}^2 < \overline{\sigma}^2 \le 0.903\overline{b}^2, \\ f^*_5(x), & \overline{\sigma}^2 > 0.903\overline{b}^2, \end{cases} \tag{3.2.25}$$

where the parameters of the density $\overline{f}^*_{25}(x)$ (3.2.18) *are determined from equations* (3.2.19)–(3.2.23) *with* $\beta = 1/2$.

We now observe the minimax algorithms of data processing generated from the least informative distribution density $\widetilde{f}^*_{25}(x)$. The shapes of score functions $\widetilde{\psi}^*_{25}(z)$ for the three branches of solution (3.2.25) are shown in Fig. 3.10.

All qualitative features of the algorithm optimal in the class $\mathcal{F}_{12}$ are preserved here. From (3.2.25) and Fig. 3.10 we obtain

- in the first zone with relatively small variances ($\overline{\sigma}^2 \le 0.548\overline{b}^2$), the LS method is optimal;
- in the third zone with relatively large variances ($\overline{\sigma}^2 > 0.903\overline{b}^2$), the Huber estimator similar to a trimmed mean (with the rejective threshold $|x| = \overline{b}/2$) is optimal;
- in the middle zone, the algorithms based on the Weber–Hermite functions provide smaller statistical weights of the observed data if their absolute deviations from the estimated parameter of location exceed $\overline{b}/2$.

3.2.5. The L_p-norm estimators of location

In applications, we recommend to use the L_p-norm approximations to the obtained explicit minimax estimators with the score functions $\psi^*_{12}(z)$ and $\psi^*_{25}(z)$, since the latter algorithms are difficult to calculate because of a complicated analytical structure of the Weber–Hermite functions.

In Chapter 2, we have established the importance of scale equivariancy for estimators of location. The minimax Huber M-estimators of location in ε-contaminated models are not scale equivariant. Simultaneous M-estimators of location and scale and M-estimators of location with a preliminary estimator of scale provide this property but they do not possess minimax properties (Huber, 1981).

Here we propose another approach to designing minimax scale equivariant estimators of location. It follows from the results of Chapter 2 that the requirement of scale equivariancy implies the use of the L_p-estimators

$$\widehat{\theta}_n = \arg\min_\theta \sum_{i=1}^n |x_i - \theta|^p, \qquad p \ge 1. \tag{3.2.26}$$

These estimators with $1 < p < 2$ were first used in (Forsythe, 1968) for robust estimation of location.

The corresponding minimax problem can be written as

$$(p^*, f^*) = \inf_{p\ge 1} \sup_{f\in\mathcal{F}} \operatorname{Var} \widehat{\theta}(p, f), \tag{3.2.27}$$

where

$$\operatorname{Var} \widehat{\theta}(p, f) = \frac{\int_0^\infty x^{2p-2} f(x)\,dx}{2n\left(\int_0^\infty x^{p-1} f'(x)\,dx\right)^2}.$$

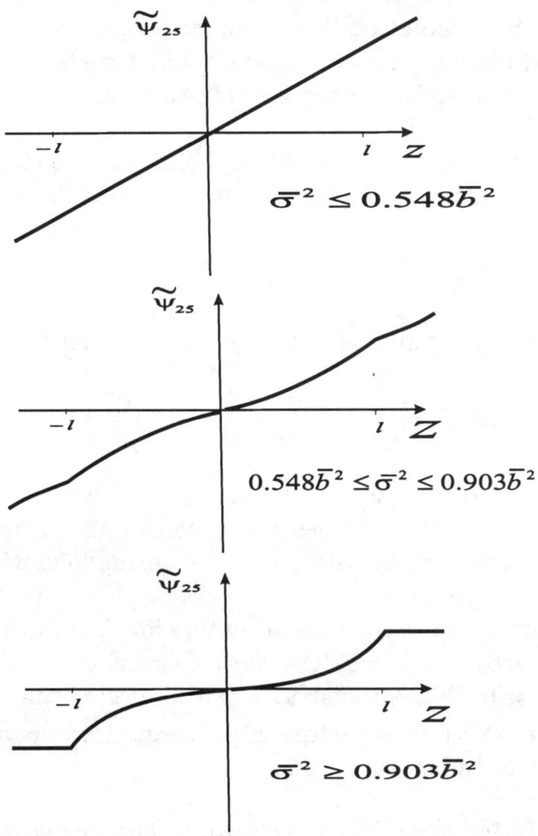

Figure 3.10. The score functions for the class $\widetilde{\mathcal{F}}_{25}$

The solution of problem (3.2.27) faces some inconvenience because of the narrowness of the parametric class of L_p-estimators, which, in general, does not include the maximum likelihood estimators. Nevertheless, it follows from the obtained results that the solution of problem (3.2.27) in the class $\mathcal{F}_1$ of nondegenerate distributions is given by the L_1-norm estimator, and in the class $\mathcal{F}_2$ with a bounded variance it is given by the L_2-norm estimator.

The following obvious assertion solves the minimax problem in the class $\mathcal{F}_3$ of ε-contaminated normal distributions (Shevlyakov, 1991).

THEOREM 3.2.6. *Let the conditions of consistency and asymptotic normality for the L_p-norm estimators hold. Then the minimax estimator is the L_1-norm estimator.*

The above conditions are just the general conditions of consistency and asymptotic normality for M-estimators formulated in Section 1.2. The straightforward checking by calculating the asymptotic variance of the L_p-norm estimators (3.2.26) at the densities with the Cauchy-type behavior of tails shows that the bounded value of the supremum of the asymptotic variance is provided only at $p = 1$.

Consider now the solution of problem (3.2.27) in the class $\mathscr{F}_{12}$ (3.2.8). It becomes much simpler if we introduce the parametric subclass of the class $\mathscr{F}_{12}$

$$\widetilde{\mathscr{F}}_{12} = \left\{ f\colon f(0) \geq \frac{1}{2a} > 0, \sigma^2(f) \leq \overline{\sigma}^2 \right\}, \tag{3.2.28}$$

where $f = f_q(x; \beta)$ is the family of exponential-power densities

$$f_q(x;\beta) = \frac{q}{2\beta\Gamma(1/q)} \exp\left(-\frac{|x|^q}{\beta^q}\right). \tag{3.2.29}$$

In formula (3.2.29), β is the scale parameter, q is the distribution shape parameter. Family (3.2.29) describes a wide collection of symmetric unimodal densities: the Laplace density with $q = 1$, the normal one with $q = 2$, and the rectangular one with $q \to \infty$.

As the L_p-estimators are the ML estimators for the location parameter θ of the density $f(x - \theta; \beta)$ with $p = q$, (the class of estimators entirely corresponds to the class of distributions), the structure of the solution of problem (3.2.27) in the class $\widetilde{\mathscr{F}}_{12}$ repeats the structure of the solution to the minimax problem in the class $\mathscr{F}_{12}$ (see Subsection 3.2.2).

THEOREM 3.2.7. *In the class $\widetilde{\mathscr{F}}_{12}$, the least informative density is of the form*

$$\widetilde{f}^*_{12}(x) = \begin{cases} \mathscr{N}(x; 0, \overline{\sigma}), & \overline{\sigma}^2/a^2 \leq 2/\pi, \\ f_{q^*}(x; \beta^*), & 2/\pi < \overline{\sigma}^2/a^2 \leq 2, \\ \mathscr{L}(x; 0, a), & \overline{\sigma}^2/a^2 > 2, \end{cases} \tag{3.2.30}$$

where q^* and β^* are determined from the equations

$$\frac{q^{*2}\Gamma(3/q^*)}{\Gamma^3(1/q^*)} = \frac{\overline{\sigma}^2}{a^2}, \qquad \beta^* * = \frac{aq^*}{\Gamma(1/q^*)}. \tag{3.2.31}$$

COROLLARY 3.2.1. *In the class $\widetilde{\mathscr{F}}_{12}$, the minimax estimator is given by the L_p-estimators* (3.2.26) *with $p = p^*$ defined by*

$$p^* = \begin{cases} 2, & \overline{\sigma}^2/a^2 \leq 2/\pi, \\ q^*, & 2/\pi < \overline{\sigma}^2/a^2 \leq 2, \\ 1, & \overline{\sigma}^2/a^2 > 2, \end{cases} \tag{3.2.32}$$

where q^* is determined from the former equation of (3.2.31).

Table 3.2. The parameters of optimal solutions in the classes $\mathscr{F}_{12}$ and $\widetilde{\mathscr{F}}_{12}$

$\overline{\sigma}^2/a^2$	$2/\pi \approx 0.637$	1.04	1.35	1.45	1.76	1.88	1.95	2
ν	0	−0.4	−0.8	−1	−2	−3	−4	$-\infty$
p^*	2	1.37	1.20	1.15	1.06	1.03	1.01	1
$I(\nu,\overline{\sigma})\,\overline{\sigma}^2$	1	1.16	1.38	1.48	1.76	1.88	1.94	2
$I(p^*)\,\overline{\sigma}^2$	1	1.17	1.40	1.49	1.76	1.88	1.94	2

The piece-wise linear approximation to the solution of equation (3.2.31) in the interval $2/\pi < \overline{\sigma}^2/a^2 \le 2$ is described by

$$p^* = \begin{cases} 2.71 - 1.12(\overline{\sigma}^2/a^2), & 2/\pi < \overline{\sigma}^2/a^2 \le 1.35, \\ 1.62 - 0.31(\overline{\sigma}^2/a^2), & 1.35 < \overline{\sigma}^2/a^2 \le 2. \end{cases} \tag{3.2.33}$$

Now we find out to which extent the solution in the parametric class of L_p-estimators is inferior to the solution based on the Weber–Hermite functions defined by Theorem 3.2.2.

In Table 3.2, the results of numerical computations of Fisher information in the classes $\mathscr{F}_{12}$ and $\widetilde{\mathscr{F}}_{12}$ are displayed for different values of $\overline{\sigma}^2/a^2$. In addition, the corresponding optimal values of the parameters p^* and ν defined by equations (3.2.31) and (5.4.10) are presented.

It is seen from Table 3.2 that the values of the Fisher information in the classes $\mathscr{F}_{12}$ and $\widetilde{\mathscr{F}}_{12}$ differ from each other at most for 2%, and so for the supremum of asymptotic variance.

3.2.6. Asymptotic relative efficiency

In studies on robustness, the Tukey contamination scheme is widely used (Tukey, 1960)

$$f(x) = (1-\varepsilon)\mathscr{N}(x;0,\sigma) + \varepsilon\mathscr{N}(x;0,k\sigma), \tag{3.2.34}$$

where ε and k are the parameters of contamination (usually, it is assumed that $\varepsilon < 0.2$).

The contamination scheme describes the case where, with large probability $1-\varepsilon$, the data occur with variance σ^2, and, with small probability ε, the outliers occur with variance $k^2\sigma^2$ ($k \gg 1$). The Huber model of ε-contaminated distributions (approximately normal) generalizes scheme (3.2.34).

Apparently, the obtained minimax algorithms having as the limiting cases robust solutions (for example, the LAV method) possess the appropriate robust properties in their entirety. Fig. 3.11 shows the dependence of the asymptotic relative efficiency (ARE) of the minimax M-estimator in the class $\mathscr{F}_{12}$ to the LS

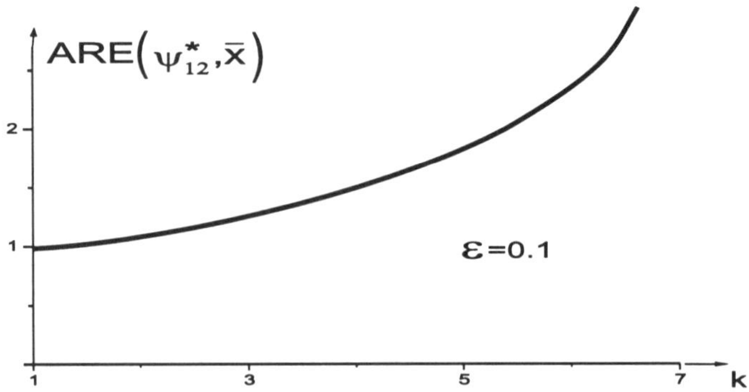

Figure 3.11. ARE of the minimax M-estimators to the sample mean under the contamination scheme

estimator (the sample mean) on the contamination parameter k with $\varepsilon = 0.1$. ARE is obtained from the following formula (we assume that $\sigma = 1$)

$$\mathrm{ARE} = \frac{\mathrm{Var}\,\bar{x}}{\mathrm{Var}\,\widehat{\theta}_n(\psi_{12}^*, f)} = \frac{(1 - \varepsilon + \varepsilon k^2)\left[\int_{-\infty}^{\infty}(\psi_{12}^*(x))' f(x)\,dx\right]^2}{\int_{-\infty}^{\infty}(\psi_{12}^*(x))^2 f(x)\,dx}, \quad (3.2.35)$$

where $f(x)$ is the distribution density (3.2.34).

The integrals in (3.2.35) are evaluated analytically with the use of the piece-wise linear approximation of the expression for ψ_{12}^* (see Fig. 3.8)

$$\psi_{12}^*(x) = \begin{cases} \psi_{12}^*(0)\,\mathrm{sgn}\,x + (\psi_{12}^*)'(0)x, & |x| \le \Delta, \\ kx, & |x| > \Delta, \end{cases}$$

where

$$k = \lim_{x\to\infty} \frac{\psi_{12}^*(x)}{x}, \qquad \Delta = \frac{\psi_{12}^*(0)}{k - (\psi_{12}^*)'(0)}.$$

The parameter ν of the optimal score function $\psi_{12}^*(x)$ is evaluated numerically from equation (5.4.10) which, in this case, can be rewritten as

$$\frac{\overline{\sigma}}{a} = 2\left(\frac{1-\varepsilon}{\sqrt{2\pi}} + \frac{\varepsilon}{\sqrt{2\pi}k}\right)(1 - \varepsilon + \varepsilon k^2)^{1/2} = \frac{\sqrt{2\nu + 1 + 1/S(\nu)}\Gamma^2(-\nu/2)}{\sqrt{2\pi}\,2^{\nu+1}\,S(\nu)\,\Gamma(-\nu)}.$$

From Fig. 3.11 we see that the estimator optimal in the class $\mathscr{F}_{12}$ is always (with the only exception of the case $k = 1$ where the estimators coincide) better than the sample mean. Here we observe the effect of the non-robustness of the LS estimators and their high sensitivity to outliers.

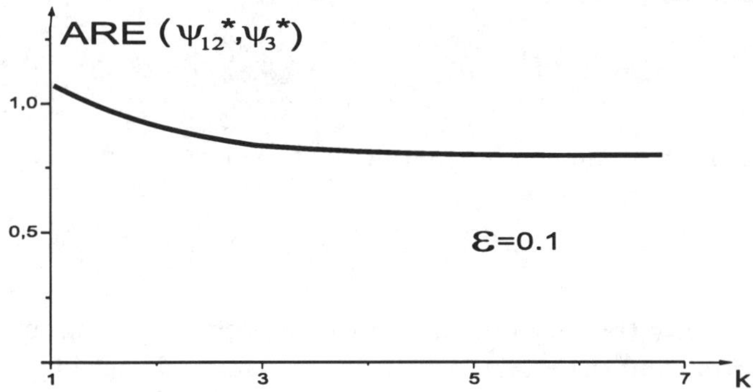

Figure 3.12. ARE of the minimax M-estimators in the classes $\mathscr{F}_{12}$ and $\mathscr{F}_3$ under the contamination scheme

We now compare the efficiencies of the qualitatively different robust algorithms, namely Huber's ψ_3^* and our ψ_{12}^*, under the contamination scheme. The graph of ARE(ψ_{12}^*, ψ_3^*) is displayed in Fig. 3.12, where ARE is determined from the relation

$$\mathrm{ARE}\,(\psi_{12}^*, \psi_3^*) = \frac{\mathrm{Var}\ \widehat{\theta}_n(\psi_3^*, f)}{\mathrm{Var}\ \widehat{\theta}_n(\psi_{12}^*, f)}$$
$$= \frac{\int_{-\infty}^{\infty}(\psi_3^*(x))^2 f(x)\,dx\left[\int_{-\infty}^{\infty}(\psi_{12}^*(x))' f(x)\,dx\right]^2}{\int_{-\infty}^{\infty}(\psi_{12}^*(x))^2 f(x)\,dx\left[\int_{-\infty}^{\infty}(\psi_3^*(x))' f(x)\,dx\right]^2}.$$

From Fig. 3.12 it follows that the estimator with the score function ψ_{12}^* is slightly inferior to the Huber estimator with large values of k but is better in the nearest neighborhood of the normal distribution. This is natural, since, first, the Huber estimator is just optimal in the class of ε-contaminated distributions (approximately normal) that contains model (3.2.34), and, second, the minimax solution ψ_{12}^* gives the sample mean being optimal for the normal density with $k = 1$.

The results of the Monte Carlo study of these estimators are represented in Chapter 8.

3.2.7. Proofs

PROOF OF LEMMA 3.2.1. Set $h(x) = g^2(x)$. Then variational problem (3.2.4) can be rewritten in the form

$$g* = \arg\min_{g} \int_0^{\infty} (g'(x))^2\,dx \tag{3.2.36}$$

under the conditions

$$\int_0^\infty g^2(x)\,dx = 1/2, \qquad \int_0^\infty x^2 g^2(x)\,dx = d^2/2.$$

For problem (3.2.36), the Euler equation is of the form

$$\frac{d^2 g}{dx^2} - \frac{1}{4}(\lambda + \mu x^2)g = 0, \tag{3.2.37}$$

where λ and μ are the Lagrangian multipliers corresponding to the normalization condition and the restriction on the variance. By setting $x = Bz$, equation (3.2.37) takes the standard form of the equation for the functions of the parabolic cylinder (the Weber–Hermite functions) (Abramowitz and Stegun, 1972; Bateman and Erdélyi, 1953)

$$\frac{d^2 g_1}{dz^2} + \left(\nu + \frac{1}{2} - \frac{z^2}{4}\right) g_1 = 0, \qquad -\infty < \nu < \infty, \tag{3.2.38}$$

where $\nu + 1/2 = -\lambda B^2/4$, $\mu B^4 = 1$, $g_1(z) = g(Bz)$.

We now rewrite the restrictions and the functional of problem (3.2.38) using the substitution $x = Bz$

$$\begin{aligned} \int_0^\infty g_1^2(z)\,dz &= \frac{1}{2B}, \\ \int_0^\infty z^2 g_1^2(z)\,dz &= \frac{d^2}{2B^3}, \\ \frac{8}{B}\int_0^\infty (g'(z))^2\,dz &= I. \end{aligned} \tag{3.2.39}$$

The linear independent real solutions of equation (3.2.38) are given by the Weber–Hermite functions $\mathcal{D}_\nu(z)$ and $\mathcal{D}_\nu(-z)$, hence the general solution of equation (3.2.38) takes the form (Abramowitz and Stegun, 1972)

$$g_1(z) = C_1\mathcal{D}_\nu(z) + C_2\mathcal{D}_\nu(-z),$$

and the requirement of boundedness leads to the choice of the branch $\mathcal{D}_\nu(z)$

$$g_1(z) = C\mathcal{D}_\nu(z). \tag{3.2.40}$$

We now substitute solution (3.2.40) into (3.2.39) and obtain the dependence of the parameters B and C on d and ν:

$$\int_0^\infty C^2\mathcal{D}_\nu^2(z)\,dz = \frac{1}{2B}, \qquad \int_0^\infty z^2 C^2\mathcal{D}_\nu^2(z)\,dz = \frac{d^2}{2B^3}. \tag{3.2.41}$$

First we use the normalization condition for which we have (Abramowitz and Stegun, 1972)

$$\int_0^\infty \mathcal{D}_\nu^2(z)\,dz = \pi^{1/2}2^{-3/2}\frac{\psi(1/2-\nu/2)-\psi(-\nu/2)}{\Gamma(-\nu)}, \tag{3.2.42}$$

where $\psi(x) = d\ln\Gamma(x)/dx$ is the digamma function and

$$C^2 = \frac{\Gamma(-\nu)}{B\pi^{1/2}2^{-1/2}[\psi(1/2-\nu/2)-\psi(-\nu/2)]}. \tag{3.2.43}$$

Representing the expression in the square brackets in the series form, we obtain

$$\psi(x)-\psi(y) = \sum_{k=0}^\infty \left(\frac{1}{y+k}-\frac{1}{x+k}\right),$$

$$\psi(1/2-\nu/2)-\psi(-\nu/2) = 2\sum_{k=0}^\infty \frac{1}{(2k-\nu)(2k-\nu+1)} = 2S(\nu),$$

$$S(\nu) = \sum_{k=0}^\infty \frac{1}{(2k-\nu)(2k-\nu+1)} = \frac{1}{2}[\psi(1/2-\nu/2)-\psi(-\nu/2)].$$

Taking the above-said into account, we rewrite formula (3.2.43) as

$$C^2 = \frac{\Gamma(-\nu)}{B\sqrt{2\pi}S(\nu)}. \tag{3.2.44}$$

In order to evaluate the second integral in (3.2.41), we use the recurrent relation for the Weber–Hermite functions

$$\mathcal{D}_{\nu+1}(z) - z\mathcal{D}_\nu(z) + \nu\mathcal{D}_{\nu-1}(z) = 0, \tag{3.2.45}$$

which yields

$$z^2\mathcal{D}_\nu^2(z) = \mathcal{D}_{\nu+1}^2(z) + 2\nu\mathcal{D}_{\nu+1}(z)\mathcal{D}_\nu(z) + \nu^2\mathcal{D}_{\nu-1}^2(z). \tag{3.2.46}$$

Substitute formula (3.2.46) into (3.2.41) and evaluate it using the relation (Bateman and Erdélyi, 1953)

$$\int_0^\infty \mathcal{D}_\mu(z)\mathcal{D}_\nu(z)\,dz = \frac{\pi 2^{\mu/2+\nu/2+1/2}}{\mu-\nu} \times \left[\frac{1}{\Gamma(1/2-\mu/2)\Gamma(-\nu/2)} - \frac{1}{\Gamma(1/2-\nu/2)\Gamma(-\mu/2)}\right]. \tag{3.2.47}$$

Using the formulas for the gamma and digamma functions

$$\Gamma(2z) = (2\pi)^{-1/2}2^{2z-1/2}\Gamma(z)\Gamma(z+1/2),$$

$$\Gamma(z+1) = z\Gamma(z), \quad \Gamma(z)\Gamma(1-z) = -z\Gamma(-z)\Gamma(z), \quad \psi(z+1) = \psi(z)+1/z,$$

after rather cumbersome transformations we obtain the relation for the distribution variance

$$2\nu + 1 + 1/S(\nu) = d^2/B^2.$$

From it and (3.2.44) we express the parameters B and C^2 via d and ν:

$$B = d/\sqrt{2\nu + 1 + 1/S(\nu)}, \qquad C^2 = \frac{\Gamma(-\nu)\sqrt{2\nu + 1 + 1/S(\nu)}}{d2^{-3/2}\pi^{1/2}S(\nu)}. \tag{3.2.48}$$

Substituting them into the expression for the density

$$h(x) = g^2(x) = g_1^2(Bx) = C^2\mathscr{D}_\nu^2(Bx),$$

we arrive at (3.2.5).

We now derive formula (3.2.7) for the functional of Fisher information at the extremals $h(x)$.

The latter integral in (3.2.39) is evaluated by differentiating the Weber–Hermite functions

$$\frac{d^m}{dz^m}\left[\exp -z^2/4\mathscr{D}_\nu(z)\right] = (-1)^m \exp -z^2/4\mathscr{D}_{\nu+m}(z), \qquad m = 1, 2, \dots,$$
$$\mathscr{D}'_\nu(z) = z\mathscr{D}_\nu(z)/2 - \mathscr{D}_{\nu+1}(z) \tag{3.2.49}$$

and using recurrent relations (3.2.45) and (3.2.46). As the result, the functional of Fisher information takes the form

$$I = \frac{2}{B}\int_0^\infty C^2[\mathscr{D}_{\nu+1}^2(z) - 2\nu\mathscr{D}_{\nu+1}(z)\mathscr{D}_\nu(z) + \nu^2\mathscr{D}_{\nu-1}^2(z)]\,dz. \tag{3.2.50}$$

The final expression for $I(\nu, d)$ is derived by substituting (3.2.48) into (3.2.50) with the use of integrals (3.2.42) and (3.2.47)

$$I(\nu, d) = [(2\nu + 1)^2 + 4(2\nu + 1)/S(\nu) + 3/S^2(\nu)]/d^2,$$

which completes the proof of Lemma 3.2.1. □

PROOF OF THEOREM 3.2.1. The validity of theorem immediately follows from Lemma 3.2.1. □

PROOF OF THEOREM 3.2.2. Here we directly check the optimality of f_{12}^* with the use of condition (3.1.28) (see Section 3.1)

$$\left[\frac{d}{dt}I(f_t)\right]_{t=0} \ge 0, \tag{3.2.51}$$

where $f_t = (1 - t)f^* + tf$ and f is an arbitrary density such that $I(f) < \infty$.

Recall that inequality (3.2.51) can be rewritten as

$$\int_{-\infty}^{\infty} (2\psi^{*\prime} - \psi^{*2})(f - f^*)\,dx \ge 0, \tag{3.2.52}$$

where $\psi^*(x) = -(f^*(x))'/f^*(x)$ is the score function.

In view of the structure of the optimal solution, it suffices to consider $f_{12}^* = f_\nu((x;\overline{\sigma}) = C^2\mathscr{D}_\nu^2(B|x|)$, since this family of extremals contains both cases: the normal density with $\nu = 0$ and the Laplace density as $\nu \to -\infty$.

Taking these considerations into account, we transform the left-hand side of inequality (3.2.52) as

$$\int_{-\infty}^{\infty} (2\psi^{*\prime} - \psi^{*2})(f - f^*)\,dx$$

$$= \frac{4\mathscr{D}_{\nu+1}(0)}{\mathscr{D}_\nu(0)} B[f(0) - f_\nu(0;\overline{\sigma})] + B^4[\overline{\sigma}^2 - \sigma^2(f)], \tag{3.2.53}$$

where $\sigma^2(f) = \int_{-\infty}^{\infty} x^2 f(x)\,dx$.

We now check the sign of both summands in (3.2.53). For $\nu \le 0$, we have $\mathscr{D}_\nu(0) > 0$ and $B = \sqrt{2\nu + 1 + 1/S(\nu)}/\sigma > 0$, hence the expression in the square brackets is nonnegative since it is one of the restrictions of the class $\mathscr{F}_{12}$:

$$f(0) - f_\nu(0;\overline{\sigma}) \ge 0 \iff f(0) \ge f_\nu(0;\overline{\sigma}) = \frac{1}{2a}.$$

Observe that equation (5.4.10) defining the optimal value of the parameter ν is the rewritten restriction of the class $\mathscr{F}_{12}$

$$f_{12}^*(0) = f_\nu(0;\overline{\sigma}) = \frac{1}{2a},$$

that holds as the equality in this case.

Further, the sign of the second summand in the right-hand side of (3.2.53) is determined by the second restriction of the class $\mathscr{F}_{12}$: $\overline{\sigma}^2 - \sigma^2(f) \ge 0$. Thus we arrive at inequality (3.2.52). □

REMARK 3.2.3. It can be also seen that the proofs of optimality of the Laplace density in the class $\mathscr{F}_1$ and the normal density in the class $\mathscr{F}_2$ follow directly from (3.2.53) after checking the signs of the first and second summands respectively.

PROOF OF THEOREM 3.2.3. The proof of Theorem 3.2.3 is analogous to that of Theorem 3.2.2, and it is performed by verifying inequality (3.2.52).

In this case, the left-hand side of inequality (3.2.52) takes the form

$$\int_{-\infty}^{\infty} (2\psi^{*\prime} - \psi^{*2})(f - f^*)\,dx = A_1 \int_{-\infty}^{\infty} [f(x) - (1-\varepsilon)\mathscr{N}(x;0,\sigma_N)]\,dx$$

$$+ A_2[\overline{\sigma}^2 - \sigma^2(f)], \quad A_1, A_2 \ge 0,$$

and it is nonnegative due to the restrictions of the class $\mathscr{F}_{23}$

$$f(x) \ge (1-\varepsilon)\mathscr{N}(x; 0, \sigma_N)$$

and

$$\sigma^2(f) \le \overline{\sigma}^2.$$

□

PROOF OF THEOREM 3.2.4. The proof is in two stages. First we obtain the structure of the optimal solution. Second, we check inequality (3.2.52) in the class $\mathscr{F}_{25}$.

Consider the following variational problems connected with each other in the domains $0 \le x \le l$ and $x > l$:

$$I_1 = \int_0^l {g_1'}^2(x)\,dx \to \min, \quad \int_0^l {g_1}^2(x)\,dx = (1-\beta)/2, \quad \int_0^l x^2 {g_1}^2(x)\,dx = d_1^2/2,$$
$$I_2 = \int_l^\infty {g_1'}^2(x)\,dx \to \min, \quad \int_l^\infty {g_1}^2(x)\,dx = \beta/2, \quad \int_l^\infty x^2 {g_1}^2(x)\,dx = (\overline{\sigma}^2 - d_1^2)/2. \tag{3.2.54}$$

By Lemma 3.2.1, we see that the general solution of Euler equation (3.2.38) takes the following forms for each problem:

$$g_1(x) = C_{11}\mathscr{D}_{\nu_1}(B_1 x) + C_{21}\mathscr{D}_{\nu_1}(-B_1 x), \quad 0 \le x \le l,$$
$$g_2(x) = C_{12}\mathscr{D}_{\nu_2}(B_2 x) + C_{22}\mathscr{D}_{\nu_2}(-B_2 x), \quad x > l.$$

The condition of optimality on the free boundary at $x = 0$: $g_1'(0) = 0$, and the boundedness of the solution as $x \to \infty$ imply the relations

$$C_{11} = C_{21} = C_1 \quad \text{and} \quad C_{22} = 0.$$

Thus, for seven unknown parameters C_1, C_{12}, B_1, B_2, ν_1, ν_2, and d_1^2, we have four equations (3.2.54), two equations of continuity of the least favorable density and its derivative at $x = l$:

$$g_1(l-0) = g_1(l+0), \qquad g_1'(l-0) = g_1'(l+0),$$

and the condition of the optimal choice of the parameter d_1^2

$$d_1^{*2} = \arg \min_{0 \le d_1^2 \le \overline{\sigma}^2} (I_1 + I_2).$$

Taking the condition of symmetry into account, we arrive at the expression for the least favorable density

$$\overline{f}_{25}^*(x) = \begin{cases} g_1^2(x) = A_1[\mathscr{D}_{\nu_1}(B_1 x) + \mathscr{D}_{\nu_1}(-B_1 x)]^2, & |x| \le l, \\ g_2^2(x) = A_2 \mathscr{D}_{\nu_2}^2(B_2 |x|), & |x| > l, \end{cases}$$

where $A_1 = C_1^2$ and $A_2 = C_{12}^2$.

We now verify inequality (3.2.52). In this case, the integrand in (3.2.52) is of the form

$$2{\psi^*}'_{25} - {\psi^*}^2_{25} = \begin{cases} 2B_1^2(2\nu_1 + 1) - B_1^4 x^2, & |x| \le l, \\ 2B_2^2(2\nu_2 + 1) - B_2^4 x^2, & |x| > l, \end{cases}$$

where $\psi_{25}^* = -\overline{f}_{25}^{*\,\prime}/\overline{f}_{25}^*$.

Integrating the above expression and extracting the summands with the restrictions of the class $\mathscr{F}_{25}$, we obtain

$$\begin{aligned} \int_{-\infty}^{\infty} (2\psi^{*\prime} - \psi^{*2})(f - f^*)\,dx &= B_1^4 \left(\int_{-\infty}^{\infty} x^2 f^*(x)\,dx - \int_{-\infty}^{\infty} x^2 f(x)\,dx \right) \\ &+ \left[B_1^2(2\nu_1 + 1) - B_2^2(2\nu_2 + 1) \right] \left(\int_{-l}^{l} f(x)\,dx - \int_{-l}^{l} f^*(x)\,dx \right) \\ &+ \left(B_2^4 - B_1^4 \right) \int_{|x|>l} x^2 [f^*(x) - f(x)]\,dx. \qquad (3.2.55) \end{aligned}$$

Now we establish the sign of each summand in the right-hand side of equality (3.2.55). The first summand is nonnegative, as it is the restriction on variance of the class $\mathscr{F}_{25}$

$$\int_{-\infty}^{\infty} x^2 f^*\,dx = \int_{-\infty}^{\infty} x^2 f(x)\,dx = \overline{\sigma}^2 - \sigma^2(f) \ge 0.$$

The second factor in the second summand is also nonnegative, since it is the restriction of approximate finiteness

$$\int_{-l}^{l} f\,dx - \int_{-l}^{l} f^*\,dx = \int_{-l}^{l} f\,dx - (1 - \beta) \ge 0. \qquad (3.2.56)$$

Now consider the third summand. Inequality (3.2.56) can be rewritten as

$$\int_{|x|\le l} (f^* - f)\,dx \le 0.$$

From the above and $\int_{-\infty}^{\infty} (f^* - f)\,dx = 0$ it follows that

$$\int_{|x|>l} (f^* - f)\,dx \ge 0.$$

By the mean value reasoning, we obtain

$$\int_{|x|>l} x^2 (f^* - f)\,dx = \xi^2 \int_{|x|>l} (f^* - f)\,dx \ge 0, \qquad l < \xi < \infty.$$

Thus, the sign of the last two summands in (3.2.55) is nonnegative if

$$B_2 > |B_1| \quad \text{and} \quad B_1^2(2\nu_1 + 1) - B_2^2(2\nu_2 + 1) \le 0.$$

We check the latter inequalities with the use of numerical calculations. The sign of the modulus for the parameter B_1 is explained by the fact that with $\nu_1 < 0$ it takes imaginary values. This entirely agrees with with the limiting case of the optimal solution $f_{25}^* = f_5^*$, whose *cosine* branch is given by the sum of the Weber–Hermite functions of imaginary arguments

$$\cos z \propto \lim_{\nu_2 \to -\infty} [\mathcal{D}_{\nu_2}(iz) + \mathcal{D}_{\nu_2}(-iz)],$$

as $e^{-z} \propto \lim_{\nu \to -\infty} \mathcal{D}_\nu(z)$. In the domain $\nu_1 \ge -1/2$, the parameter B_1 takes real values, in its turn, the values of B_2 are only real. □

3.3. Robust estimation of location in models with bounded subranges

3.3.1. The least informative density in the class $\mathcal{F}_{55}$

Consider the class of densities with the restriction on their mass in the central zone or the class of approximately finite densities

$$\mathcal{F}_5 = \left\{ f: \int_{-l}^{l} f(x)\,dx \ge 1 - \beta, 0 < \beta \le 1 \right\}. \tag{3.3.1}$$

The constraint on the distribution mass can be rewritten as the constraint on the distribution subrange

$$\mathcal{F}_5 = \left\{ f: F^{-1}(1 - \beta/2) - F^{-1}(\beta/2) \le b, 0 < \beta \le 1 \right\}, \tag{3.3.2}$$

where $b = 2l$.

We recall (see Section 3.1) that in this case the least informative density has the *cosine*-central part and the exponential tails

$$f^*(x; A, B, C, D, b) = \begin{cases} A\cos^2(Bx), & |x| \le b/2, \\ C\exp(-D|x|), & |x| > b/2, \end{cases} \tag{3.3.3}$$

where the values $A = A(\beta, b)$, $B = B(\beta, b)$, $C = C(\beta, b)$, and $D = D(\beta, b)$ are chosen to satisfy the conditions

- the normalization condition

$$\int_{-\infty}^{\infty} f^*(x; A, B, C, D, b)\,dx = 1;$$

- the characterization condition of the class $\mathscr{F}_5$

$$\int_{-b/2}^{b/2} f^*(x; A, B, C, D, b)\, dx = 1 - \beta;$$

- the conditions of smoothness at $x = b/2$

$$f^*(b/2 - 0; A, B, C, D, b) = f^*(b/2 + 0; A, B, C, D, b),$$
$$f^{*\prime}(b/2 - 0; A, B, C, D, b) = f^{*\prime}(b/2 + 0; A, B, C, D, b).$$

The exponential tails of the least informative density and the corresponding form of the robust minimax contrast function $\rho^* = |x|$ for $|x| > b/2$ imply that the observed data with $|x| > b/2$ are simply ignored ('rejected') when we apply this method. The smaller b, the more data is rejected.

We now consider the class $\mathscr{F}_{55}$ with the inequality constraints on distribution subranges

$$\mathscr{F}_{55} = \{f : F^{-1}(1 - \beta_1/2) - F^{-1}(\beta_1/2) \le b_1, F^{-1}(1 - \beta_2/2) - F^{-1}(\beta_2/2) \le b_2\} \tag{3.3.4}$$

with $0 \le \beta_2 \le \beta_1 \le 1, b_1 \le b_2$. The following result holds in this case (Shevlyakov, 1995).

THEOREM 3.3.1. *In the class $\mathscr{F}_{55}$, the least informative density is of the form*

$$f_{55}^*(x) = \begin{cases} f^*(x; A_2, B_2, C_2, D_2, b_2), & b_2/b_1 \le k_1, \\ f^*(x; A^*, B^*, C^*, D^*, b^*), & k_1 < b_2/b_1 \le k_2, \\ f^*(x; A_1, B_1, C_1, D_1, b_1), & b_2/b_1 > k_2, \end{cases} \tag{3.3.5}$$

where

- *the function $f^*(x; A, B, C, D, b)$ is defined by equation* (3.3.3)*;*

- *the values of the parameters $A_1, \dots, D_1$ are set to $A_1 = A(\beta_1, b_1)$, $B_1 = B(\beta_1, b_1)$, $C_1 = C(\beta_1, b_1)$, $D_1 = D(\beta_1, b_1)$;*

- *the values of the parameters $A_2, \dots, D_2$ are set to $A_2 = A(\beta_2, b_2)$, $B_2 = B(\beta_2, b_2)$, $C_2 = C(\beta_2, b_2)$, $D_2 = D(\beta_2, b_2)$;*

- *the values of the parameters A^*, B^*, C^*, D^*, and b^*, $b_1 < b^* < b_2$, are determined from the equations including:*

  - *the normalization condition*

$$\int_{-\infty}^{\infty} f^*(x; A^*, B^*, C^*, D^*, b^*)\, dx = 1;$$

– *the characterization conditions of the class*

$$\int_{-b_1/2}^{b_1/2} f^*(x;\, A^*, B^*, C^*, D^*, b^*)\, dx = 1 - \beta_1,$$
$$\int_{-b_2/2}^{b_2/2} f^*(x;\, A^*, B^*, C^*, D^*, b^*)\, dx = 1 - \beta_2;$$

– *the conditions of smoothness at* $x = b^*$

$$f^*(b^* - 0; A^*, B^*, C^*, D^*, b^*) = f^*(b^* + 0; A^*, B^*, C^*, D^*, b^*),$$
$$f^{*\prime}(b^* - 0; A^*, B^*, C^*, D^*, b^*) = f^{*\prime}(b^* + 0; A^*, B^*, C^*, D^*, b^*);$$

- *the switching parameters* k_1 *and* k_2 *of solution* (3.3.5) *are derived from the equations*

$$\int_0^{b_2/2k_2} f^*(x;\, A_1, B_1, C_1, D_1, b_1)\, dx = (1 - \beta_1)/2,$$
$$\int_0^{k_1 b_1/2} f^*(x;\, A_2, B_2, C_2, D_2, b_2)\, dx = (1 - \beta_2)/2.$$

Three branches of solution (3.3.5) are connected with the degree in which the constraints are taken into account:

- in the first zone ($b_2/b_1 \le k_1$), only the second restriction matters;
- in the third zone ($b_2/b_1 > k_2$), only the first restriction is substantial;
- in the intermediate zone, both restrictions are used.

From (3.3.5) we can conclude that

- for relatively small distribution dispersion (in the first zone), the 'mild' robust algorithm based on $f^*(x;\, A_2, B_2, C_2, D_2, b_2)$ is optimal;
- for relatively large distribution dispersion (in the third zone), the hard robust algorithm (with the hard rejection of sample elements) based on $f_1^*(x; A_1, B_1, C_1, D_1, b_1)$ is optimal,
- in the middle zone, a compromise between these algorithms is the best solution.

3.3.2. The least informative density in the class $\mathscr{F}_{15}$

Now we consider the intersection of the classes $\mathscr{F}_1$ and $\mathscr{F}_5$ with the constraints on the value of a density at the center of symmetry and on the distribution subrange:

$$\mathscr{F}_{15} = \left\{ f\colon f(0) \geq \frac{1}{2a}, F^{-1}(1-\beta/2) - F^{-1}(\beta/2) \leq b, 0 < \beta \leq 1 \right\}. \quad (3.3.6)$$

The following result is true in this case.

THEOREM 3.3.2. *In the class $\mathscr{F}_{15}$, the least informative density is of the form*

$$f_{15}^*(x) = \begin{cases} f^*(x; A1, B1, C1, D1, b1), & b/a \leq k, \\ f^*(x; A^*, B^*, C^*, D^*, b^*), & k < b/a \leq 2, \\ \mathscr{L}(x; 0, a), & b/a > 2, \end{cases} \quad (3.3.7)$$

where

- *the function $f^*(x; A, B, C, D, b)$ is defined by equations (3.3.3);*
- *the values of the parameters $A1, \dots, D1$ are set to $A1 = A(\beta, b)$, $B1 = B(\beta, b)$, $C1 = C(\beta, b)$, $D1 = D(\beta, b)$;*
- *the values of the parameters A^*, B^*, C^*, D^*, and b^*, $2a < b^* < b$, are determined from the equations including*
  - *the normalization condition*

$$\int_{-\infty}^{\infty} f^*(x; A^*, B^*, C^*, D^*, b^*)\, dx = 1;$$

  - *the characterization conditions of the class*

$$f^*(x; A^*, B^*, C^*, D^*, b^*) = \frac{1}{2a},$$

$$\int_{-b/2}^{b/2} f^*(x; A^*, B^*, C^*, D^*, b^*)\, dx = 1 - \beta;$$

  - *the conditions of smoothness at $x = b^*$*

$$f^*(b^* - 0; A^*, B^*, C^*, D^*, b^*) = f^*(b^* + 0; A^*, B^*, C^*, D^*, b^*),$$
$$f^{*\prime}(b^* - 0; A^*, B^*, C^*, D^*, b^*) = f^{*\prime}(b^* + 0; A^*, B^*, C^*, D^*, b^*);$$

- *the switching parameter k of solution (3.3.7) is given by*

$$\int_0^{ka} f^*(x; A1, B1, C1, D1, b)\, dx = (1 - \beta)/2.$$

Here the conclusions are similar to those of Theorem 3.3.1, only with the relatively large distribution dispersion in the tail domain, the robust method is the L_1-norm method, i.e., the sample median for location. The proofs of the both theorems are carried out by direct checking optimality condition (3.1.28), which gives the the characterization inequalities of the class of densities.

3.4. Robust estimators of multivariate location

3.4.1. Preliminaries

In this section we consider the problems of robust minimax estimation of a multivariate location parameter.

In the literature, the estimation of multivariate location is usually examined in the context of a much more general and difficult problem in robust statistics: the simultaneous estimation of location and shape of the data (Campbell, 1980; Campbell, 1982; Davies, 1987; Devlin *et al.*, 1981; Donoho, 1982; Hampel *et al.*, 1986; Huber, 1981; Lopuhaä, 1989; Maronna, 1976; Meshalkin, 1971; Rocke and Woodruff, 1993; Rousseeuw, 1984; Rousseeuw and Leroy, 1987; Shurygin, 1994a).

We recall the precise formulation of this problem.

Let $\mathbf{x}_1, \dots, \mathbf{x}_n$ be a sample from an m-variate elliptical distribution with a density f of the form

$$f(\mathbf{x}) = (\det \mathbf{C})^{-1/2} h\left[(\mathbf{x}-\mathbf{t})^T \mathbf{C}^{-1}(\mathbf{x}-\mathbf{t})\right], \qquad (3.4.1)$$

where $\mathbf{t} = (t_1, \dots, t_m)^T$ is an m-variate location vector, $\mathbf{C}$ is an $(m\times m)$-covariance matrix and $h(|\mathbf{x}|)$ is a density in $\mathbf{R}^m$ ($|\cdot|$ stands for the Euclidean norm).

The problem is to estimate the location vector $\mathbf{t}$ and covariance matrix $\mathbf{C}$ when h is only approximately known.

Meshalkin estimators. In (Meshalkin, 1971), this problem was first considered for the important particular case of an m-variate normal distribution. Meshalkin proposes exponential weighting and proves the consistency of estimators $\widehat{\mathbf{t}}$ and $\widehat{\mathbf{C}}$, which are solutions of simultaneous matrix equations

$$\sum_{i=1}^{n} (\mathbf{x}_i - \widehat{\mathbf{t}}) \exp(-\lambda d_i/2) = 0,$$

$$\sum_{i=1}^{n} \left[(\mathbf{x}_i - \widehat{\mathbf{t}})(\mathbf{x}_i - \widehat{\mathbf{t}})^T + (1+\eta)^{-1}\widehat{\mathbf{C}}\right] \exp(-\lambda d_i/2) = 0,$$

where $d_i^2 = (\mathbf{x}_i - \widehat{\mathbf{t}})^T \widehat{\mathbf{C}}^{-1} (\mathbf{x}_i - \widehat{\mathbf{T}})$, and $\lambda, \eta > 0$ are some suitable constants. In the univariate case $m = 1$, Meshalkin recommends the values $\lambda = \eta = 1/2$.

M-estimators. In (Maronna, 1976), robust M-estimators were introduced for multivariate location and covariance, their consistency and asymptotic normality was proved, and qualitative robustness properties were studied. In (Huber, 1981), Maronna's definition was extended by defining M-estimators as solutions of the simultaneous matrix equations

$$\frac{1}{n}\sum_{i=1}^{n} v_1(d_i)(\mathbf{x}_i - \widehat{\mathbf{t}}) = 0, \tag{3.4.2}$$

$$\frac{1}{n}\sum_{i=1}^{n} \left[v_2(d_i)(\mathbf{x}_i - \widehat{\mathbf{t}})(\mathbf{x}_i - \widehat{\mathbf{t}})^T - v_3(d_i)\widehat{\mathbf{C}}\right] = 0, \tag{3.4.3}$$

where v_1, v_2 and v_3 are real-valued functions on $[0, \infty)$.

In particular, in (Huber, 1964) it was suggested to take $v_3(y) = 1$, $v_1(y) = \psi_1(y)/y$ and $v_2(y) = \psi_2(y)/y$, where $\psi_1(y) = \psi_H(y, k)$ and $\psi_2(y) = \psi_H(y, k^2)$. The function $\psi_H(y) = \min\{k, \max\{y, -k\}\}$ is the Huber ψ-function.

Obviously, M-estimators generalize the Meshalkin estimators.

REMARK 3.4.1. Equations (3.4.2) and (3.4.3) determine the location estimator $\mathbf{t}$ as the weighted mean

$$\widehat{\mathbf{t}} = \frac{\sum_{i=1}^{n} v_1(d_i)\mathbf{x}_i}{\sum_{i=1}^{n} v_1(d_i)} \tag{3.4.4}$$

with weights $v_1(d_i)$ depending on the estimators $\widehat{\mathbf{t}}$ and $\widehat{\mathbf{C}}$ sought for. Also, the estimator $\widehat{\mathbf{C}}$ can be written in a similar way (see (Huber, 1981)). These representations are the basis for iterative procedures of calculating simultaneous estimators of location and covariance (see (Huber, 1981) and Section 8.1.1).

Maximum likelihood estimators. M-estimators (3.4.2) and (3.4.3) embrace the maximum likelihood estimators as a particular case (Huber, 1981)

$$\frac{1}{n}\sum_{i=1}^{n} v_1(d_i)(\mathbf{x}_i - \widehat{\mathbf{t}}) = 0, \tag{3.4.5}$$

$$\frac{1}{n}\sum_{i=1}^{n} \left[v_2(d_i)(\mathbf{x}_i - \widehat{\mathbf{t}})(\mathbf{x}_i - \widehat{\mathbf{t}})^T - \widehat{\mathbf{C}}\right] = 0, \tag{3.4.6}$$

where

$$v_1(r) = v_2(r) = -\frac{f'(r)}{rf(r)}.$$

S-estimators. In (Rousseeuw and Yohai, 1984), S-estimators were defined in the regression context as the solution to the problem of minimization of σ under the condition

$$\frac{1}{n}\sum_{i=1}^{n}\rho\left(\frac{y_i - \theta^T \mathbf{x}_i}{\sigma}\right) = b_0 \tag{3.4.7}$$

over all $(\theta, \sigma) \in \mathbf{R}^m \times (0, \infty)$, where $0 < b_0 < \sup \rho$. The particular case $\rho(y) = y^2$ in (5.5.3) obviously gives the LS estimators.

In (Lopuhaä, 1989), this definition was extended to S-estimators of multivariate location and covariance as the solutions $\theta_n = (\mathbf{t}_n, \mathbf{C}_n)$ to the problem of minimization of $\det(\mathbf{C})$ provided that

$$\frac{1}{n}\sum_{i=1}^{n}\rho\left([(\mathbf{x}_i - \mathbf{t})^T \mathbf{C}^{-1}(\mathbf{x}_i - \mathbf{T})]^{1/2}\right) = b_0. \tag{3.4.8}$$

It was also shown that S-estimators satisfy conditions (3.4.2) and (3.4.3) for M-estimators, and obtains that S-estimators have a limiting normal distribution which is similar to the limiting normal distribution of M-estimators.

Shurygin estimators. In (Shurygin, 1994a; Shurygin, 1995; Shurygin, 2000), the so-called *stoikii* (sturdy) estimators were designed of multivariate location and covariance optimizing complex criteria of efficiency and stability. The derived estimators are similar in their structure to the Meshalkin estimators with the weights depending on the form of an underlying distribution.

3.4.2. Least informative distributions

Here we are mainly interested in robust minimax estimation of multivariate location and therefore in the structure of least informative distributions determining the structure of robust minimax estimators.

Huber solution in ε-contaminated models. In (Huber, 1981), the least informative distribution was given over the class of spherically symmetric ε-contaminated normal distributions in $\mathbf{R}^3$. It is of the following form (see (Huber, 1981, p. 230)):

$$f^*(r) = \begin{cases} a \exp(-r^2/2), & r \le r^*, \\ b r^{-2} \exp(-cr), & r > r^*, \end{cases}$$

where

$$a = (1 - \varepsilon)(2\pi)^{-3/2},$$
$$b = (1 - \varepsilon)(2\pi)^{-3/2} r^{*2} \exp(r^{*2}/2 - 2),$$
$$c = r^* - 2/r^*.$$

The constants r^* and ε are related by the condition of normalization for f^*

$$4\pi^2 \int f^*(r) r^2 dr = 1.$$

Then the minimax estimator for location is given by the maximum likelihood principle

$$-\frac{f^{*\prime}(r)}{f^*(r)} = \begin{cases} r, & r \le r^*, \\ c + 2/r, & r > r^*; \end{cases}$$

(cf. (3.4.5)).

The solution in the class with a bounded covariance matrix. In (Luneva, 1983), the problem was considered to minimize the Fisher information under distributions symmetric about zero with bounded covariance matrices

$$\mathscr{F} = \left\{ f \colon \int \cdots \int \mathbf{x}\mathbf{x}^T f(x_1, \dots, x_m)\, dx_1 \cdots dx_m \le \overline{\mathbf{C}} \right\},$$

where $\mathbf{C}$ is a given $m \times m$ positive definite matrix.

In this case, the least informative distribution is normal

$$f^*(\mathbf{x}) = \mathscr{N}_m(\mathbf{x}; \mathbf{0}, \overline{\mathbf{C}}) \tag{3.4.9}$$

with the corresponding minimax LS estimator of location in the form of the sample mean

$$\widehat{\mathbf{t}} = \frac{1}{n} \sum_{i=1}^{n} \mathbf{x}_i.$$

Bokk solution in ε-contaminated models. In (Bokk, 1990), the above Huber solution was extended to the case of arbitrary dimension m. For the class of spherically symmetric ε-contaminated normal distributions

$$\mathscr{F} = \{ f \colon f(r) \ge (1 - \varepsilon)\, \mathscr{N}_m(r; \sigma), 0 \le \varepsilon < 1 \}, \tag{3.4.10}$$

$$\mathscr{N}_m(r; \sigma) = \frac{1}{(2\pi)^{m/2} \sigma^m} \exp\left(-\frac{r^2}{2\sigma^2} \right),$$

the least informative distribution minimizing the Fisher information

$$I(f) = \int_0^\infty \left(\frac{f'(r)}{f(r)} \right)^2 f(r) r^{m-1} dr$$

is given by

$$f^*(r) = \begin{cases} (1-\varepsilon)\mathcal{N}_m(r;\sigma), & r \le r^*, \\ (1-\varepsilon)\dfrac{\mathcal{N}_m(r^*;\sigma)}{K_\nu^2(\lambda^{1/2}r^*)}\dfrac{r^{*2\nu}}{r^{2\nu}}K_\nu^2(\lambda^{1/2}r), & r > r^*, \end{cases} \tag{3.4.11}$$

where

- K_ν is the modified Macdonald function of order ν (Abramowitz and Stegun, 1972);
- $\nu = m/2 - 1$;
- λ satisfies the equation
$$\lambda^{1/2}\frac{K_{\nu+1}(\lambda^{1/2}r^*)}{K_\nu(\lambda^{1/2}r^*)} = \frac{r^*}{2\sigma^2};$$
- and r^* is determined from the normalization condition
$$\frac{2\pi^{m/2}}{\Gamma(m/2)}\int_0^\infty r^{m-1}f^*(r)\,dr = 1.$$

The minimax estimator for location $\mathbf{t}$ is obtained from the maximum likelihood equation (3.4.5).

3.4.3. The L_p-norm estimators of multivariate location

In this subsection we apply the results of Chapter 2 on orthogonal and scale equivariancy of the L_p-norm estimators of multivariate location in order to obtain relatively simple and efficient estimators. In Subsection 3.2 we use the L_p-norm estimators with $1 < p < 2$ to approximate the precise solutions based on the Weber–Hermite functions. Now we demonstrate that those results can be partly generalized in the multivariate case.

The least informative distributions minimizing the Fisher information for a multivariate location parameter are derived in the parametric classes of the exponential-power spherically symmetric distributions with the following characterizing restrictions:

- a bounded variance;
- a bounded value of a distribution density at the center of symmetry;
- the intersection of these restrictions.

For the first two cases, the least informative distributions are normal and Laplace, respectively. In the latter case, the optimal solution has three branches:

- with relatively small variances, it is normal;
- with relatively large variances, it is the Laplace;
- and it is a compromise between them with intermediate variances.

The corresponding robust minimax M-estimators of location are given by the L_2-norm, L_1-norm and L_p-norm methods respectively.

Let $\mathbf{x}_1, \dots, \mathbf{x}_n$ be a sample from an m-variate spherically symmetric density

$$f(\mathbf{x} - \mathbf{t}) = f(|\mathbf{x} - \mathbf{t}|), \qquad \mathbf{x}, \mathbf{t} \in \mathbf{R}^m,$$

with f belonging to the parametric class of exponential-power distributions

$$\mathscr{F}_q = \left\{ f : f_q(r; \beta) = \frac{q\Gamma(m/2)}{2\pi^{m/2}\beta^m\Gamma(m/q)} \exp\left(-\frac{r^q}{\beta^q}\right) \right\}, \tag{3.4.12}$$

where

$$q \ge 1, \qquad r = |\mathbf{x} - \mathbf{t}| = \left(\sum_{j=1}^{m} (x_j - t_j)^2 \right)^{1/2},$$

and β is the scale parameter.

The L_p-norm estimator of a location parameter $\mathbf{t} = (t_1, \dots, t_m)$ is defined as

$$\widehat{\mathbf{t}}_{L_p} = \arg\min_{\mathbf{t}} \sum_{i=1}^{n} r_i^p, \qquad p \ge 1, \tag{3.4.13}$$

$$r_i = \left(\sum_{j=1}^{m} (x_{ij} - t_j)^2 \right)^{1/2}. \tag{3.4.14}$$

We use the L_p-norm estimators, since they are the maximum likelihood estimators of location for densities (3.4.12) when $p = q$.

Now we search for the minimax variance L_p-norm estimators of multivariate location in the class $\mathscr{F}_q$. From spherical symmetry it follows that the saddle point (p^*, q^*) of the covariance matrix $\mathbf{C}(p, q)$ of the L_p-norm estimator (3.4.13)

$$\mathbf{C}(p^*, q) \le \mathbf{C}(p^*, q^*) = \mathbf{C}(q^*, q^*) = \mathbf{I}^{-1}(q^*),$$

where $\mathbf{I}$ is the Fisher information matrix, is determined from the solution of the variational problem

$$f^*(r) = \arg\min_{f \in \mathscr{F}} \int_0^\infty \left[\frac{f'(r)}{f(r)} \right]^2 f(r) r^{m-1}\, dr. \tag{3.4.15}$$

Hence, for the class $\mathscr{F}_q$, we have the simplest problem of the parameter minimization

$$(q^*, \beta^*) = \arg\min_{q,\beta} \frac{q^2\Gamma\left(\frac{m-2}{q} + 2\right)}{\beta^2\Gamma(m/q)}. \tag{3.4.16}$$

Least informative distribution in the class $\mathscr{F}_q$. Using additional restrictions on densities (3.4.12), we obtain the multivariate analogs of the univariate least informative densities described in Subsection 3.2 (Shevlyakov, 1991)

PROPOSITION 3.4.1. *In the class of nondegenerate densities*

$$\mathscr{F}_{1q} = \left\{ f_q : f_q(0; \beta) \geq \frac{1}{2a^m} > 0 \right\}, \tag{3.4.17}$$

the least informative density is the multivariate analog of the Laplace density

$$f_1^*(r) = \mathscr{L}_m(r; \beta^*) = \frac{1}{2a^m} \exp\left(-\frac{r}{\beta^*}\right), \tag{3.4.18}$$

where

$$\beta^* = \frac{a}{2^{(m-1)/m}\pi^{(m-1)/(2m)}\Gamma^{1/m}((m+1)/2)}.$$

In this case, the minimization problem (3.4.16) is immediately solved by excluding the parameter β from the equation $f_q(\mathbf{0}; \beta) = 1/(2x^m)$ with $x \leq a$ followed by substituting it into the expression for Fisher information. Thus we have $q^* = 1$, and the following minimization with respect to x yields $x^* = a$.

PROPOSITION 3.4.2. *In the class with bounded component variances*

$$\mathscr{F}_{2q} = \left\{ f_q : \sigma_k^2(f_q) = \int \cdots \int x_k^2 f_q(r)\, dx_1 \cdots dx_m \leq \overline{\sigma}^2,\ k = 1, \dots, m \right\}, \tag{3.4.19}$$

the least informative density is normal

$$f_2^*(r) = \mathscr{N}_m(r; \overline{\sigma}) = \frac{1}{(2\pi)^{m/2}\overline{\sigma}^m} \exp\left(-\frac{r^2}{2\overline{\sigma}^2}\right). \tag{3.4.20}$$

This assertion immediately follows from the above-mentioned general result in (Luneva, 1983): the multivariate normal density $\mathscr{N}_m(\mathbf{x}; \boldsymbol{\theta}, \overline{\mathbf{C}})$ is the least informative in the class of multivariate distributions with a bounded covariance matrix: $\mathbf{C}(f) \leq \overline{\mathbf{C}}$.

THEOREM 3.4.1. *In the intersection of the classes $\mathscr{F}_{1q}$ and $\mathscr{F}_{2q}$*

$$\mathscr{F}_{12q} = \left\{ f_q : f_q(0; \beta) \geq \frac{1}{2a^m} > 0,\ \sigma_k^2(f_q) \leq \overline{\sigma}^2,\ k = 1, \dots, m \right\}, \tag{3.4.21}$$

the least informative density is of the form

$$f_{q^*}(r) = \begin{cases} \mathscr{N}_m(r; \overline{\sigma}), & \overline{\sigma}^2/a^2 \leq b_1(m), \\ f_\alpha(r; \beta^*), & b_1(m) < \overline{\sigma}^2/a^2 \leq b_2(m), \\ \mathscr{L}_m(r; \beta^*), & \overline{\sigma}^2/a^2 > b_2(m), \end{cases} \tag{3.4.22}$$

where

$$b_1(m) = \frac{2^{2/m}}{2\pi}, \qquad b_2(m) = \frac{m+1}{(4\pi)^{(m-1)/m}\,\Gamma^{2/m}((m+1)/2)},$$

and the parameters α *and* β^* *are determined from the equations*

$$\frac{\overline{\sigma}}{a} = \frac{\alpha^{1/m}\Gamma^{1/m}(m/2)\Gamma^{1/2}((m+2)/\alpha)}{(\pi m)^{1/2}\Gamma^{1/2}(m/\alpha)},$$
$$\beta^* = m^{1/2}\overline{\sigma}\Gamma^{1/2}(m/\alpha)\Gamma^{1/2}\left(\frac{m+2}{\alpha}\right).$$

Three branches of solution (3.4.22) appear due to the degree in which the restrictions are taken into account:

- in the first domain $\overline{\sigma}^2/a^2 \le b_1(m)$, it is just the restriction on a variance that matters: $\sigma_k^2(\widetilde{f}_2) = \overline{\sigma}^2, k = 1,\dots,m$; the restriction on the value of a density at the center of symmetry has the form of the strict inequality: $f_2(0) > 1/(2a^m)$;
- in the third domain $\overline{\sigma}^2/a^2 > b_2(m)$, the restriction on the value of a density is substantial: $f_1(0) = 1/(2a^m)$, $\sigma_k^2(f_1) < \overline{\sigma}^2$, $k = 1,\dots,m$;
- in the middle domain both restrictions hold as the equalities: $f_\alpha(0) = 1/(2a^m)$, $\sigma_k^2(f_\alpha) = \overline{\sigma}^2$, $k = 1,\dots,m$,, thus they determine the unknown parameters α and β.

Theorem 3.4.1 is an analog of Theorem 3.2.1 in the case of the multivariate exponential-power distributions (3.4.12).

COROLLARY 3.4.1. *The minimax variance estimator of location is the multivariate L_p-norm estimator with $p = q^*$: thus, in the first domain with relatively small variances, the L_2-norm method is optimal; in the third domain with relatively large variances, the L_1-norm method is optimal; in the middle domain, the L_p-norm estimators with $1 < p < 2$ are the best.*

REMARK 3.4.2. It can be seen from Theorem 3.4.1 that the optimal value of q^* is determined independently of β^* due to the scale equivariancy of L_p-norm estimators.

The switching bounds for the minimax algorithm from the L_1-norm estimator to the L_p-norm with $1 < p < 2$ and to the L_2-norm estimator are given by the functions $b_1(m)$ and $b_2(m)$. The values of these bounds are given in Table 3.3.

It can be seen from Table 3.3 that, first, the asymptotic values of the bounds are being attained rather rapidly as $m \to \infty$, and, second, with m increasing, these values become smaller, in asymptotics approximately three times less

Table 3.3. The switching bounds of the L_p-norm estimators

m	1	2	3	4	5	∞
$b_1(m)$	$2/\pi$	$1/\pi$	$1/(2^{1/3}\pi)$	$1/(2^{1/2}\pi)$	$1/(2^{2/3}\pi)$	$1/(2\pi)$
$b_2(m)$	2	$3/\pi$	$(2/\pi)^{2/3}$	$5/(6^{1/2}\pi)$	$3/(2\pi^{4/5})$	$e/(2\pi)$

than with $m = 1$. This notice is confirmed by the behavior of the robust minimax variance multivariate L_p-norm estimators under ε-contaminated normal distributions

$$f(r) = (1-\varepsilon)\mathcal{N}_m(r;1) + \varepsilon\mathcal{N}_m(r;k), \qquad 0 \le \varepsilon < 1, \tag{3.4.23}$$

$$\mathcal{N}_m(r;k) = \frac{1}{(2\pi)^{m/2}k^m}\exp\left(-\frac{r^2}{2k^2}\right), \qquad k > 1.$$

The asymptotic relative efficiency of the L_1 and L_2-norm estimators under distributions (3.4.23) is given by

$$\mathrm{ARE}(L_1, L_2) = b(m)(1-\varepsilon+\varepsilon k^2)(1-\varepsilon+\varepsilon/k)^{-2},$$

where

$$b(m) = \frac{(m-1)^2\Gamma^2((m-1)/2)}{2m\Gamma^2(m/2)}, \qquad m \ge 2.$$

The behavior of ARE is presented in Figure 3.13. Here we display some values of $b(m)$, for example, $b(1) = 2/\pi$, $b(2) = \pi/4$, $b(3) = 8/(3\pi)$, and $b(\infty) = 1$. We can see that under the normal distribution, the superiority of the L_2-estimator vanishes fast as $m \to \infty$. In other words, all estimators become catastrophically bad with high dimension (see also (Maronna, 1976; Huber, 1981; Shurygin, 2000).

3.4.4. Some remarks on the general nonparametric case

We now consider the character of minimax solutions in the general case of spherically symmetric distributions.

For the variational problem (3.4.15), the Euler equation takes the form (Huber, 1981)

$$u'' + [(m-1)/r]u' - \lambda u = 0, \tag{3.4.24}$$

where $u(r) = \sqrt{f(r)}$, λ is the Lagrange multiplier corresponding to the normalization condition.

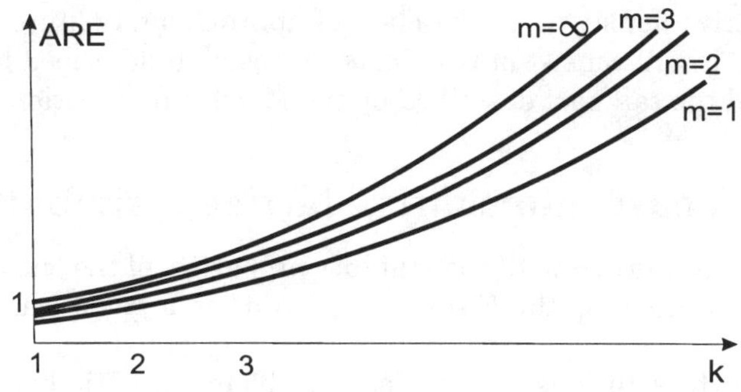

Figure 3.13. The behavior of ARE(L_1, L_2) under ε-contaminated normal distributions

Setting $w(r) = r^\nu u(r)$, $\nu = m/2 - 1$, and $z = \sqrt{|\lambda|}r$, we obtain the equation for the Bessel functions

$$z^2 w''(z) + zw'(z) - (z^2 \operatorname{sgn} \lambda + \nu^2)w(z) = 0.$$

Its solutions can be written as

$$w(z) = \begin{cases} J_\nu(z) \text{ or } N_\nu(z), & \lambda < 0, \\ I_\nu(z) \text{ or } K_\nu(z), & \lambda \geq 0, \end{cases} \tag{3.4.25}$$

where $J_\nu(z)$ and $N_\nu(z)$ are the Bessel and Neyman functions of order ν, $I_\nu(z)$ and $K_\nu(z)$ are the modified Bessel and Macdonald functions (Abramowitz and Stegun, 1972).

Using solutions (3.4.25), we can describe the multivariate analogs of the univariate least informative densities.

The first is the Bokk generalization of Huber least informative density under ε-contaminated distributions (see Subsection 3.4.1).

The second generalizes the *cosine*-type density minimizing Fisher information over the class of finite distributions, and this result also belongs to (Bokk, 1990).

Consider the class of finite spherically symmetric densities in $\mathbf{R}^m$

$$\mathcal{F}_m = \left\{ f\colon \frac{2\pi^{m/2}}{\Gamma(m/2)} \int_0^R r^{m-1} f(r)\, dr = 1,\ f(R) = f'(R) = 0 \right\}.$$

The least informative density is of the form

$$f^*(r) = \text{const} \cdot r^{-2\nu} J_\nu^2(r_0), \qquad 0 \leq r_0 \leq R,$$

where r_0 is the first root of the equation $J_\nu(r) = 0$.

REMARK 3.4.3. Finally note that it is possible to write out the structure of the least informative density over the class of approximately finite multivariate distributions: it will consist of two parts, the central described by the Bessel functions, and the tail part described by the Macdonald functions.

3.5. Least informative lattice distributions

This section is concerned with the stability properties of the least informative distributions minimizing the Fisher information in a given class of distributions.

Generally, the solutions of variational problems essentially depend on the regularity conditions of the functional class. The stability of these optimal solutions with respect to the violations of regularity conditions is studied under lattice distributions. The discrete analogs of the Fisher information are obtained in these cases. They have the form of the Hellinger metrics while estimating a real continuous location parameter and the form of the χ^2 metrics while estimating an integer discrete location parameter. The analytical expressions for the corresponding least informative discrete distributions are derived in some classes of lattice distributions by means of generating functions and the Bellman recursive functional equations of dynamic programming. These classes include the class of nondegenerate distributions with a restriction on the value of the density at the center of symmetry, the class of finite distributions, and the class of contaminated distributions. The obtained least informative lattice distributions preserve the form of their prototypes in the continuous case. These results show the stability of robust minimax structures under different types of transitions from the continuous distribution to the discrete one (Shevlyakov, 1991; Vilchevski and Shevlyakov, 1997).

3.5.1. Preliminaries

As shown before, the form of the solution obtained by the minimax approach substantially depends on the characteristics of the distribution class. As a rule, the classes of continuous and symmetric distributions are considered. In many real-life problems of data processing, the results of measurements include groups of equal values. Furthermore, the results of measurements usually come rounded in accordance with the scale of the measurement device playing the role of a discretizer. Thus, in these cases, the use of continuous distribution models does not seem adequate to the original problem of data processing, and it is quite important for applications to design robust methods for discrete distribution models corresponding to the real nature of data.

Here we describe the analogs of Fisher information for the discrete distribution classes while considering

- the direct discretization procedure of the Fisher information functional in the problem of estimation of a continuous location parameter; and

- the discrete analog of the Rao–Cramér inequality in the problem of estimation of a discrete location parameter.

In the latter case, the obtained form of the Rao–Cramér inequality is similar to the Chapman–Robbins inequality (Chapman and Robbins, 1951).

The derived terms corresponding to the Fisher information functional are quite different in the above cases, but the solutions of the variational problems of minimization of these functionals (the least informative distributions) are the same.

Moreover, they demonstrate a remarkable correspondence with their continuous analogs. Thus we can conclude that the structure of robust minimax procedures is rather stable to deviations from the assumptions of regularity of the distribution classes.

3.5.2. Discrete analogs of the Fisher information

Consider the class of lattice distributions

$$f_l(x) = \sum_i p_i \delta(x - i\Delta), \qquad \sum_i p_i = 1, \tag{3.5.1}$$

where $\delta(\cdot)$ is the Dirac delta-function, Δ is the step of discretization.

We consider two different cases

- the location parameter is continuous with $\theta \in \mathbf{R}$;
- the location parameter is discrete with $\theta \in \mathbf{Z}$.

In the first case, the following result is true.

THEOREM 3.5.1. *In the class of lattice distributions with continuous parameter* θ (3.5.1), *the variational problem of minimization of the Fisher information for the location parameter is equivalent to the optimization problem*

$$\sum \left(\sqrt{p_{i+1}} - \sqrt{p_i}\right)^2 \to \min. \tag{3.5.2}$$

In the second case with discrete location parameter θ, the following analog of the Rao–Cramér inequality holds.

THEOREM 3.5.2. *Let* $x_1, \dots, x_n$ *be independent identically distributed random variables with distribution density* $f(x - \theta)$ (3.5.1), *and* $p_i > 0$, $i \in \mathbf{Z}$, $x_1, \dots, x_n, \theta \in \mathbf{Z}$. *Let* $\widehat{\theta}_n = \widehat{\theta}_n(x_1, \dots, x_n)$ *be a discrete unbiased estimator of the discrete location parameter*

$$\widehat{\theta}_n \in \mathbf{Z}, \qquad \mathsf{E}\widehat{\theta}_n = \theta.$$

Then the variance of this estimator satisfies the inequality

$$\operatorname{Var}\widehat{\theta}_n \geq \left[\left(\sum_{i\in\mathbf{Z}} \frac{(p_{i-1}-p_i)^2}{p_i} + 1\right)^n - 1\right]^{-1}. \tag{3.5.3}$$

REMARK 3.5.1. The key feature of the obtained result is that in the discrete case the lower boundary of the estimator's variance decreases exponentially as $n \to \infty$ providing the corresponding efficiency of estimation to be much greater than in the continuous case.

COROLLARY 3.5.1. *In the class of lattice distributions* (3.5.1) *with a discrete parameter* θ, *the problem of minimization of Fisher information is equivalent to the optimization problem*

$$\sum_{i\in\mathbf{Z}} \frac{p_{i-1}^2}{p_i} \to \min. \tag{3.5.4}$$

3.5.3. Least informative lattice distributions

Now we consider the discrete analogs of the least informative distributions for the classes of continuous distributions $\mathcal{F}_1$, $\mathcal{F}_3$, and $\mathcal{F}_4$ considered in Section 3.1.

Let $\mathcal{P}_1$ be the class of lattice symmetric nondegenerate distributions

$$\mathcal{P}_1 = \left\{p_i,\ i \in \mathbf{Z}\colon p_i > 0, p_0 \geq \gamma_0 > 0, p_{-i} = p_i, \sum p_i = 1\right\}.$$

THEOREM 3.5.3. *In the class* $\mathcal{P}_1$ *of lattice distributions, the solution of optimization problem* (3.5.2) *is of the form*

$$p_{-i}^* = p_i^* = \alpha^i \gamma_0, \quad \alpha = \frac{1-\gamma_0}{1+\gamma_0}, \quad i = 0, 1, \dots,. \tag{3.5.5}$$

THEOREM 3.5.4. *In the class* $\mathcal{P}_1$ *of lattice distributions, the solution of optimization problem* (3.5.4) *is the same as in Theorem* 3.5.3.

REMARK 3.5.2. The least informative lattice distribution f_{l1}^* (3.5.1) with the geometric progression of p_i^* is the discrete analog of the least informative Laplace density f_1^* for the distribution class $\mathcal{F}_1$ of nondegenerate distributions.

Consider the discrete analog of the class $\mathcal{F}_3$ of ε-contaminated distributions with the restrictions on the values of p_i in the central zone:

$$\mathcal{P}_3 = \left\{p_i, i \in \mathbf{Z}\colon p_i > 0,\ p_{-i} = p_i \geq \gamma_i > 0, i = 0, 1, \dots, k; \sum p_i = 1\right\}. \tag{3.5.6}$$

THEOREM 3.5.5. *In the class* $\mathscr{P}_3$ *of lattice distributions with the additional restrictions on* γ_i

$$\gamma_i^{1/2} - \gamma_{i-1}^{1/2} \le \frac{(1-\alpha^{1/2})^2}{2\alpha^{1/2}} \sum_{j=0}^{i-1} \gamma_j^{1/2},$$

the solution of variational problem (3.5.2) *is of the form*

$$p_{-i}^* = p_i^* = \begin{cases} \gamma_i, & i = 0, 1, \dots, s^*, s^* \le k, \\ \alpha^{i-s^*}\gamma_{s^*}, & i > s^*, \end{cases} \tag{3.5.7}$$

where

$$\alpha = \frac{1 - \gamma_0 - 2\sum_{i=0}^{s^*} \gamma_i}{1 - \gamma_0 - 2\sum_{i=0}^{s^*} \gamma_i + 2\gamma_{s^*}};$$

the sewing number s^* *is determined by the maximum value of* s *satisfying*

$$2(\gamma_{s-1}\gamma_s)^{1/2} + \left(1 - \gamma_0 - 2\sum_{i=0}^{s-1} \gamma_i\right)^{1/2}$$
$$\times \left(\left(1 - \gamma_0 - 2\sum_{i=0}^{s} \gamma_i\right)^{1/2} - \left(1 - \gamma_0 - 2\sum_{i=0}^{s-2} \gamma_i\right)^{1/2} \right) > 0.$$

The connection of this result with the Huber least informative density f_3^* (see Section 3.1) is obvious.

Finally, consider the discrete analog of the class of finite distributions $\mathscr{F}_4$:

$$\mathscr{P}_4 = \left\{p_i, i \in \mathbf{Z}\colon p_{-i} = p_i > 0 \text{ for } i = 0, 1, \dots, n; p_i = 0 \text{ for } i > n; \sum p_i = 1\right\}.$$

THEOREM 3.5.6. *In the class* $\mathscr{P}_4$ *of lattice distributions, the solution of optimization problem* (3.5.2) *is of the form*

$$p_{-i}^* = p_i^* = \frac{1}{n+1}\cos^2\left(\frac{i\pi}{2(n+1)}\right), \qquad i = 0, \dots, n. \tag{3.5.8}$$

The results of Theorems 3.5.3–3.5.6 show the stability of robust minimax solutions under the violations of regularity conditions caused by different types of transitions from the continuous to the discrete case.

3.5.4. Proofs

PROOF OF THEOREM 3.5.1. For the variational problem of minimization of the Fisher information, the condition of non-negativeness for the density $f \ge 0$ is satisfied with the following change of variables $f = g^2$:

$$I(g) = \int_{-\infty}^{\infty} (g'(x))^2 dx \to \min_g, \qquad \int_{-\infty}^{\infty} g^2(x) dx = 1. \tag{3.5.9}$$

Consider the δ_h-sequence approximation to expression (3.5.1) with $\Delta = 1$

$$f_h(x) = g_h^2(x), \qquad g_h(x) = \sum_i \frac{p_i^{1/2}}{2\pi h^2} \exp\left\{-\frac{(x-i)^2}{4h^2}\right\}.$$

In this case, functional (3.5.9) and the normalization condition are written as

$$I_h = \frac{1}{h^2} - \frac{1}{4h^4}\sum_i\sum_j \sqrt{p_i}\sqrt{p_j}(i-j)^2 \exp\left\{-\frac{(i-j)^2}{8h^2}\right\},$$

$$\sum_i\sum_j \sqrt{p_i}\sqrt{p_j} \exp\left\{-\frac{(i-j)^2}{8h^2}\right\} = 1.$$

The main part of the functional I_h is $-\sum \sqrt{p_i}\sqrt{p_{i+1}}$ as $h \to 0$. Recalling the normalization condition for p_i, we arrive at the assertion of Theorem 3.5.1. □

PROOF OF THEOREM 3.5.2. Consider the likelihood

$$L(x_1,\dots,x_n|\theta) = p_{x_1-\theta}\cdots p_{x_n-\theta}.$$

In this case, the normalization and unbiasedness conditions are

$$\sum_{x_1,\dots,x_n\in\mathbf{Z}} L(x_1,\dots,x_n \mid \theta) = 1, \tag{3.5.10}$$

$$\sum_{x_1,\dots,x_n\in\mathbf{Z}} \widehat{\theta}_n(x_1,\dots,x_n)L(x_1,\dots,x_n \mid \theta) = \theta. \tag{3.5.11}$$

Considering the unbiasedness condition for the parameter value $\theta+1$

$$\sum_{x_1,\dots,x_n\in\mathbf{Z}} \widehat{\theta}_n(x_1,\dots,x_n)L(x_1,\dots,x_n|\theta+1) = \theta+1$$

and subtracting (3.5.11) from it, we obtain

$$\sum_{x_1,\dots,x_n\in\mathbf{Z}} \widehat{\theta}_n(x_1,\dots,x_n)\left[L(x_1,\dots,x_n|\theta+1) - L(x_1,\dots,x_n|\theta)\right] = 1. \tag{3.5.12}$$

Set $\widehat{\theta}_n(x_1,\dots,x_n) = \widehat{\theta}_n$ and $L(x_1,\dots,x_n|\theta) = L(\theta)$. Then by the normalization condition (3.5.10), from (3.5.12) we obtain

$$\sum_{x_1,\dots,x_n\in\mathbf{Z}} (\widehat{\theta}_n - \theta)\left[\frac{L(\theta+1)-L(\theta)}{L(\theta)}\right] L(\theta) = 1. \tag{3.5.13}$$

Finally, the Cauchy–Bunyakovskii inequality and (3.5.13) yield

$$\sum_{x_1,\dots,x_n\in\mathbf{Z}} (\widehat{\theta}_n - \theta)^2 L(\theta) \sum_{x_1,\dots,x_n\in\mathbf{Z}} \left[\frac{L(\theta+1)-L(\theta)}{L(\theta)}\right]^2 L(\theta) \ge 1$$

and the Rao–Cramér type inequality in the form

$$\operatorname{Var}\widehat{\theta}_n \ge \left(\sum_{x_1,\dots,x_n\in\mathbf{Z}} \left[\frac{L(\theta+1)-L(\theta)}{L(\theta)}\right]^2 L(\theta)\right)^{-1}. \tag{3.5.14}$$

Theorem 3.5.2 immediately follows from (3.5.14). □

PROOF OF THEOREM 3.5.3. Set $\lambda_i = \sqrt{p_i}$, $i \in \mathbf{Z}$. Let the parameter $\lambda_0 = \sqrt{p_0} \ge \sqrt{\gamma_0} > 0$ be free; it will be optimized at the final stage.

Variational problem (3.5.2) can be reformulated as

$$\sum_{i\in\mathbf{Z}} \lambda_i\lambda_{i+1} \to \max_{\Lambda}, \tag{3.5.15}$$

where $\Lambda = \{\lambda_i, i \in \mathbf{Z}\}$. In this case, the Lagrange functional is of the form

$$2\left(\sqrt{p_0}\lambda_1 + \sum_{i=1}^{\infty}\lambda_i\lambda_{i+1}\right) - \mu\left(p_0 + 2\sum_{i=1}^{\infty}\lambda_i^2 - 1\right) \to \max_{\mu,\Lambda}, \tag{3.5.16}$$

where μ is the Lagrangian multiplier corresponding to the normalization condition.

The extremum conditions for problem (3.5.16) are given by the simultaneous equations

$$\begin{aligned} \sqrt{p_0} - 2\mu\lambda_1 + \lambda_2 &= 0, \\ \lambda_1 - 2\mu\lambda_2 + \lambda_3 &= 0, \\ &\dots \\ \lambda_{k-1} - 2\mu\lambda_k + \lambda_{k+1} &= 0, \\ &\dots \end{aligned} \tag{3.5.17}$$

To solve (3.5.17), we make use of the generating function in the form

$$F(x) = \sum_{i=0}^{\infty} \lambda_{i+1}x^i, \qquad |x| < 1. \tag{3.5.18}$$

We obtain the obvious expression for (3.5.18) by multiplying equations (3.5.17) by x^i, $i = 0, 1, \dots$, and summing them, which yields

$$F(x) = \frac{\lambda_1 - \sqrt{p_0}x}{x^2 - 2\mu x + 1}. \tag{3.5.19}$$

Set $\lambda_1 = t\sqrt{p_0}$ in (3.5.19), hence

$$F(x) = \frac{t - x}{x^2 - 2\mu x + 1}\sqrt{p_0}. \qquad (3.5.20)$$

The denominator of (3.5.20) can be written as

$$x^2 - 2\mu x + 1 = (x - x_0)(x - 1/x_0), \qquad x_0 = \mu - \sqrt{\mu^2 - 1},$$

with $x_0 = t$. Therefore (3.5.20) takes the form

$$F(x) = \frac{t}{1 - tx}\sqrt{p_0} = t\sqrt{p_0}\sum_{i=0}^{\infty} t^i x^i. \qquad (3.5.21)$$

Comparing series (3.5.18) and (3.5.21), we obtain

$$\lambda_i = t^i\sqrt{p_0}, \qquad i \in \mathbf{N}.$$

The value of t is determined from the normalization condition

$$p_0 + 2p_0\sum_{i=1}^{\infty} t^{2i} = 1$$

yielding

$$t = \frac{1 - p_0}{1 + p_0}.$$

Functional (3.5.15) depends of the free parameter p_0 as follows:

$$2\left(\sqrt{p_0}\lambda_1 + \sum_{i=1}^{\infty}\lambda_i\lambda_{i+1}\right) = \sqrt{1 - p_0^2}.$$

By virtue of the condition $p_0 \ge \gamma_0 > 0$, we obtain the optimal solution

$$p_0^* = \arg\max_{p_0 \ge \gamma_0 > 0}\sqrt{1 - p_0^2} = \gamma_0.$$

It remains to set $\alpha = t^2(p_0^*) = (1 - \gamma_0)/(1 + \gamma_0)$. □

REMARK 3.5.3. If the parameter λ_0 is not free, then the following equation must be added to (3.5.17):

$$-\mu\lambda_0 + \lambda_1 = 0.$$

In our case, it holds as the strict inequality: $-\mu\sqrt{\gamma_0} + \lambda_1 < 0$.

PROOF OF THEOREM 3.5.4. In the symmetric case $p_{-i} = p_i$ with the free parameter $p_0 \geq \gamma_0 > 0$, optimization problem (3.5.4) can be represented as

$$\left.\begin{aligned} I = \min_{p_1,\dots}\left[\sum_{i=0}^{\infty}\left(\frac{p_i^2}{p_{i+1}} + \frac{p_{i+1}^2}{p_i}\right) - 1\right] \\ \text{under the condition } \sum_{i=1}^{\infty} p_i = \frac{1-p_0}{2}. \end{aligned}\right\} \tag{3.5.22}$$

Consider the following auxiliary optimization problem:

$$\text{minimize } \sum_{i=0}^{\infty}\left(\frac{p_i^2}{p_{i+1}} + \frac{p_{i+1}^2}{p_i}\right) \quad \text{under the condition } \sum_{i=1}^{\infty} p_i = b.$$

Set the optimal value of the functional as

$$\Phi(p_0, b) = \min_{p_1,\dots}\left[\frac{p_0^2}{p_1} + \frac{p_1^2}{p_0} + \sum_{i=1}^{\infty}\left(\frac{p_i^2}{p_{i+1}} + \frac{p_{i+1}^2}{p_i}\right), \sum_{i=1}^{\infty} p_i = b, p_i \geq 0\right]$$

or

$$\min_{0\leq p_1 \leq b}\left[\frac{p_0^2}{p_1} + \frac{p_1^2}{p_0} + \min_{p_2,\dots}\left[\frac{p_1^2}{p_2} + \frac{p_2^2}{p_1} + \sum_{i=2}^{\infty}\left(\frac{p_i^2}{p_{i+1}} + \frac{p_{i+1}^2}{p_i}\right), \sum_{i=2}^{\infty} p_i = b - p_1, p_i \geq 0\right]\right]$$
$$= \min_{0\leq p_1\leq b}\left[\frac{p_0^2}{p_1} + \frac{p_1^2}{p_0} + \Phi(p_1, b - p_1)\right].$$

Consider the function

$$\psi(y) = \min_{z_1,\dots}\left[\frac{y^2}{z_1} + \frac{z_1^2}{y} + \sum_{i=1}^{\infty}\left(\frac{z_i^2}{z_{i+1}} + \frac{z_{i+1}^2}{z_i}\right), \sum_{i=1}^{\infty} z_i = 1, z_i \geq 0\right].$$

Then the relation

$$\Phi(p_0, b) = b\psi\left(\frac{p_0}{b}\right)$$

holds. Therefore, we arrive at the recursive Bellman equations

$$\Phi(p_0, b) = \min_{0\leq p_1\leq b}\left[\frac{p_0^2}{p_1} + \frac{p_1^2}{p_0} + \Phi(p_1, b - p_1)\right]$$

or

$$b\psi\left(\frac{p_0}{b}\right) = b\min_{0\leq z\leq 1}\left[\frac{p_0^2}{b^2}\frac{1}{z} + \frac{z^2}{p_0/b} + (1-z)\psi\left(\frac{z}{1-z}\right)\right].$$

The Bellman function $\psi(y)$ satisfies the functional equation

$$\psi(y) = \min_{0 \le z \le 1} \left[\frac{y^2}{z} + \frac{z^2}{y} + (1-z)\psi\left(\frac{z}{1-z}\right) \right]. \tag{3.5.23}$$

It can be directly checked that the solution of (3.5.23) is

$$\psi(y) = \frac{1}{1+y} + (1+y)^2. \tag{3.5.24}$$

Thus,

$$\min_{0 \le z \le 1} \left[\frac{y^2}{z} + \frac{z^2}{y} + (1-z)\left[(1-z) + \frac{1}{(1-z)^2} \right] \right] = \min_{0 \le z \le 1} \left[\frac{y^2}{z} + \frac{z^2}{y} + (1-z)^2 + \frac{1}{(1-z)} \right].$$

Differentiating, we obtain

$$-\frac{y^2}{z^2} + 2\frac{z}{y} - 2(1-z) + \frac{1}{(1-z)^2} = (z - y(1-z)) \left[\frac{2}{y} + \frac{z + y(1-z)}{(1-z)^2 z^2} \right] = 0.$$

The derivative equals zero with $z = y/(1+y)$, which implies (3.5.24).

It follows from (3.5.22) that the Fisher information is of the form

$$I = \frac{1-p_0}{2}\psi\left(2\frac{p_0}{1-p_0}\right) - 1 = \frac{4p_0^2}{1-p_0^2}$$

and

$$\min_{p_0 \ge \gamma_0 > 0} I = \frac{4\gamma_0^2}{1-\gamma_0^2}$$

with

$$p_i = \left(\frac{1-\gamma_0}{1+\gamma_0}\right)^i \gamma_0,$$

which completes the proof. □

PROOF OF THEOREM 3.5.5. The proof is based on

- the solution of the infinite system of equations similar to (3.5.17) with the first equation

$$\lambda_{s^*+1} - 2\mu\lambda_{s^*+2} + \lambda_{s^*+3} = 0;$$

- the maximization of functional (3.5.15) with checking the restrictions

$$p_i \ge \gamma_i > 0, \qquad i = 0, 1, \dots, k,$$

- and the inequalities of gradient type

$$\lambda_k - 2\mu\lambda_{k+1} + \lambda_{k+2} < 0, \qquad 0 \le k \le s^*.$$

□

PROOF OF THEOREM 3.5.6. In this case, the Lagrange functional is of the form

$$2\sum_{i=0}^{n-1} \lambda_i \lambda_{i+1} - \mu \left(\lambda_0^2 + 2\sum_{i=1}^{n} \lambda_i^2 - 1 \right) \to \max_{\mu,\lambda_0,\dots,\lambda_n}, \tag{3.5.25}$$

where μ is the Lagrangian multiplier corresponding to the normalization condition. The extremum conditions for problem (3.5.25) are given by

$$\begin{aligned} -\mu\lambda_0 + \lambda_1 &= 0, \\ \lambda_0 - 2\mu\lambda_1 + \lambda_2 &= 0, \\ &\dots \\ \lambda_{n-2} - 2\mu\lambda_{n-1} + \lambda_n &= 0, \\ \lambda_{n-1} - 2\mu\lambda_n &= 0. \end{aligned} \tag{3.5.26}$$

Simultaneous equations (3.5.26) yield the recursive equations for the Chebyshev polynomials of the first kind T_i. Thus,

$$\begin{aligned} &\lambda_1 = \mu\lambda_0, \ \lambda_2 = (2\mu^2 - 1)\lambda_0, \ \dots, \\ &\lambda_i = T_i(\mu)\lambda_0, \qquad i = 0, 1, \dots, n. \end{aligned}$$

It remains to recall the normalization condition. □

4

Robust estimation of scale

In this chapter the problem of robust estimation of scale is mainly treated as the problem subordinate to robust estimation of location. Special attention is paid to the optimization approach to constructing the measures of spread in the data analysis setting and to the Huber minimax variance estimator of the scale parameter under ε-contaminated normal distributions, since the latter estimator is applied to the problem of robust estimation of the correlation coefficient.

4.1. Introductory remarks

4.1.1. Preliminaries

Following (Huber, 1981), we define the *scale estimator* as a positive statistic S_n that is equivariant under scale transformations

$$S_n(\lambda x_1, \dots, \lambda x_n) = \lambda S_n(x_1, \dots, x_n) \qquad \lambda > 0. \tag{4.1.1}$$

Moreover, its invariance under changes of sign and shifts is also desirable:

$$S_n(-x_1, \dots, -x_n) = S_n(x_1, \dots, x_n), \tag{4.1.2}$$

$$S_n(x_1 + \mu, \dots, x_n + \mu) = S_n(x_1, \dots, x_n). \tag{4.1.3}$$

In actual practice, scale problems, as a rule, do not occur independently of location (or regression) problems, in which the scale usually is a nuisance parameter. Such problems of scale estimation are thoroughly studied in (Huber, 1981) with the use of M-, L- and R-estimators. In what follows, we describe the main representatives of these classes.

However, we distinguish the data analysis probability-free setting for constructing the measures of the data spread from the statistical setting where the scale parameter of the underlying distribution is estimated.

With an optimization approach in data analysis, the role of scale is secondary: the scale estimator is subordinated to the location estimator.

4.1.2. Scale estimation in data analysis via the optimization approach to location

Consider the optimization approach to constructing the estimator $\widehat{\theta}_n$ of location (central tendency) for the one-dimensional data $x_1, \dots, x_n$

$$\widehat{\theta}_n = \arg\min_{\theta} J(\theta), \tag{4.1.4}$$

where $J(\theta)$ is the goal function.

In data analysis, the scale estimator S_n (the measure of spread of the data about the location estimator) can be naturally defined as an appropriately transformed value of the optimization criterion $J(\theta)$ at the optimal point $\widehat{\theta}_n$ (Orlov, 1976)

$$S_n \sim J(\widehat{\theta}_n). \tag{4.1.5}$$

In the case of M-estimators of location where $J(\theta) = n^{-1}\sum_1^n \rho(x_i - \theta)$,

$$S_n =_{\propto} \rho^{-1}\left(\frac{1}{n}\sum_{i=1}^{n} \rho(x_i - \widehat{\theta}_n)\right),$$

with

$$\widehat{\theta}_n = \arg\min_{\theta} \sum_{i=1}^{n} \rho(x_i - \theta).$$

In particular,

- the standard deviation

$$S_n = s = \sqrt{\frac{1}{n}\sum_{i=1}^{n}(x_i - \overline{x})^2}$$

for the LS or L_2-norm method with $\rho(u) = u^2$ and $\widehat{\theta}_{LS} = \overline{x}$;

- the mean absolute deviation

$$S_n = d = \frac{1}{n}\sum_{i=1}^{n} |x_i - \operatorname{med} x|$$

for the L_1-norm method with $\rho(u) = |u|$ and $\widehat{\theta}_{L_1} = \operatorname{med} x$;

- and, more generally, the pth-power deviations (Gentleman, 1965)

$$S_n = S_{L_p} = c_n \left(\sum_{i=1}^{n} |x_i - \widehat{\theta}_{L_p}|^p\right)^{1/p},$$

for the L_p-norm method with $\rho(u) = |u|^p$,

$$\widehat{\theta}_{L_p} = \arg\min_\theta \sum_{i=1}^n |x_i - \theta|^p, \qquad p \geq 1,$$

where c_n is a normalization constant chosen, say, from the condition of asymptotic unbiasedness under the standard normal $\mathsf{E}_\Phi S_n = \sigma$;

- half of the sample range $S_n = R/2 = (x_{(n)} - x_{(1)})/2$ for the L_∞ or Chebyshev metric with $\rho(u) = \max|u|$ and $\widehat{\theta}_{L_\infty} = (x_{(1)} + x_{(n)})/2$.

The following estimator is also of interest: the least median squares (LMS) deviation

$$S_n = S_{\text{LMS}} = \operatorname{med}|x - \widehat{\theta}_{\text{LMS}}|$$

for the LMS method with $J(\theta) = \operatorname{med}(x_i - \theta)^2$ and $\widehat{\theta}_{\text{LMS}}$ given by

$$\widehat{\theta}_{\text{LMS}} = \arg\min_\theta \operatorname{med}(x_i - \theta)^2.$$

Apparently, the estimator S_{LMS} is close to the median absolute deviation $\text{MAD}\, x = \operatorname{med}|x - \operatorname{med} x|$.

REMARK 4.1.1. Summarizing the above, we say that any location and, more generally, regression estimator obtained with the optimization approach generates the corresponding scale estimator. Thus, for the collection of location estimators of Chapter 2 and Chapter 3, we have the appropriate collection of scale estimators.

REMARK 4.1.2. The above-formulated optimization approach to designing scale estimators is close to the scale estimators obtained from S-estimators for location (Hampel *et al.*, 1986, p. 115), where the S-estimator of location defined by

$$\widehat{\theta}_n = \arg\min_\theta s(x_1 - \theta, \dots, x_n - \theta)$$

simultaneously yields the scale estimator $S_n = s(x_1 - \widehat{\theta}_n, \dots, x_n - \widehat{\theta}_n)$.

4.2. Measures of scale defined by functionals

In this section we consider the case where the measure of spread for a random variable ξ with distribution function F is defined by means of some functional $S(F)$.

Let ξ be a random variable with some symmetric and absolutely continuous distribution function F. Denote the center of symmetry as θ and define the measure of spread of ξ about θ in the terms of the distance of ξ from θ, namely $|\xi - \theta|$.

The requirements of scale equivariancy and monotonicity of stochastic ordering imposed on the functional $S(F)$ to be a measure of spread are formulated in (Bickel and Lehmann, 1973): $S(F_{a\xi+b}) = |a|S(F_\xi)$ for all a and b, with $S(F_\xi) \le S(F_\eta)$ if $F_\xi <_{st} F_\eta$, where F_ξ and F_η are the distribution functions of $|\xi - \theta_\xi|$ and $|\eta - \theta_\eta|$ respectively.

One may consider the following groups of functionals.

- The first group includes the functionals constructed with the use of the deviation of each element of a population from some typical (central) element θ. Usually the expectation $\theta(F) = \mu(F) = \int x\,dF(x)$, or the median $\theta(F) = \mathrm{Med}(F) = F^{-1}(1/2)$ are used. Denote the distribution functions of $|\xi - \theta(F)|$ and $|\xi_1 - \xi_2|$ as F_1 and F_2, where the random variables ξ_1 and ξ_2 are independent with common distribution F.

  Now define the class of scale functionals as

$$S(F) = \left\{ \int_0^1 [F_1^{-1}(t)]^p \, dK(t) \right\}^{1/p}, \tag{4.2.1}$$

  where $K(t)$ is some distribution function on $[0, 1]$ and $p > 0$. For example, if $\theta(F) = \mu(F)$ and $K(t) = t$, $0 < t < 1$, in (4.2.1), then this formula yields the mean absolute deviation with $p = 1$ and the standard deviation with $p = 2$. Furthermore, if $K(t) = t/(1-\alpha)$, $0 \le t \le 1-\alpha$, $0 \le \alpha \le 1/2$, then we arrive at the α-trimmed variants of the above-mentioned measures. The other part of this group is defined by the functional $F_1^{-1}(1/2)$, and, in particular, it yields the median absolute deviation functional with $\theta(F) = \mathrm{Med}(F)$.

- The second group comprises the functionals constructed with the use of the deviations between all the elements of a population, and it is of the form

$$S(F) = \left\{ \int_0^1 [F_2^{-1}(t)]^p \, dK(t) \right\}^{1/p}. \tag{4.2.2}$$

  For instance, if $p = 1$ and $K(t) = t$, $0 \le t \le 1$, then we arrive at the Gini mean difference from (4.2.2). For $p = 2$ we obtain the standard deviation multiplied by $\sqrt{2}$. The median absolute deviation functional can also be described by (4.2.2) if we set $S(F) = F_2^{-1}(1/2)$.

- The third group consists of the functionals defined by the distances between the characteristic points of F, for example, between the quantiles of given levels

$$S(F) = \left\{ \int_0^1 |F^{-1}(1-\alpha) - F^{-1}(\alpha)|^p \, dK(t) \right\}^{1/p}. \tag{4.2.3}$$

In particular, the inter-α-quantile ranges are related to this group:

$$S(F) = F^{-1}(1-\alpha) - F^{-1}(\alpha), \qquad 0 < \alpha < 1/2.$$

A general scheme of the construction of scale functionals can be described as follows:

(i) the initial random variable ξ is transformed into $|\xi - \theta(F)|^p$, $|\xi_1 - \xi_2|^p$, etc.;

(ii) then those transformed random variables are processed by the operations of averaging, or of 'median', or of 'Hodges–Lehmann', etc.

In other words, the scale functional is defined via some location functional for the transformed variables. In this case, the variety of scale measures is determined by both the varieties of transformations and of location measures, and the rich experience obtained with location studies can be applied to the case of scale estimation.

Thus some new variants of scale measures can be proposed (Shulenin, 1993), for example, applying the 'median' operation to $|\xi_1 - \xi_2|$ leads to the median of absolute differences

$$S(F) = \mathrm{Med}(F_2) \qquad S_n = \mathrm{med}\{|x_i - x_j|,\ 1 \le i < j \le n\}; \tag{4.2.4}$$

the 'operation of Hodges–Lehmann' yields such scale estimators as

$$S_n = \mathrm{med}\{(|x_i - \mathrm{med}\, x| + |x_j - \mathrm{med}\, x|)/2,\ 1 \le i < j \le n\},$$

or

$$S_n = \mathrm{med}\{(|x_i - x_j| + |x_k - x_l|)/2,\ 1 \le i,j,k,l \le n\},$$

along with their trimmed variants.

The choice of a concrete functional among the above can be made on the basis of the comparison of their estimation accuracy in the chosen distribution model.

4.3. *M*-, *L*-, and *R*-estimators of scale

4.3.1. *M*-estimators of the scale parameter

Now we consider the problem of estimating the scale parameter σ for the family of densities

$$f(x;\sigma) = \frac{1}{\sigma} f\left(\frac{x}{\sigma}\right), \qquad \sigma > 0. \tag{4.3.1}$$

Any M-estimator of σ is defined as the solution of the equation

$$\sum_{i=1}^{n} \chi\left(\frac{x_i}{S_n}\right) = 0, \tag{4.3.2}$$

where χ is the score function, usually even: $\chi(-x) = \chi(x)$.

The estimator S_n corresponds to the functional $S(F)$ defined by

$$\int \chi\left(\frac{x}{S(F)}\right) dF(x) = 0.$$

The influence function is of the following form (Huber, 1981, p. 109)

$$IF(x; F, S) = \frac{\chi(x/S(F))S(F)}{\int (x/S(F))\chi'(x/S(F))\, dF(x)}. \tag{4.3.3}$$

The breakdown point for ε-contamination is given by $\varepsilon^* = -\chi(0)/\|\chi\| \le 1/2$, where $\|\chi\| = \chi(\infty) - \chi(0)$ (Huber, 1981, p. 110).

The following particular cases are of interest:

- the standard deviation $s = (n^{-1}\sum x_i^2)^{1/2}$ with $\chi(x) = x^2 - 1$;
- the mean absolute deviation $d = n^{-1}\sum |x_i|$ with $\chi(x) = |x| - 1$;
- the pth-power deviation $S_{L_p} = (n^{-1}\sum |x_i|^p)^{1/p}$ with $\chi(x) = |x|^p - 1$;
- the median absolute deviation MAD = med $|x_i|$ with $\chi(x) = \operatorname{sgn}(|x| - 1)$.

Figures 4.1–4.3 illustrate the above cases.

REMARK 4.3.1. All the above estimators are the absolute deviations from zero, since the location is assumed given in this setting.

Like for location, M-estimators (4.3.2) yield the maximum likelihood estimators of the scale parameter σ for the family of densities $\sigma^{-1}f(x/\sigma)$ with

$$\chi(x) = -x\frac{f'(x)}{f(x)} - 1. \tag{4.3.4}$$

The sufficient conditions of regularity providing the Fisher consistency and asymptotic normality of S_n are imposed on the densities f and the score functions χ (Hampel *et al.*, 1986, pp. 125, 139):

(F1) f is twice continuously differentiable and satisfies $f(x) > 0$ $\forall x \in \mathbf{R}$.

(F2) The Fisher information for scale

$$I(f; \sigma) = \frac{1}{\sigma^2}\int \left[-x\frac{f'(x)}{f(x)} - 1\right]^2 f(x)\, dx \tag{4.3.5}$$

satisfies $0 < I(f; \sigma) < \infty$.

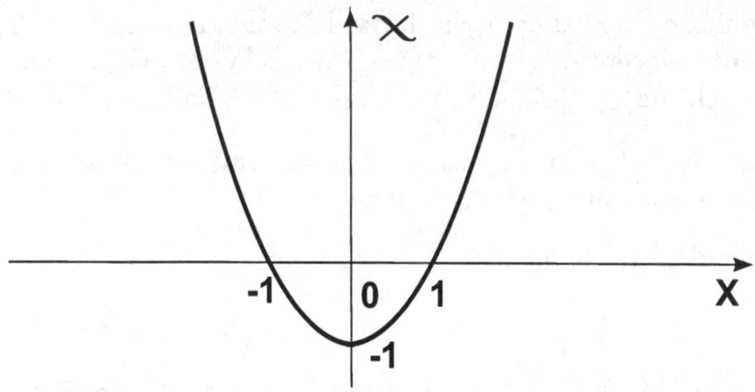

Figure 4.1. The score function for the standard deviation

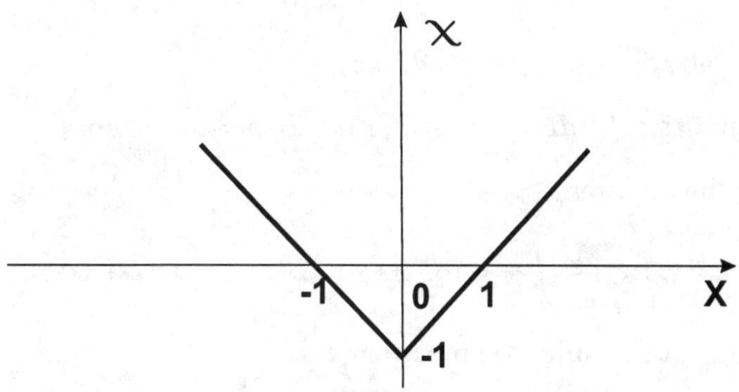

Figure 4.2. The score function for the mean absolute deviation

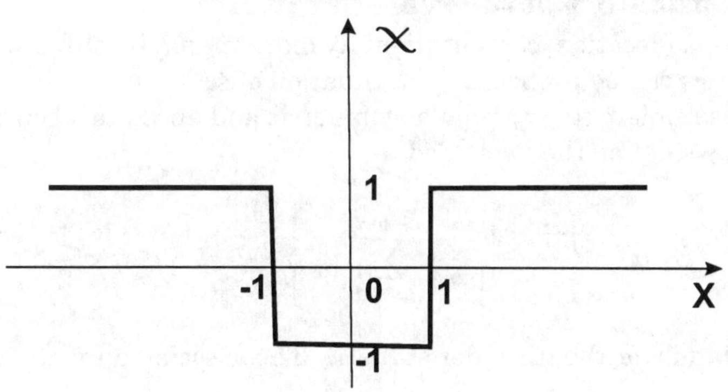

Figure 4.3. The score function for the median absolute deviation

(χ1) χ is well-defined and continuous on $\mathbf{R} \setminus C(\chi)$, where $C(\chi)$ is finite. At each point of $C(\chi)$ there exist finite left and right limits of χ, which are different. Moreover, $\chi(-x) = \chi(x)$ if $(-x, x) \subset \mathbf{R} \setminus C(\chi)$, and there exists $d > 0$ such that $\chi(x) \le 0$ on $(0, d)$ and $\chi(x) \ge 0$ on (d, ∞).

(χ2) The set $D(\chi)$ of points at which χ is continuous but at which χ' is not defined or not continuous is finite.

(χ3) $\int \chi \, dF = 0$ and $\int \chi^2 \, dF < \infty$.

(χ4) $0 < \int x\chi'(x) \, dF(x) < \infty$.

Under conditions (F1), (F2), (χ1)–(χ4), $\sqrt{n}(S_n - \sigma)$ is asymptotically normal with asymptotic variance (Hampel *et al.*, 1986)

$$V(f, \chi) = \frac{\int \chi^2(x) \, dF(x)}{\left(\int x\chi'(x) \, dF(x)\right)^2}. \tag{4.3.6}$$

Let us briefly discuss these conditions.

- The condition $\int \chi \, dF = 0$ provides the Fisher consistency.
- Using the notation

$$A(\chi) = \int \chi^2(x) \, dF(x), \qquad B(\chi) = \int x\chi'(x) \, dF(x),$$

we have for the influence function (4.3.3)

$$IF(x; F, S) = \frac{\chi(x)}{B(\chi)}.$$

4.3.2. L-estimators of the scale parameter

As in the case of location, computationally more simple L-estimators based on order statistics can be proposed for estimation of scale.

Given a sample $x_1, \dots, x_n$ from a symmetric and absolutely continuous distribution F, we define the two-sided α- as

$$\widehat{S}_1(\alpha) = \left\{ \frac{1}{n - 2[\alpha n]} \sum_{i=[\alpha n]}^{n-[\alpha n]} x_{(i)}^2 \right\}^{1/2}, \qquad 0 \le \alpha < \frac{1}{2}, \tag{4.3.7}$$

where $x_{(i)}$ stands for the ith order statistic. The associated functional is of the form

$$S_1(F, \alpha) = \left\{ \frac{1}{1 - 2\alpha} \int_{F^{-1}(\alpha)}^{F^{-1}(1-\alpha)} x^2 \, dF(x) \right\}^{1/2}, \qquad 0 \le \alpha < \frac{1}{2}.$$

For $\alpha = 0$, formula (4.3.7) yields the

$$\widehat{S}_1(0) = s = \left\{ \frac{1}{n} \sum_{i=1}^{n} x_i^2 \right\}^{1/2}.$$

The two-sided α-trimmed mean absolute deviation is defined by

$$\widehat{S}_2(\alpha) = \frac{1}{n - 2[\alpha n]} \sum_{i=[\alpha n]}^{n-[\alpha n]} |x_{(i)}|, \qquad 0 \le \alpha < \frac{1}{2} \tag{4.3.8}$$

with the functional of the form

$$S_2(F, \alpha) = \frac{1}{1 - 2\alpha} \int_{F^{-1}(\alpha)}^{F^{-1}(1-\alpha)} |x| \, dF(x), \qquad 0 \le \alpha < \frac{1}{2}.$$

For $\alpha = 0$, formula (4.3.8) yields the

$$\widehat{S}_2(0) = d = \frac{1}{n} \sum_{i=1}^{n} |x_i|.$$

The limiting cases $\alpha \to 1/2$ give the median absolute deviation med $|x|$ for both estimators.

The expressions for the influence functions of L-estimators can be found in (Huber, 1981, pp. 111–113).

4.3.3. R-estimators of the scale parameter

The relative scale between two samples can be estimated by rank tests for scale. Following (Huber, 1981), we describe such an approach to constructing scale estimators.

Given the samples $(x_1, \dots, x_m)$ and $(y_1, \dots, y_n)$ from populations with the distribution functions F and G, let R_i be the rank of x_i in the sample of size $N = m + n$. Then the test statistic $\sum_1^m a(R_i)$ with $a_i = a(i)$ is defined by

$$a_i = N \int_{(i-1)/N}^{i/N} J(t) \, dt$$

for some score-generating function J satisfying

$$J(1 - t) = J(t), \qquad \int J(t) \, dt = 0.$$

The functional $S = S(F, G)$ estimating relative scale between F and G is given by

$$\int J \left(\frac{m}{N} F(x) + \frac{n}{N} G \left(\frac{x}{S} \right) \right) dF(x) = 0.$$

Such a measure of relative scale satisfies $S(F_{aX}, F_X) = a$, where F_{aX} stands for the distribution of the random variable aX.

The efficient score-generating function $J(t)$ is given by

$$J(t) = -F^{-1}(t)\,\frac{f'[F^{-1}(t)]}{f[F^{-1}(t)]} - 1,$$

(cf. (4.3.4)).

EXAMPLE 4.3.1. For the standard normal $F = \Phi$, the choice

$$J(t) = [\Phi^{-1}(t)]^2 - 1$$

leads to the efficient R-estimator (Huber, 1981).

4.4. Huber minimax estimator of scale

In this section we give well-known formulas for the Huber minimax solution under ε-contaminated normal distributions (Huber, 1964; Huber, 1981), because this solution is essential for constructing the minimax estimator of the correlation coefficient in Chapter 7.

As the problem of estimating the scale parameter for the random variable ξ can be reduced to that of estimating the location parameter $\tau = \log\sigma$ for the random variable $\eta = \log|\xi|$, where σ is a scale parameter for ξ, the minimax solution for scale can be obtained by rewriting the minimax solution for location. This approach is realized in (Huber, 1964; Huber, 1981). Here we follow a straightforward way of minimizing the Fisher information for scale.

4.4.1. The least informative distribution

For the family of densities

$$p(x;\sigma) = \frac{1}{\sigma} f\left(\frac{x}{\sigma}\right),$$

the Fisher information for scale is

$$\begin{aligned} I(f;\sigma) &= \int \left[\frac{\partial \log p(x;\sigma)}{\partial\sigma}\right]^2 p(x;\sigma)\,dx \\ &= \int \left[-\frac{f'(x/\sigma)}{f(x/\sigma)}\,\frac{x}{\sigma^2} - \frac{1}{\sigma}\right]^2 \frac{1}{\sigma} f\left(\frac{x}{\sigma}\right)\,dx, \\ &= \frac{1}{\sigma^2}\int \left[-\frac{f'(x)}{f(x)}x - 1\right]^2 f(x)\,dx. \end{aligned} \qquad (4.4.1)$$

Without loss of generality we can assume that the true scale is $\sigma = 1$.

Consider the variational problem of minimizing the Fisher information (4.4.1)

$$f^* = \arg\min_{f \in \mathscr{F}_\varepsilon} I(f; 1) \tag{4.4.2}$$

in the class of ε-contaminated normal densities

$$\mathscr{F}_\varepsilon = \{f: f(x) \geq (1 - \varepsilon)\,\phi(x),\ 0 \leq \varepsilon < 1\}, \tag{4.4.3}$$

where $\phi(x)$ is the standard normal density.

First, we show that the problem of minimization of (4.4.1) is equivalent to that of minimization of the functional

$$J(f) = \int x^2 \left[-\frac{f'(x)}{f(x)}\right]^2 f(x)\,dx.$$

Obviously, from (4.4.1) it follows that

$$I(f; 1) = J(f) + 2 \int x f'(x)\,dx + 1.$$

Assume that $\lim_{x \to \infty} = 0$. Then, integrating $\int x f'(x)\,dx$ by parts, we obtain

$$I(f; 1) = J(f) - 1. \tag{4.4.4}$$

Then, by the standard substitution $f(x) = g^2(x)$ and the symmetry condition, the variational problem of minimizing the Fisher information for scale with the side normalization condition takes the form

$$\text{minimize} \quad \int_0^\infty x^2 g'(x)^2\,dx$$

under the condition

$$\int_0^\infty g^2(x)\,dx = \frac{1}{2}.$$

The Lagrange functional for this problem is

$$L(g, \lambda) = \int_0^\infty x^2 g'(x)^2\,dx + \lambda \left(\int_0^\infty g^2(x)\,dx - 1/2\right),$$

and the Euler equation can be represented as

$$x^2 g''(x) + 2x g'(x) - \lambda g(x) = 0.$$

Its suitable solutions are of the *t-distribution* type forms

$$g(x) = C_1 x^{k_1} + D_2 x^{k_2}, \qquad k_{1,2} = (-1 \pm \sqrt{1 + 4\lambda}). \tag{4.4.5}$$

Hence the optimal solution f^* of (4.4.2) is constructed by smooth 'glueing' the 'free' extremals (4.4.5) with the constraint on the density $f(x) = (1-\varepsilon)\phi(x)$ as follows:

$$f^*(x) = \begin{cases} A_0|x|^{p_0}, & |x| < x_0, \\ (1-\varepsilon)\phi(x), & x_0 \le |x| \le x_1, \\ A_1|x|^{p_1}, & |x| > x_1. \end{cases} \tag{4.4.6}$$

The parameters of 'glueing' A_0, A_1, x_0, x_1, p_1 and p_2 in (4.4.6) are determined from the equations equations which comprise the conditions of normalization, continuity and differentiability of the solution at $x = x_0$ and $x = x_1$ (see the conditions of regularity (F1) and (F2) in Subsection 4.3.1), and the relation between the exponents p_0 and p_1:

$$\begin{gathered} \int_{-\infty}^{\infty} f^*(x)\,dx = 1, \\ f^*(x_i - 0) = f^*(x_i + 0), \quad f^{*\prime}(x_i - 0) = f^{*\prime}(x_i + 0), \quad i = 0, 1; \\ p_0 + p_1 = -2. \end{gathered} \tag{4.4.7}$$

By substituting the solution of system (4.4.7) into (4.4.6), we obtain the least informative density (Huber, 1981, p. 120)

$$f^*(x) = \begin{cases} (1-\varepsilon)\,\phi(x_0)\left(\dfrac{x_0}{|x|}\right)^{(x_0^2)}, & |x| < x_0, \\ (1-\varepsilon)\phi(x), & x_0 \le |x| \le x_1, \\ (1-\varepsilon)\,\phi(x_1)\left(\dfrac{x_1}{|x|}\right)^{(x_1^2)}, & |x| > x_1, \end{cases} \tag{4.4.8}$$

where the parameters x_0 and x_1 satisfy the equations

$$\begin{gathered} x_0^2 = (1-k)^+, \qquad x_1^2 = 1 + k, \\ 2\int_{x_0}^{x_1} \phi(x)\,dx + \frac{2x_0\phi(x_0) + 2x_1\phi(x_1)}{x_1^2 - 1} = \frac{1}{1-\varepsilon}. \end{gathered} \tag{4.4.9}$$

In the case of sufficiently small ε ($\varepsilon < 0.205$, $x_0 = 0$, $x_1 > \sqrt{2}$), the least informative density f^* corresponds to a distribution that is normal in the central zone and is like a t-distribution with $k = x_1^2 - 1 \ge 1$ degrees of freedom in the tails. For $\varepsilon > 0.205$ and $x_0 > 0$, an additional t-distribution part of f^* appears about $x = 0$: actually, this effects in trimming some smallest data values along with greatest ones.

4.4.2. Efficient M- and L-estimators

For the least informative density (4.4.8), the efficient M-estimator of scale is defined by the score function

$$\chi^*(x) = -x\,\frac{f^{*\prime}(x)}{f^*(x)} - 1 = \begin{cases} x_0^2 - 1, & |x| < x_0, \\ x^2 - 1, & x_0 \le |x| \le x_1, \\ x_1^2 - 1, & |x| > x_1, \end{cases} \tag{4.4.10}$$

Now we are able to check the optimality of the obtained solution for f^*. As shown in Section 3.1 (see also (Huber, 1981, p. 82)), f^* minimizes the Fisher information over the class $\mathscr{F}$ if and only if

$$\left[\frac{d}{dt}I(f_t)\right]_{t=0} \ge 0, \tag{4.4.11}$$

where $f_t = (1-t)f^* + tf$ and f is a density providing $0 < I(f) < \infty$. Inequality (4.4.11) can be rewritten as

$$\int [2x\,\chi^{*\prime}(x) - \chi^{*2}(t)][f(x) - f^*(x)]\,dx \ge 0, \tag{4.4.12}$$

where $\chi^*(x)$ is given by (4.4.10).

Substituting (4.4.10) into (4.4.12), we obtain

$$f(x) - (1-\varepsilon)\phi(x) \ge 0.$$

Thus (4.4.12) is equivalent to the restriction of the class of ε-contaminated normal densities, which confirms the validity of the expression for f^*.

The efficient L-estimator is the trimmed standard deviation

$$S_n = \left\{\frac{1}{n - [\alpha_1 n] - [\alpha_2 n]} \sum_{i=[\alpha_1 n]+1}^{n-[\alpha_2 n]} x_i^2\right\}^{1/2}, \tag{4.4.13}$$

where

$$\alpha_1 = F^*(x_0) - 1/2, \qquad \alpha_2 = 1 - F^*(x_1).$$

Fig. 4.4 and 4.5 illustrate the possible forms of the score function.

The limiting case $\varepsilon \to 1$ gives the median absolute deviation: the limiting M- and L-estimators of τ coincide with the median of $\{\log|x_i|\}$, hence the corresponding estimator is the median of $\{|x_i|\}$.

The above M- and L-estimators are biased at the standard normal distribution Φ. To make them asymptotically unbiased in this case, one should divide them by an appropriate constant. The values of these constants are given in (Huber, 1981, pp. 125–126).

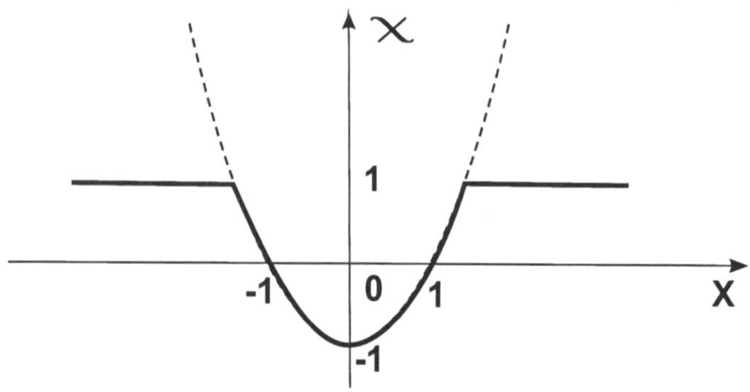

Figure 4.4. The score function for the one-sided trimmed standard deviation

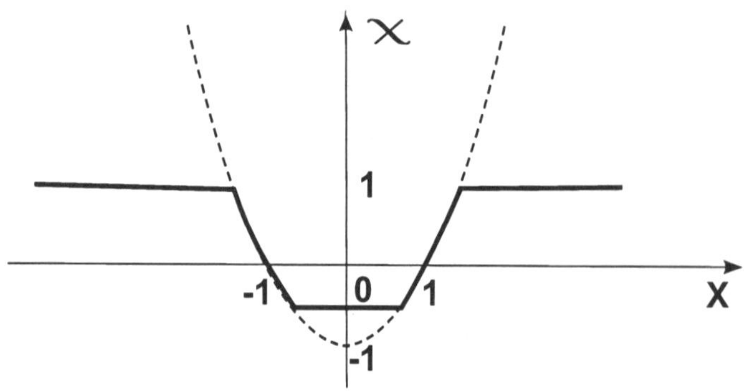

Figure 4.5. The score function for the two-sided trimmed standard deviation

4.4.3. Remark on minimax aspects

It follows from the general results of Section 1.2 on minimax estimation of the location parameter that the above M-estimator of scale is minimax with regard to the asymptotic variance

$$V(f,\chi^*) \le V(f^*,\chi^*) = \frac{1}{I(f^*;1)}$$

under ε-contaminated normal distributions satisfying $S(F) = 1$ or, in other words, the Fisher consistency condition

$$\int \chi(x)\,dF(x) = 0. \qquad (4.4.14)$$

It is possible to ignore this rather restrictive condition for sufficiently small

ε ($\varepsilon < 0.04$ and $x_1 > 1.88$) using a kind of standardized variance instead of the asymptotic variance $V(f, \chi)$ (for details, see (Huber, 1981, pp. 122–126)).

4.5. Final remarks

In this section we briefly discuss some scale estimators mainly belonging to two groups, which generate scale estimators subordinated to location ones. The first group contains the maximum likelihood scale estimators for given densities, in particular, from the parametric family of exponential-power distributions. The estimators of the second group are obtained by reducing the problem of scale estimation to the problem of estimating the location parameter, for example, the well-known median absolute deviation appears to be a minimax variance estimator in the class of distribution densities analogous to the class of nondegenerate distributions.

4.5.1. Estimating the scale parameter of exponential-power densities

As before, we assume that the location parameter is known: $\theta = 0$. Consider the family of exponential-power densities

$$f_q(x;\sigma) = \frac{q}{2\sigma\Gamma(1/q)} \exp\left(-\frac{|x|^q}{\sigma^q}\right), \qquad q \geq 1.$$

Then the maximum likelihood estimator of the scale parameter σ is

$$S_n = \left(\frac{q}{n}\right)^{1/q} \left(\sum_{i=1}^{n} |x_i|^q\right)^{1/q}.$$

The minimum of the Fisher information $I(q;\sigma) = q/\sigma^2$ is attained at $q^* = 1$, i.e., the least informative density is the Laplace and the corresponding scale estimator is the mean absolute deviation $d = n^{-1}\sum_i |x_i|$.

In the multivariate case of spherically symmetric exponential-power densities

$$f_q(r;\sigma) = \frac{q\Gamma(m/2)}{2\pi^{m/2}\sigma^m\Gamma(m/q)} \exp\left(-\frac{r^q}{\sigma^q}\right), \quad q \geq 1, \quad r = \left(\sum_{j=1}^{m} u_j^2\right)^{1/2},$$

for $\widehat{\sigma}$ we obtain, in addition,

$$S_n = \left(\frac{q}{mn}\right)^{1/q} \left(\sum_{i=1}^{n} r_i^q\right)^{1/q}.$$

Similarly, the Fisher information is $I(q; m, \sigma) = (mq)/\sigma^2$, and its minimum is also attained at $q = 1$. Thus the corresponding estimator of σ is given by the multivariate analog of the mean absolute deviation

$$S_n = d_n = \frac{1}{mn}\sum_{i=1}^{n} r_i, \qquad r_i = \left(\sum_{j=1}^{m} x_{ij}^2\right)^{1/2}.$$

4.5.2. Scale analogs to minimax variance location estimators

Following (Huber, 1981), we rewrite the least informative density for location into the corresponding least informative density for scale using the change of variables $\eta = \log|\xi|$ and $\tau = \log\sigma$. Denote the distribution function of η by $G(y - \tau)$. Then

$$G(y) = F(e^y) - F(-e^y), \qquad g(y) = 2e^y f(e^y).$$

Without loss of generality we assume that $\sigma = 1$ and $\tau = 0$. Now we consider the Laplace density $g^*(y) = L(y; 0, a)$ minimizing the Fisher information for location over the class of nondegenerate distributions with the restriction $g(0) \geq 1/(2a) > 0$. Hence the corresponding restriction on the density f is of the form $f(1) \geq 1/(4a) > 0$, and the least informative density minimizing the Fisher information for scale is

$$f^*(x) = \begin{cases} \dfrac{1}{4a|x|^{1-1/a}}, & |x| \leq 1, \\ \dfrac{1}{4a|x|^{1+1/a}}, & |x| > 1. \end{cases}$$

Therefore the score function is

$$\chi^*(x) = -x\frac{f^{*\prime}(x)}{f^*(x)} - 1 = \begin{cases} -1/a, & |x| \leq 1, \\ 1/a, & |x| > 1 \end{cases}$$

with the Fisher information

$$I(f^*) = \int_{-\infty}^{\infty}\left(-x\frac{f^{*\prime}(x)}{f^*(x)} - 1\right)^2 f^*(x)\,dx = \frac{1}{a^2}.$$

The above score function $\chi^*(x)$ corresponds to the median absolute deviation $S_n = \operatorname{med}|x|$. Thus this estimator minimizes the asymptotic variance over the class of distributions with the restriction on the value of the density at $x = 1$. Certainly, such a restriction does not seem very natural, because it means a bounded above dispersion in the distribution zones about the points $x = \pm 1$.

It is possible to present some other examples of applying the above approach to the least informative distributions, in particular, to the *cosine-exponential* distribution optimal over the class with a bounded distribution subrange, but their substantial interpretation is rather embarrassing.

4.5.3. Two examples of scale estimators possessing both high global robustness and efficiency in the normal case

The median absolute deviation is the 'most robust estimator of scale' (Huber, 1981, p. 122), but it has a very low efficiency $4/\pi^2 \approx 0.367$ under the normal distribution. Thus the problem of designing robust estimators with high breakdown points (close to 50%) and efficiency about 0.9–0.95 remains open yet.

Here we represent two close to each other estimators which partly satisfy the above conditions.

The first is called the median of absolute differences

$$S_n = \operatorname{med} |x_i - x_j|, \qquad 1 \le i < j \le n$$

with the efficiency 0.864 in the normal case (Shulenin, 1993).

The second, the Q_n-estimator (Rousseeuw and Croux, 1993), is defined by the 0.25-quantile of absolute differences

$$S_n = Q_n = f_n \cdot 2.2219 \{|x_i - x_j| : i < j\}_{(k)},$$

where $k = \binom{h}{2}$ and $h = [n/2] + 1$. The constant f_n is a small sample factor. Under the normal distribution, the efficiency of Q_n is 0.82.

5

Robust regression and autoregression

In this chapter we extend the results of Chapter 3 to the problems of robust linear regression.

Most attention is paid to minimax variance estimation in autoregression and regression–autoregression models which the solutions of Chapter 3 based on the Weber–Hermite functions are applied to.

5.1. Introductory remarks

In statistics as a whole, regression problems are related to most important for theory and applications. Besides robustness, they cause a lot of specific questions (Hampel *et al.*, 1986; Huber, 1981); we recall only three of them:

- the choice of a model (linear or nonlinear, parametric, nonparametric or semiparametric) and its order;
- the choice of a criterion of goodness-of-fit;
- and the choice of a computational algorithm.

Fig. 5.1 contains the data and its fits illustrating the abovesaid. At least three choices of the model (two straight lines and the parabola) are obviously possible with the data presented, and it is quite difficult to make the final decision without additional information.

Here we consider the classical linear regression model in matrix notation

$$\mathbf{x} = \mathbf{\Phi}\boldsymbol{\theta} + \mathbf{e}, \tag{5.1.1}$$

or in scalar notation

$$x_i = \sum_{j=1}^{m} \phi_{ij}\theta_j + e_i, \qquad i = 1, \dots, n, \tag{5.1.2}$$

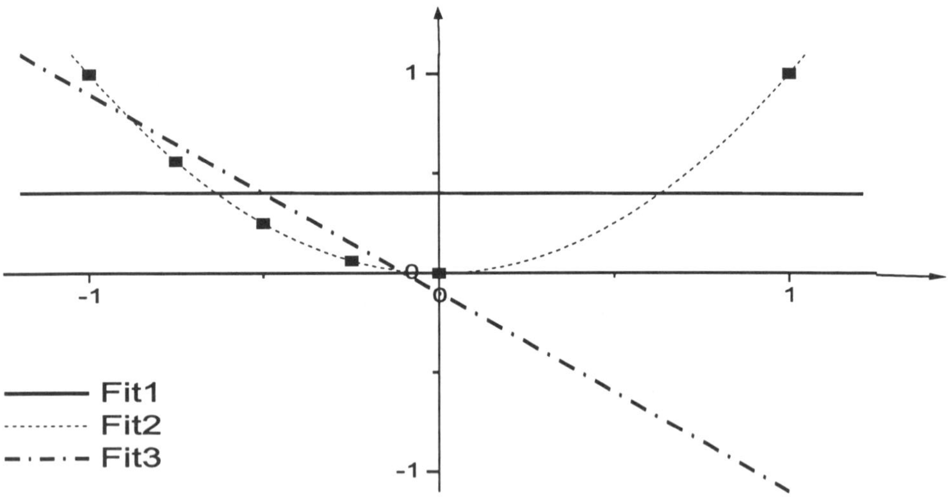

Figure 5.1. The possible fits to the data

where

- $\mathbf{x} = (x_1, \dots, x_n)^T$ is the vector of observations or *response variables*;
- $\boldsymbol{\theta} = (\theta_1, \dots, \theta_m)^T$ is the vector of unknown parameters to be estimated;
- $\boldsymbol{\Phi} = (\phi_{ij})_{n,m}$ is the given design matrix, and the variables $\phi_{i1}, \dots, \phi_{im}$ are called the *explanatory variables* or *carriers*;
- $\mathbf{e} = (e_1, \dots, e_n)^T$ is the vector of independent random errors with common symmetric density f belonging to a certain class $\mathscr{F}$.

Furthermore, applying the regression estimator to the data

$$\begin{pmatrix} \phi_{11} & \phi_{12} & \cdots & \phi_{1m} & x_1 \\ \phi_{21} & \phi_{22} & \cdots & \phi_{2m} & x_2 \\ \cdots & \cdots & \cdots & \cdots & \cdots \\ \phi_{n1} & \phi_{n2} & \cdots & \phi_{nm} & x_n \end{pmatrix}$$

yields $\widehat{\boldsymbol{\theta}} = (\widehat{\theta}_1, \dots, \widehat{\theta}_m)^T$, where the estimators $\widehat{\theta}_j$ are called the *regression coefficients*. Substituting these estimators for θ_j into (5.1.2), we obtain

$$\widehat{x}_i = \sum_{j=1}^{m} \phi_{ij} \widehat{\theta}_j,$$

where $\widehat{x}_i$ is called the *predicted* or *estimated* value of x_i. The difference between the actually observed and estimated values

$$r_i = x_i - \widehat{x}_i$$

is called the *residual* r_i.

The classical approach originates from Gauss and Legendre (see (Stigler, 1981) for historical remarks), and it suggests minimizing the sum of squares of residuals

$$\text{minimize} \quad \sum_{i=1}^{n} r_i^2, \tag{5.1.3}$$

or, which is equivalent, solving m simultaneous equations obtained by differentiation of (5.1.3),

$$\sum_{i=1}^{n} r_i \phi_{ij} = 0, \qquad j = 1, \dots, m.$$

In matrix notation, it is of the form

$$\mathbf{\Phi}^T \mathbf{\Phi} \boldsymbol{\theta} = \mathbf{\Phi}^T \mathbf{x}.$$

If $\mathbf{\Phi}$ has the full rank m and the independent errors e_i have zero means $\mathsf{E}e_i = 0$ with common variance $\mathsf{E}e_i^2 = \sigma^2 < \infty$, then the solution can be written as

$$\widehat{\boldsymbol{\theta}} = (\mathbf{\Phi}^T \mathbf{\Phi})^{-1} \mathbf{\Phi}^T \mathbf{x} \tag{5.1.4}$$

with the covariance matrix of estimators in the form

$$\mathbf{V} = \frac{\sigma^2}{n - m} (\mathbf{\Phi}^T \mathbf{\Phi})^{-1}. \tag{5.1.5}$$

This least squares estimator (5.1.4) has many remarkable properties (see, e.g. (Kendall and Stuart, 1963; Rao, 1965)). The main of them are: first, the LS estimator is optimal in the class of all linear unbiased estimators (Gauss–Markov theorem), and second, under normal error distributions, the LS estimator is optimal in the class of all unbiased estimators.

The optimality in Gauss–Markov theorem can be understood with regard to any of the following four criteria:

- The variance of an arbitrary linear combination $\sum_{j=1}^{m} \lambda_j \widehat{\theta}_j$ is minimum in the class of all linear unbiased estimators of $\sum_{j=1}^{m} \lambda_j \theta_j$, in particular, the variance of each component θ_j is minimum.

- $\det \mathbf{V}(\boldsymbol{\theta})$ is minimum (the determinant of a covariance matrix is called the generalized variance).

- $\mathbf{V}(\boldsymbol{\theta})$ is minimum, that is, the difference between the covariance matrices of any estimator of a given class and the LS estimator is a positive definite matrix.

- $\operatorname{tr}\mathbf{V}(\boldsymbol{\theta})$ is minimum.

Recall that the LS estimator is extremely unstable under outliers and gross errors in the data, in other words, it is completely non-robust.

5.2. The minimax variance regression

Robust versions of the LS procedure are given by M-estimators that provide a straightforward generalization for the linear regression problem (5.1.1). In this case, the estimator $\widehat{\boldsymbol{\theta}}$ is obtained by minimizing the goodness-of-fit criterion

$$\text{minimize} \quad \sum_{i=1}^{n} \rho(r_i), \tag{5.2.1}$$

or, to ensure scale invariance for $\widehat{\boldsymbol{\theta}}$,

$$\text{minimize} \quad \sum_{i=1}^{n} \rho\left(\frac{r_i}{S_n}\right), \tag{5.2.2}$$

where S_n is some robust estimator for the scale of residuals.

For the differentiable and convex contrast functions ρ, the above relations can be replaced by the simultaneous equations

$$\sum_{i=1}^{n} \psi(r_i)\phi_{ij} = 0, \qquad j = 1, \dots, m, \tag{5.2.3}$$

or

$$\sum_{i=1}^{n} \psi\left(\frac{r_i}{S_n}\right)\phi_{ij} = 0, \qquad j = 1, \dots, m, \tag{5.2.4}$$

where $\psi = \rho'$ is the score function.

In (Huber, 1973), it was suggested to estimate σ by solving (5.2.4) simultaneously with the equation

$$\sum_{i=1}^{n} \psi^2\left(\frac{r_i}{S_n}\right) = (n-m)A. \tag{5.2.5}$$

The constant A is chosen so that S_n converges to σ when the e_i have the normal distribution with mean zero and variance σ^2.

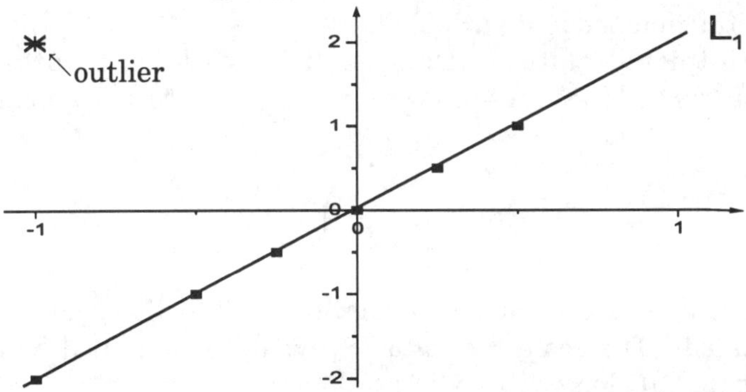

Figure 5.2. Robustness of the L_1-regression with respect to an outlier in the x-direction

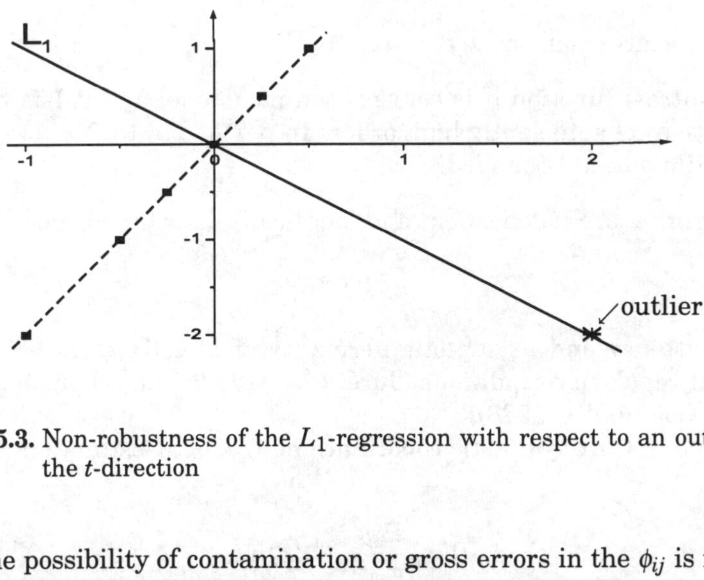

Figure 5.3. Non-robustness of the L_1-regression with respect to an outlier in the t-direction

If the possibility of contamination or gross errors in the ϕ_{ij} is ignored and only contamination of x_i is allowed, then the regression M-estimators are robust, as long as ψ is bounded. However, allowing for the realistic possibility of gross errors in the ϕ_{ij} may yield an unsatisfactory situation even for monotone and bounded ψ: outliers in independent variables (leverage points) can have a considerable influence on estimators (Rousseeuw and Leroy, 1987). Fig. 5.2 and 5.3 illustrate this effect.

One of the most robust estimators, the L_1-estimator minimizing the sum of the absolute values of residuals $\sum_i |r_i|$ (see Chapter 6 for general properties of L_1-approximations), protects the estimator of a straight line against gross errors in the x-direction, and it completely fails with leverage points caused by gross errors in the independent variable t.

To deal with this difficulty, in (Hampel, 1974; Mallows, 1975) the *generalized M-estimators* (*GM*-estimators) were suggested, which are defined as a solution of

$$\sum_{i=1}^{n} w(\boldsymbol{\phi}_i)\boldsymbol{\phi}_i \psi\left(\frac{r_i}{S_n}\right) = 0, \tag{5.2.6}$$

where $w(u_1, \dots, u_m)$ is a continuous scalar weight function chosen so that $w(\boldsymbol{\phi}_i)\boldsymbol{\phi}_i$ is bounded. The scale estimator S_n would be obtained by simultaneously solving (5.2.6) along with (5.2.5).

As above in the case of estimating the location parameter, under certain regularity conditions (Huber, 1981, p. 165), M-estimators (5.2.4) are consistent and asymptotically normal. The main of those conditions are:

(R1) $\mathbf{\Phi}$ has full rank m, and the diagonal elements of the hat matrix

$$\mathbf{H} = \mathbf{\Phi}(\mathbf{\Phi}^T\mathbf{\Phi})^{-1}\mathbf{\Phi}^T$$

are uniformly small: $\max_i h_i = h \ll 1$.

(R2) The contrast function ρ is convex, non-monotone, and it has bounded derivatives of sufficiently high order. In particular, $\psi(x) = \rho'(x)$ should be continuous and bounded.

(R3) The errors e_i are independent and identically distributed such that

$$\mathsf{E}\psi(e_i) = 0.$$

The consistency and asymptotic normality of M-estimators hold under more general regularity conditions (Jurečkovà, 1977; Onishchenko and Tsybakov, 1987; Zolotukhin, 1988).

Furthermore, the asymptotic covariance matrix of M-estimators is of the form

$$\mathbf{V}(f, \psi) = \frac{\mathsf{E}\psi^2}{(\mathsf{E}\psi')^2}(\mathbf{\Phi}^T\mathbf{\Phi})^{-1}. \tag{5.2.7}$$

Thus, the elements of the covariance matrix (5.2.7) are proportional to the scalar factor

$$v(f, \psi) = \frac{\int \psi^2(x)f(x)\,dx}{\left(\int \psi'(x)f(x)\,dx\right)^2},$$

which is the asymptotic variance $\mathsf{Var}\,\theta_n(\psi, f)$ of M-estimators of location being the functional of the pair (ψ, f) (see Section 1.2).

The matrix factor depends only on the positive definite matrix $\mathbf{\Phi}^T\mathbf{\Phi}$. Therefore the minimax properties of M-estimators of location are directly extended to the linear regression model (5.1.2) (Huber, 1972)

$$\mathbf{V}(\psi^*, f) \le \mathbf{V}(\psi^*, f^*),$$
$$f^* = \arg\min_{f\in\mathcal{F}} I(f), \qquad \psi^* = -f^{*\prime}/f^*, \tag{5.2.8}$$

where $I(f)$ is the Fisher information for location, and the inequality $\mathbf{A} \ge \mathbf{B}$ is understood in the sense of non-negative definiteness of the matrix $(\mathbf{A} - \mathbf{B})$.

Finally we observe that all the results on the minimax variance estimators of location obtained in Chapter 3 can be applied here with obvious modifications.

5.3. Robust autoregression

5.3.1. Preliminaries

Autoregressive models are widely used in theory and applications for description of time series (Anderson, 1971; Kendall and Stuart, 1968). This is due to the two of their specific features. First, the autoregressive model represents a stochastic equation in differences, and so it can be used for description of the output of dynamic systems (Astrom and Eykhoff, 1971; Eykhoff, 1974; Ljung, 1987; Ljung, 1995; Tsypkin, 1984; Walter and Pronzato, 1997). Second, the process of autoregression is one of the simplest models for stochastic processes, and in a certain sense (by the criterion of maximum entropy), it is the best approximation to an arbitrary stationary stochastic process (Kleiner *et al.*, 1979; Martin, 1981).

In this section we consider some extensions of M-estimators to the problem of robust estimation of autoregressive parameters and formulate minimax variance robust algorithms of estimation over some classes of error distributions.

We now pose the estimation problem with the autoregressive model.

DEFINITION 5.3.1. A sequence $x_1, \dots, x_n, \dots$, is said to be a linear autoregressive model if

$$x_n = \sum_{j=1}^{m} \beta_j x_{n-j} + \xi_n, \qquad n = j+1, j+2, \dots, \tag{5.3.1}$$

where

- x_n are the observations;
- $\boldsymbol{\beta} = (\beta_1, \dots, \beta_m)$ is the vector of autoregressive parameters;

- ξ_n are independent identically distributed random variables referred to as the *innovations*.

Then the problem is formulated as follows: given a sample $x_1, \dots, x_n$, estimate the parameter vector $\boldsymbol{\beta} \in \mathbf{R}^m$.

The classical approach suggests normality of the innovations ξ_i. As robust statistics mainly aims at the struggle against outliers in the data, we consider the following two basic and simple outlier generating mechanisms for autoregression (Fox, 1972):

- the *innovations outliers* (*IO*) model

$$x_n = \sum_{j=1}^{m} \beta_j x_{n-j} + \xi_n, \tag{5.3.2}$$

  where the distribution F_ξ of the innovations ξ_i is heavy-tailed;

- the *additive outliers* (*AO*) model is of the form

$$y_n = x_n + \eta_n, \tag{5.3.3}$$

  where x_n is the Gaussian autoregression and η_n are independent and identically distributed with common distribution

$$F_\eta = (1 - \varepsilon)\delta_0 + \varepsilon H;$$

  δ_0 is the degenerate distribution having its whole mass at zero, and H is a heavy-tailed symmetric distribution.

If ε is small, then the x_n are observed perfectly most of the time with probability $P(\eta_n = 0) = 1 - \varepsilon$.

Though the *AO*-model seems more realistic than the *IO*-model, here we are mainly interested in the *IO*-model of outliers, because the methods of protection from additive outliers are as a whole the same as in regression problems.

In general, time series models generate a wider variety of cases where outliers may occur than, for example, regression models.

Consider the class of M-estimators of autoregressive parameters

$$\widehat{\boldsymbol{\beta}}_n = \arg\min_{\boldsymbol{\beta}} \sum_{i=m+1}^{n} \rho\left(x_i - \sum_{j=1}^{m} \beta_j x_{i-j}\right), \tag{5.3.4}$$

where $\rho(u)$ is a given contrast function. For instance, setting $\rho(u) = u^2$, we have the LS estimators; with $\rho(u) = -\log f_\xi(u)$, we have the ML estimators.

Introducing the score function $\psi(u) = \rho'(u)$, we write out the simultaneous equations giving the M-estimators sought for in the implicit form

$$\sum_{i=m+1}^{n} \psi\left(x_i - \sum_{j=1}^{m} \widehat{\beta}_j x_{i-j}\right) x_{i-j} = 0, \qquad j = 1, \dots, m.$$

Just these estimators are the goal of our further analysis.

5.3.2. The properties of M-estimators for autoregression

Like M-estimators of regression parameters, under certain regularity conditions, M-estimators of autoregressive parameters are consistent and asymptotically normal (Martin, 1979; Martin, 1981; Polyak and Tsypkin, 1983).

Assume the following.

(AR1) Autoregression is stable, i.e., all roots of the characteristic equation

$$q^m = \sum_{j=1}^{m} \beta_j q^{m-j}$$

lie outside the interval $[-1, 1]$.

(AR2) The errors are independent with common density f symmetric about zero and bounded variance

$$\mathsf{E}\xi_n^2 = \sigma^2 < \infty.$$

(AR3) The contrast function $\rho(u)$ is nonnegative, symmetric, convex, and twice differentiable.

Under assumptions (AR1), (AR2), (AR3), and some additional regularity conditions, the estimator (5.3.4) is consistent (Polyak and Tsypkin, 1983): $\widehat{\boldsymbol{\beta}}_n$ tends to $\boldsymbol{\beta}$ in probability, and it is asymptotically normal

$$\sqrt{n}(\boldsymbol{\beta}_n - \boldsymbol{\beta}) \sim \mathcal{N}(\mathbf{0}, \mathbf{V}),$$

where the covariance matrix $\mathbf{V} = \mathbf{V}(f, \psi)$ is of the form

$$\mathbf{V}(f, \psi) = v(f, \psi)\mathbf{R}^{-1}, \qquad v(f, \psi) = \frac{\int \psi^2 f\, dx}{\sigma^2 \left(\int \psi' f\, dx\right)^2},$$

$$\mathbf{R} = \begin{pmatrix} \rho_0 & \rho_1 & \cdots & \rho_{m-1} \\ \rho_1 & \rho_2 & \cdots & \rho_{m-2} \\ \cdots & \cdots & \cdots & \cdots \\ \rho_{m-1} & \rho_{m-2} & \cdots & \rho_0 \end{pmatrix}. \tag{5.3.5}$$

Here $\mathbf{R}$ is the matrix of autocorrelations $\rho_i = \mu_i/\sigma^2$, where

$$\mu_i = \lim_{n\to\infty} \mathsf{E} x_{n-i} x_n$$

are the autocovariances of some autoregressive process.

It is essential that the covariance matrix $\mathbf{V}(f, \psi)$ depends on the shape of f and ψ only through the scalar factor $v(f, \psi)$ equal to the asymptotic variance of M-estimators of location divided by the variance of errors.

The matrix $\mathbf{R}$ does not depend on the shape of the density, and it is defined only by the coefficients $\boldsymbol{\beta}$.

If we assume that the distribution density f is not completely known but belongs to a given class $\mathcal{F}$, then M-estimators of autoregressive parameters possess the minimax property (Martin, 1979)

$$\mathbf{V}(\psi^*, f) \le \mathbf{V}(\psi^*, f^*)$$
$$f^* = \arg\min_{f\in\mathcal{F}} \sigma^2(f)I(f), \qquad \psi^* = -f^{*\prime}/f^*. \tag{5.3.6}$$

It follows from (5.3.6) that the minimax variance estimator is the maximum likelihood one with the score function $\psi^* = -f^{*\prime}/f^*$ for the least favorable density f^* minimizing the functional $J(f) = \sigma^2(f)I(f)$, which is the product of the distribution variance and the Fisher information. Hence, in the case of autoregression, the results are qualitatively different from the case of regression.

5.3.3. The minimax variance estimators of autoregressive parameters under distributions with bounded variance

Here an exact result concerning minimax properties of the LS estimators of autoregressive parameters is formulated, and as its corollary, exact solutions of the variational problem (5.3.6) are given for the classes of distribution densities $\mathcal{F}_{12}$, $\mathcal{F}_{23}$ and $\mathcal{F}_{25}$.

The following important result on the least favorable distribution is true for autoregression (Whittle, 1962).

THEOREM 5.3.1. *If the class $\mathcal{F}$ contains the normal density $f(x) = \mathcal{N}(x; 0, \sigma_N)$, then the latter is a solution of the variational problem* (5.3.6)*:*

$$f^*(x) = \mathcal{N}(x; 0, \sigma_N).$$

COROLLARY 5.3.1. *The minimax estimator of autoregressive parameters is given by the LS method.*

Comparing this solution with the solutions for location and regression, where it holds only for the class $\mathcal{F}_2$ (in particular cases, for the classes $\mathcal{F}_{12}$, $\mathcal{F}_{23}$ and $\mathcal{F}_{25}$), here the LS method is optimal for the classes $\mathcal{F}_1$, $\mathcal{F}_2$, $\mathcal{F}_3$, and $\mathcal{F}_5$, and also for the combinations of these classes $\mathcal{F}_{12}$, $\mathcal{F}_{23}$, and $\mathcal{F}_{25}$.

For the sake of completeness, we now formulate these results for the classes $\mathscr{F}_{12}$, $\mathscr{F}_{23}$, and $\mathscr{F}_{25}$.

THEOREM 5.3.2. *In the classes*

$$\mathscr{F}_{12} = \{f : f(0) \ge 1/(2a) > 0, \sigma^2(f) \le \overline{\sigma}^2\},$$
$$\mathscr{F}_{23} = \{f : f(x) \ge (1-\varepsilon)\mathscr{N}(x; 0, \sigma_N), \sigma^2(f) \le \overline{\sigma}^2\},$$
$$\mathscr{F}_{25} = \{f : F^{-1}(3/4) - F^{-1}(1/4) \le b, \sigma^2(f) \le \overline{\sigma}^2\},$$

the solution of variational problem (5.3.6) *is the normal density*

$$f^*(x) = \mathscr{N}(x; 0, \sigma_N),$$

where σ_N takes any value satisfying the constraints of the classes, and in each case $J(f^*) = 1$.

In spite of the fact that the optimal solutions over the classes $\mathscr{F}_{12}$ and $\mathscr{F}_{25}$ may differ from each other by the standard deviation σ_N, the minimax variance estimators do not depend on those due to scale equivariancy. The estimators are given by (5.1.4)

$$\widehat{\boldsymbol{\beta}}_n = (\boldsymbol{\Phi}^T\boldsymbol{\Phi})^{-1}\boldsymbol{\Phi}\mathbf{x}, \tag{5.3.7}$$

where $\mathbf{x} = (\mathbf{x}_{m+1}, \mathbf{x}_{m+2}, \dots, \mathbf{x}_n)^T$, $\boldsymbol{\Phi} = (x_{i-j})$ is an $(n-m) \times m$ matrix. The covariance matrix of estimator (5.3.7) is of the form

$$\mathbf{V} = \mathbf{V}(f^*, \psi^*) = \mathbf{R}^{-1},$$

providing the guaranteed accuracy of estimation

$$\mathbf{V}(\psi^*, f) \le \mathbf{R}^{-1}$$

for all $f \in \mathscr{F}$.

We have established that the normal density is the least favorable over a sufficiently wide class, and so the LS method yields the optimal estimator. The other results occur if to narrow a class of distribution densities we introduce an additional lower bound on the variance of innovations $\sigma^2(f) \ge \underline{\sigma}^2$ (Polyak and Tsypkin, 1983; Tsypkin, 1984).

For the problem of autoregression, such a restriction has a plain motivation. In the case of regression, the growth of errors deteriorates the conditions of estimation. On the contrary, for autoregression, the innovations e_i generate the process itself, and the growth of their power improves estimation.

Considering the classes of distribution densities with the additional lower bound on variance $\sigma^2(f) \ge \underline{\sigma}^2$, we observe that for sufficiently small $\underline{\sigma}^2$, the effect of narrowing the class has no impact, and the optimal density remains normal. Nevertheless, with large $\underline{\sigma}^2$, the solutions become different.

The following is true (Shevlyakov, 1991).

THEOREM 5.3.3. *In the class* $\mathscr{F}_1$ *with additional restrictions* $\sigma^2(f) \geq \underline{\sigma}^2$ *and* $\underline{\sigma}^2 \leq 2a^2$,

$$\widetilde{\mathscr{F}}_1 = \left\{ f\colon f(0) \geq \frac{1}{2a} > 0, \sigma^2(f) \geq \underline{\sigma}^2, \underline{\sigma}^2 \leq 2a^2 \right\}, \tag{5.3.8}$$

the least favorable density is of the form

$$\widetilde{f}_1^*(x) = \begin{cases} \mathscr{N}(x; 0, \underline{\sigma}), & \underline{\sigma}^2/a^2 \leq 2/\pi, \\ f(x; \nu, \underline{\sigma}), & 2/\pi < \underline{\sigma}^2/a^2 \leq 2, \\ L(x; 0, a), & \underline{\sigma}^2/a^2 = 2, \end{cases} \tag{5.3.9}$$

where $f(x; \nu, \underline{\sigma})$ *are the Weber–Hermite distribution densities of the form*

$$f(x; \nu, \underline{\sigma}) = \frac{\Gamma(-\nu)\sqrt{2\nu + 1 + 1/S(\nu)}}{\sqrt{2\pi}\,\underline{\sigma}\,S(\nu)} \mathscr{D}_\nu^2 \left(\frac{|x|}{\underline{\sigma}} \sqrt{2\nu + 1 + 1/S(\nu)} \right) \tag{5.3.10}$$

with $\nu \in (-\infty, 0]$ *determined from the equation (see Section* 3.2*)*

$$\frac{\overline{\sigma}}{a} = \frac{\sqrt{2\nu + 1 + 1/S(\nu)}\,\Gamma^2(-\nu/2)}{\sqrt{2\pi}\, 2^{\nu+1}\, S(\nu)\, \Gamma(-\nu)}. \tag{5.3.11}$$

This result does not cover the case where $\underline{\sigma}^2 > 2a^2$. It is possible to describe the structure of the optimal solution in this domain: it consists of the extremals based on the Weber–Hermite functions. The unknown parameters of the solution can be determined from simultaneous equations including the restrictions of the class and transversality conditions. Nevertheless, here it is reasonable to seek for an approximate solution in the form of a linear combination of the limiting densities (Tsypkin, 1984)

$$\widetilde{f}_1^*(x) = (1 - \alpha)L(x; 0, a_1) + \alpha N(x; \mu, \sigma_N),$$

where the parameters α, a_1, μ, and σ_N are determined from the restrictions of the class

$$\widetilde{f}_1^*(0) = \frac{1}{2a_1}, \qquad \sigma^2(\widetilde{f}_1^*) \geq \underline{\sigma}^2 > 2a^2,$$

and the optimality condition $\widetilde{f}_1^* = \arg\min J(f)$.

The minimax estimator corresponding to the least favorable density of Theorem 5.3.2 coincides with that in the class $\mathscr{F}_{12}$ for the location parameter and regression whose score functions $\psi^* = -f^{*\prime}/f^*$ are intermediate between the linear (the LS method) and the sign (the LAV method) (see Fig. 3.8).

The minimax properties of the obtained estimator provide the guaranteed accuracy of estimation in the sense of the boundedness of the covariance matrix for all densities from the class $\widetilde{\mathscr{F}}_1$

$$\mathbf{V}(f, \widetilde{\psi}_1^*) \le \mathbf{V}(\widetilde{f}_1^*, \widetilde{\psi}_1^*) = v(\widetilde{f}_1^*, \widetilde{\psi}_1^*)\mathbf{R}^{-1},$$

$$v(\widetilde{f}_1^*, \widetilde{\psi}_1^*) = \begin{cases} 1, & \underline{\sigma}^2/a^2 \le 2/\pi, \\ 1/(\underline{\sigma}^2 I(\widetilde{f}_1^*)), & 2/\pi < \underline{\sigma}^2/a^2 \le 2, \\ 2, & \underline{\sigma}^2/a^2 = 2, \end{cases}$$

where $I(\widetilde{f}_1^*)$ is given by (3.2.7).

Here robustness is evidently confirmed due to the invariance of $v(\widetilde{f}_1^*, \widetilde{\psi}_1^*)$ to distribution variance.

5.3.4. Proofs

PROOF OF THEOREM 5.3.1. Observe that the least favorable density should obviously have finite variance $\sigma^2(f^*) < \infty$. We prove the assertion of this theorem using the Cauchy–Bunyakovskii inequality

$$\left(\int_{-\infty}^{\infty} xf'(x)\,dx\right)^2 = \left(\int_{-\infty}^{\infty} x\frac{f'(x)}{f(x)} f(x)\,dx\right)^2 \tag{5.3.12}$$

$$\le \int_{-\infty}^{\infty} x^2 f(x)\,dx \cdot \int_{-\infty}^{\infty} \left[\frac{f'(x)}{f(x)}\right]^2 f(x)\,dx = \sigma^2(f)I(f).$$

Indeed, integrating the left-hand side of (5.3.13) by parts, we obtain

$$\int_{-\infty}^{\infty} xf'(x)\,dx = xf(x)|_{-\infty}^{\infty} - \int_{-\infty}^{\infty} f(x)\,dx = -1, \tag{5.3.13}$$

because by virtue of boundedness of the variance, $\lim_{x\to\pm\infty} xf(x) = 0$. Thus it follows from (5.3.12) and (5.3.13) that

$$I(f)\sigma^2(f) \ge 1, \tag{5.3.14}$$

and as inequality (5.3.14) becomes the equality at the normal density

$$I(\mathscr{N}(x; 0, \sigma_N)) = \frac{1}{\sigma_N^2},$$

we arrive at the required assertion

$$f^*(x) = \mathscr{N}(x; 0, \sigma_N) \tag{5.3.15}$$

for arbitrary σ_N unless $f^* \in \mathscr{F}$. □

PROOF OF COROLLARY 5.3.1. The minimax estimator of autoregressive parameters is defined by the linear score function

$$\psi^*(z) = z/\sigma_N^2,$$

and by the LS method respectively. □

PROOF OF THEOREM 5.3.2. The validity of the theorem immediately follows from Theorem 5.3.1. □

PROOF OF THEOREM 5.3.3. In this case we use the scheme of the proof of Theorem 3.2.2.

Convexity of the Fisher information $I(f)$ implies convexity of the functional $J(f) = \sigma^2(f)I(f)$ in problem (5.3.6). Hence the density $f^* \in \widetilde{\mathscr{F}}_1$ minimizes $J(f)$ if and only if

$$\left.\frac{d}{dt}J(f_t)\right|_{t=0} \geq 0, \tag{5.3.16}$$

where $f_t = (1-t)f^* + tf$, $f \in \widetilde{\mathscr{F}}_1$ and $J(f) < \infty$.

Now we verify inequality (5.3.16) for solution (5.3.9). Rewrite the left-hand side of (5.3.16) as

$$\begin{aligned} \left.\frac{d}{dt}J(f_t)\right|_{t=0} &= \left.\frac{d}{dt}\left\{[(1-t)\sigma^2(f^*) + t\sigma^2(f)]I(f_t)\right\}\right|_{t=0} \\ &= \left.\frac{d}{dt}I(f_t)\right|_{t=0}\sigma^2(f^*) + I(f^*)[\sigma^2(f) - \sigma^2(f^*)] \geq 0. \end{aligned} \tag{5.3.17}$$

As in the proof of Theorem 3.2.2, it is suffices to check inequality (5.3.17) for the middle branch of solution (5.3.9), since the first and third branches are its limiting cases:

$$\mathscr{N}(x; 0, \underline{\sigma}) = f(x; 0, \underline{\sigma})$$

and

$$L(x; 0, a) = \lim_{\nu \to -\infty} f(x; \nu, \sigma).$$

In Section 3.2 the following expression was obtained for the derivative of the Fisher information:

$$\left.\frac{d}{dt}I(f_t)\right|_{t=0} = 4B\frac{\mathscr{D}_{\nu+1}(0)}{\mathscr{D}_\nu(0)}B[f(0) - f^*(0)] - B^4[\sigma^2(f) - \sigma^2(f^*)], \tag{5.3.18}$$

where

$$B = \frac{1}{\underline{\sigma}}\sqrt{2\nu + 1 + \frac{1}{S(\nu)}}. \tag{5.3.19}$$

Furthermore, the restrictions of the class $\widetilde{\mathscr{F}}_1$ hold as equalities

$$\sigma^2(f^*) = \underline{\sigma}^2, \qquad f^*(0) = \frac{1}{2a}. \tag{5.3.20}$$

Substituting (5.3.18) and (5.3.20) for $dI(f_t)/dt$, $\sigma^2(f^*)$, and $f^*(0)$ into the left-hand side of inequality (5.3.17), we obtain

$$\begin{aligned} \frac{d}{dt}J(f_t) &= \left\{4B\frac{\mathscr{D}_{\nu+1}(0)}{\mathscr{D}_\nu(0)}\left[f(0) - \frac{1}{2a}\right] - B^4[\sigma^2(f) - \underline{\sigma}^2]\right\}\underline{\sigma}^2 + I(f^*)[\sigma^2(f) - \underline{\sigma}^2] \\ &= 4B\underline{\sigma}^2\frac{\mathscr{D}_{\nu+1}(0)}{\mathscr{D}_\nu(0)}B\left[f(0) - \frac{1}{2a}\right] + [I(f^*) - B^4\underline{\sigma}^2][\sigma^2(f) - \underline{\sigma}^2]. \end{aligned}$$

By virtue of the restrictions of the class $\widetilde{\mathscr{F}}_1$, the terms $f(0) - 1/(2a)$ and $\sigma^2(f) - \underline{\sigma}^2$ are nonnegative. Furthermore, $\mathscr{D}_{\nu+1}(0)/\mathscr{D}_\nu(0) > 0$ (see Section 3.2). Therefore, inequality (5.3.17) holds if $I(f^*) - B^4\underline{\sigma}^2 > 0$.

Now we establish the latter. By Lemma 3.2.1, the Fisher information is of the following form at the densities $f(x; \nu, \underline{\sigma})$:

$$I(f^*) = I(f(x; \nu, \underline{\sigma})) = \frac{1}{\underline{\sigma}^2}\left[(2\nu+1)^2 + 4(2\nu+1)S(\nu) + \frac{3}{S^2(\nu)}\right]. \tag{5.3.21}$$

By (5.3.19) and (5.3.21),

$$\begin{aligned} I(f^*) - B^4\underline{\sigma}^2 &= \frac{1}{\underline{\sigma}^2}\left[(2\nu+1)^2 + 4(2\nu+1)S(\nu) + \frac{3}{S^2(\nu)}\right] - \frac{1}{\underline{\sigma}^2}\left[2\nu + 1 + \frac{1}{S(\nu)}\right]^2 \\ &= \frac{2}{S(\nu)\underline{\sigma}^2}\left[2\nu + 1 + \frac{1}{S(\nu)}\right] = \frac{2}{S(\nu)B^2} > 0. \end{aligned}$$

The positiveness of $S(\nu)$ follows from its definition (see Subsection 3.2.7), which completes the proof. □

5.4. Robust identification in dynamic models

In this section we demonstrate how to apply the Weber–Hermite densities obtained in Section 3.2 to the problems of identification in dynamic models of control theory.

Identification of parameters of a dynamic object is one of the main stages of a control process, and this topic represents a rich field for the use of statistical methods (Eykhoff, 1974; Kashyap and Rao, 1976; Lee, 1964; Ljung, 1987; Ljung, 1995; Sage and Melsa, 1971; Walter and Pronzato, 1997).

We now briefly describe a general approach to identification of dynamic objects (Tsypkin, 1984). Let the control object be given by the linear difference equation

$$x(n) = -\sum_{j=1}^{m} a_j x(n-j) + \sum_{j=0}^{m} b_j u(n-j) + \xi(n) + \sum_{j=1}^{m} d_j \xi(n-m), \quad n = 0, 1, 2, \dots, \tag{5.4.1}$$

where

- $x(n)$ are the outputs of the object;
- $u(n)$ are the control values;
- $\xi(n)$ are the stationary uncorrelated errors with symmetric distribution density f.

This equation is rather general: the particular cases of model (5.4.1) are given by the processes of regression, autoregression, moving average, and ARMA. For instance,

- the regression or R-objects

$$x(n) = \sum_{j=0}^{m} b_j u(n-j) + \xi(n), \tag{5.4.2}$$

  for $a_j = d_j = 0, j = 1, 2, \dots, m$;

- the autoregression or AR-objects

$$x(n) + \sum_{j=1}^{m} a_j x(n-j) = \xi(n), \tag{5.4.3}$$

  for $b_0 = 0$, $b_j = d_j = 0, j = 1, 2, \dots, m$;

- the moving average

$$x(n) = \sum_{j=0}^{m} d_j \xi(n-j),$$

  for $b_0 = 0$, $b_j = a_j = 0, j = 1, 2, \dots, m$;

- the ARMA-processes

$$x(n) + \sum_{j=1}^{m} a_j x(n-j) = \sum_{j=0}^{m} d_j \xi(n-m),$$

  for $b_j = 0, j = 0, 1, \dots, m$.

The objects described by the general equation (5.4.1) are called the *regression–autoregression* (RAR)-objects.

There exist various approaches to identification of the parameter vector

$$\boldsymbol{\theta} = (a_1, a_2, \dots, a_m, b_0, b_1, \dots, b_m, d_1, d_2, \dots, d_m)$$

(see, e.g. (Eykhoff, 1974; Ljung, 1987; Ljung, 1995; Walter and Pronzato, 1997)).

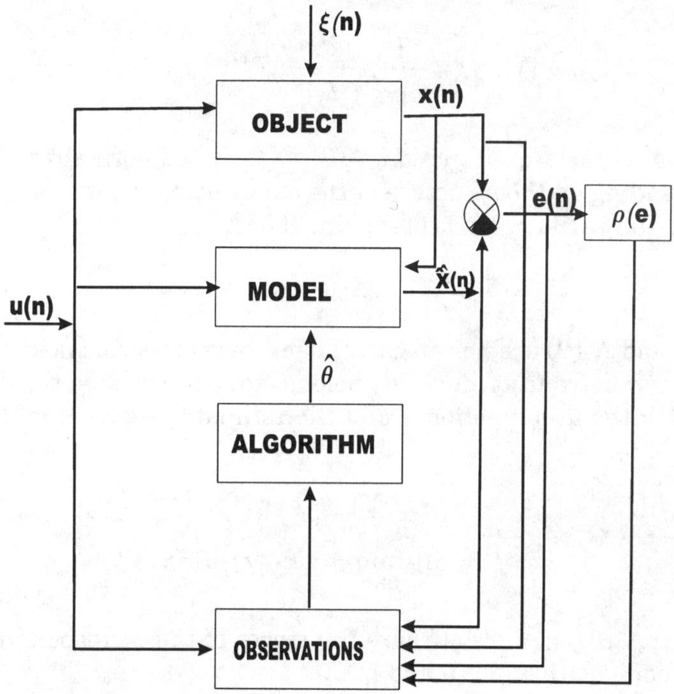

Figure 5.4. The block-scheme of identification

Here we consider that based on the use of the predicting or tuning model, in other words, the moving estimator of the output $x(n)$ of the form (Tsypkin, 1984)

$$\widehat{x}(n) = -\sum_{j=1}^{m} \widehat{a}_j x(n-j) + \sum_{j=0}^{m} \widehat{b}_j u(n-j) + \sum_{j=1}^{m} \widehat{d}_j [x(n-m) - \widehat{x}(n-m)].$$

The discrepancies $e(n) = x(n) - \widehat{x}(n)$ measured between the object model and the predicting model determine the procedure of identification based on minimization of the average losses

$$\widehat{\boldsymbol{\theta}}(n) = \arg\min_{\boldsymbol{\theta}} \frac{1}{n} \sum_{i=1}^{n} \rho[x(i) - \widehat{x}(i)],$$

where ρ is the loss function or the contrast function.

This approach to identification is illustrated in Fig. 5.4.

The asymptotic covariance matrix

$$\mathbf{V} = n\mathbf{V}_n = n\mathsf{E}(\widehat{\boldsymbol{\theta}} - \boldsymbol{\theta})(\widehat{\boldsymbol{\theta}} - \boldsymbol{\theta})^T$$

is a measure of efficiency of the form

$$\mathbf{V}(f, \psi) = \frac{\int \psi^2 f\, dx}{\left(\int \psi' f\, dx\right)^2} \mathbf{A}^{-1}(\boldsymbol{\theta}, \sigma^2(f)), \tag{5.4.4}$$

where $\psi = \rho'$ is the score function, $\mathbf{A}(\boldsymbol{\theta}, \sigma^2(f))$ is the normalized information matrix depending on the structure of the object being identified, as well as on the distribution variance $\sigma^2(f)$ (Tsypkin, 1984):

$$\mathbf{A}(\boldsymbol{\theta}, \sigma^2(f)) = \mathbf{A}_1(\boldsymbol{\theta}) + \sigma^2(f)\mathbf{A}_2(\boldsymbol{\theta}).$$

Here $\mathbf{A}_1(\boldsymbol{\theta})$ and $\mathbf{A}_2(\boldsymbol{\theta})$ are non-negative definite symmetric matrices.

If the error distribution density f belongs to some class $\mathscr{F}$, then the optimal choice of both the loss function ρ and the estimator is given by the formulas (Tsypkin, 1984)

$$\rho^*(x) = -\log f^*(x),$$
$$f^* = \arg\min_{f \in \mathscr{F}} [\mu + \eta\sigma^2(f)] I(f), \tag{5.4.5}$$

where $\mu = \operatorname{tr} \mathbf{A}_1(\boldsymbol{\theta})$, $\eta = \operatorname{tr} \mathbf{A}_2(\boldsymbol{\theta})$ are the traces of the corresponding matrices both being nonnegative: $\mu \ge 0$ and $\eta \ge 0$.

The particular cases of problem (5.4.5) are:

- identification of regression objects when $\eta = 0$ and $a_1 = a_2 = \dots = a_m = 0$ in (5.4.1);
- identification of autoregressive objects when $\mu = 0$ and $b_0 = b_1 = \dots = b_m = 0$ in (5.4.1).

In the general case of identification, such a choice of ρ provides the following minimax variance and guaranteed accuracy estimation properties:

$$\operatorname{tr} \mathbf{V}(\psi^*, f) \le \operatorname{tr} \mathbf{V}(\psi^*, f^*). \tag{5.4.6}$$

In the particular cases of regression and autoregression, the scalar minimax principle is replaced by the matrix minimax principle

$$\mathbf{V}(\psi^*, f) \le \mathbf{V}(\psi^*, f^*). \tag{5.4.7}$$

In general, it seems impossible to obtain solutions of variational problem (5.4.5) in an analytic form if both μ and η are nonzero. In (Tsypkin, 1984), numerical and also analytic methods were used to describe the approximate form of optimal solutions, and the least favorable densities are intermediate between the normal and Laplace densities, namely their linear combination (see Subsection 5.3.3).

However, just this variational problem of minimizing the informational functional $[\mu + \eta\sigma^2(f)]\, I(f)$ gives an opportunity to apply the Weber–Hermite distribution densities obtained in Section 3.2.

Consider the following particular case of the RAR-model (5.4.1) directly combining the R- and AR-models of Sections 5.1 and 5.2

$$x_n = \sum_{j=1}^{m} \beta_j x_{n-j} + \sum_{k=1}^{p} \theta_k \phi_{nk} + \xi_n, \quad n = j+1, j+2, \dots, \tag{5.4.8}$$

where the parameter vector $\mathbf{c} = (\beta_1, \beta_2, \dots, \beta_m, \theta_1, \theta_2, \dots, \theta_p)$ is determined by the optimal M-estimators of the form

$$\widehat{\mathbf{c}}_n = \arg\min_{\mathbf{c}} \sum_{i=j+1}^{n} \rho^* \left(x_i - \sum_{j=1}^{m} \beta_j x_{i-j} - \sum_{k=1}^{p} \theta_k \phi_{ik} \right),$$

$$\rho^*(x) = -\log f^*(x), \qquad f^* = \arg\min_{f\in\mathcal{F}} [\mu + \eta\sigma^2(f)]\, I(f). \tag{5.4.9}$$

Solving the latter variational problem in some class $\mathcal{F}$, we use the exact analytic expression for the Fisher information in the class $\overline{\mathcal{F}_2}$ with a given variance (Lemma 3.2.1), namely using the decomposition of the optimization problem

$$\min_{f\in\mathcal{F}} [\mu + \eta\sigma^2(f)]\, I(f) = \min_{d} \left\{ (\mu + \eta d^2) \min_{f\in\mathcal{F}\cap\overline{\mathcal{F}_2}} I(f) \right\}, \quad \overline{\mathcal{F}_2} = \{f\colon \sigma^2(f) = d^2\}.$$

Obviously, while solving the inner optimization problem in the class $\mathcal{F}\cap\overline{\mathcal{F}_2}$ with fixed variance, there necessarily occur the structures based on the Weber–Hermite functions. In particular, in the class of nondegenerate densities $\mathcal{F}_1 = \{f\colon f(0) \ge 1/(2a) > 0\}$ with the additional restrictions $2a^2/\pi \le d^2 \le 2a^2$, the solution of the inner problem is given by Theorem 3.2.1, namely, optimal solution (3.2.6) with $\overline{\sigma}$ substituted for d

$$f^* = \arg \min_{f\in\mathcal{F}\cap\overline{\mathcal{F}_2}} I(f) = f(x; \nu, d).$$

Then the optimum of the Fisher information is given by $I(f^*) = I(\nu, d)$ (3.2.7), and finally the optimal solution is obtained by minimizing $(\mu + \eta d^2)\, I(\nu, d)$ in the scalar parameter d provided that

$$\nu = \nu(d). \tag{5.4.10}$$

Thus variational problem (5.4.9) is reduced to the problem of parameter optimization. In this case, the optimal solution belongs to the family of the Weber–Hermite densities $f(x; \nu, d)$, as before for the classes $\mathcal{F}_{12}$ (Theorem 3.2.1) and $\widetilde{\mathcal{F}}_{25}$ (Theorem 3.2.4). A fundamental difficulty occurs while realizing this procedure: the optimal solution depends on the parameters μ and ν, which in turn depend on the unknown parameter vector $\mathbf{c}$ of the RAR-model (5.4.8). It is obvious how to overcome this obstacle using an iterative procedure, but these questions lie beyond the limits of this work.

5.5. Final remarks

5.5.1. Minimax variance robust regression

Minimax variance robust estimators of a location parameter are the basis for designing various extensions to more complicated problems of estimation including:

- parametric and nonparametric regression;
- autoregression and the mixed models of regression–autoregression, autoregression–moving average;
- univariate and multivariate recurrent 'online' estimation procedures;
- smoothing the data.

A great body of researches is devoted to the use and development of the minimax approach in the above-mentioned areas. Not pretending to be complete, here we only enlist (Huber, 1973; Huber, 1981; Martin, 1980; Martin, 1981; Martin and Yohai, 1984; Polyak and Tsypkin, 1978; Polyak and Tsypkin, 1980; Polyak and Tsypkin, 1983; Rousseeuw and Yohai, 1984; Cypkin, 1976; Tsypkin, 1984; Shurygin, 1994a; Shurygin, 1996; Shurygin, 2000) on robust regression and mixed models; (Katkovnik, 1979; Katkovnik, 1985; Nemirovskii, 1981; Nemirovskii, 1985; Nemirovskii *et al.*, 1983; Tsybakov, 1982; Tsybakov, 1983) on robust nonparametric regression; (Polyak and Tsypkin, 1980; Tsypkin, 1984; Tsypkin and Poznyak, 1981) on robust recurrent algorithms; (Tukey, 1977; Tukey, 1979; Huber, 1979) on robust smoothing.

The minimax variance estimators of location designed for the classes of error distributions in Chapter 3 (with bounded variances and subranges) also, with fair ease, can be reformulated for the above regression problems. The characteristic property of most of these estimators is that their structure is defined by the interrelation between distribution dispersions at the central and tail domains: with relatively light tails there occur the LS and close methods; with relatively heavy tails, various robust versions (in particular, the L_1-norm estimator) appear, and there always exists an intermediate zone of compromise between them.

The adaptive variants of such regression algorithms with 'online' estimation of characteristics of a distribution class (through estimating of residuals) are studied in Chapter 8.

The entire Chapter 6 is devoted to the important case of the L_1-norm estimation.

5.5.2. The LMS, LTS, and other regressions

Various robust estimators have been proposed to provide high global robustness of estimation, among them the median of pairwise slopes (Theil, 1950; Adichie, 1967; Sen, 1968); the resistant line (Tukey, 1977; Velleman and Hoaglin, 1981); L-estimators (Bickel, 1973; Koenker and Bassett, 1978); R-estimators (Jurečkovà, 1971; Jaeckel, 1972). None of these estimators achieve the breakdown point of 30%, and some of them are defined only for simple regression when $m = 2$.

To achieve the maximum value 50% of the breakdown point, in (Siegel, 1982) the coordinate-wise *repeated median* estimator was constructed, whose computation requires all subsets of m observations from the data and, therefore, may take a lot of time.

The LMS regression. In (Rousseeuw, 1984), both the equivariant and high-breakdown methods were introduced, and as the goal function the median of residuals is chosen instead of their sum:

$$\text{minimize} \quad \operatorname{med} r_i^2, \tag{5.5.1}$$

and call it the *least median of squares* (LMS) estimator.

It turns out that this estimator is very robust with respect to outliers in the x-direction as well as to outliers in the carriers $\boldsymbol{\phi}$. It can be shown that the LMS estimator possesses the highest possible value of the breakdown point, namely 50%, and it is equivariant under linear transformations of the explanatory variable, since (7.1.11) depends only on residuals. Unfortunately, the LMS estimator is of low asymptotic efficiency with convergence rate $n^{-1/3}$ (for details see (Rousseeuw and Leroy, 1987)).

The LTS regression. To improve the latter property of the LMS estimator, in (Rousseeuw, 1984) the *least trimmed squares* (LTS) estimator was proposed defined as the solution of the problem

$$\text{minimize} \quad \sum_{i=1}^{k} r_{(i)}^2, \tag{5.5.2}$$

where $r_{(i)}^2$ are the ordered squared residuals.

Formula (5.5.2) yields the LS method when $k = n$, and provides the best robustness properties when k is approximately $n/2$.

The S-regression. The S-regression (Rousseeuw and Yohai, 1984) belongs to the class of high-breakdown affine equivariant estimators minimizing some robust measure of the dispersion of the residuals

$$\text{minimize} \quad S(r_1(\boldsymbol{\theta}), r_2(\boldsymbol{\theta}), \dots, r_n(\boldsymbol{\theta})), \tag{5.5.3}$$

where $S(r_1(\boldsymbol{\theta}), r_2(\boldsymbol{\theta}), \dots, r_n(\boldsymbol{\theta}))$ is defined as the solution of

$$\frac{1}{n}\sum_{i=1}^{n} \rho\left(\frac{r_i}{S}\right) = K. \tag{5.5.4}$$

The constant K is usually chosen equal to $\mathsf{E}_\Phi \rho$, where Φ is the standard normal.

S-estimators are asymptotically equivalent to M-estimators, and they also have convergence rate $n^{-1/2}$.

The WLS regression. All above estimators are computed not so fast as compared to the LS estimators. In order to use the computational and statistical advantages of the well-organized procedures of the LS method, it is reasonable to apply the *weighted least squares* (WLS) estimators based on the detection of outliers and defined as the solution of the problem

$$\text{minimize} \quad \sum_{i=1}^{n} w_i r_i^2. \tag{5.5.5}$$

One may define the weights w_i as

$$w_i = \begin{cases} 1, & |r_i/S_n| \le 3, \\ 0, & |r_i/S_n| > 3, \end{cases}$$

where S_n is some robust estimator for scale of residuals, for example, based on the preliminary LMS regression

$$S_n = C\sqrt{\operatorname{med} r_i^2}$$

with the constant C providing consistency at normal error distributions: $\mathsf{E}_\Phi S_n = \sigma$.

Certainly, the rejection bound 3 is arbitrary but quite reasonable because of the so-called 3σ-rule.

The WLS estimator still possesses the high-breakdown property, but it is more accustomed to the conventional statistical tools aimed at working with normality assumptions and/or least squares.

6

Robustness of L_1-norm estimators

This chapter is about the general and fundamental property of invariance (stability) of best L_1-approximations to rare impulsive noises, which in turn implies robustness of the LAV estimators of regression parameters.

Robustness properties of L_1-norm estimators (in particular, their breakdown points) are studied in the linear case of approximation by algebraic and trigonometric polynomials, and in the nonlinear case of approximation by exponentials.

6.1. Introductory remarks

In robust statistics, L_1-norm methods play an important role for a number of related reasons. The sample median is the simplest example of the L_1-norm or least absolute values estimators defined by the property that they minimize the L_1-norm of the deviations from the fit

$$\operatorname{med} x = \arg\min_{\theta} \sum |x_i - \theta|. \tag{6.1.1}$$

First, the sample median is a limiting case of a family of minimax variance M-estimators, minimizing the maximum asymptotic variance in ε-contamination models when the parameter of contamination ε tends to its boundary value 1 (Huber, 1964).

Second, it is the minimax variance estimator under distributions with relatively large variances (Vilchevski and Shevlyakov, 1994).

Third, it minimizes the maximum asymptotic bias under asymmetric contamination (Huber, 1981; Smolyak and Titarenko, 1980).

Finally, it is the simplest estimator having the highest value of the breakdown point, namely 1/2 (Hampel *et al.*, 1986). These properties are naturally generalized for L_1-regression problems (Huber, 1981; Hampel *et al.*, 1986; Shevlyakov, 1996).

Vice versa, the topics of L_1-approximations have been intensively and independently studied in the approximation theory (Akhiezer, 1958; Rice, 1964).

In this chapter we study relations between the robustness of L_1-norm estimators and fundamental properties of best L_1-approximations.

6.1.1. The general properties of L_1-approximations

In this section we represent some basic results on approximation in the L_1-norm.

L_1-approximation problem. Throughout this chapter we use several standard notations. The function which is to be approximated is denoted by $x(t)$. The collection of parameters of an approximating function is denoted by A and the parameters by $a_1, \ldots, a_m$; thus $A = (a_1, \ldots, a_m)$.

The space in which the parameters lie is the ordinary Euclidean space E_m of dimension m.

The distance of the approximation $F(A, t)$ from $x(t)$ in the L_1-norm is measured by

$$L_1(x, A) = \int_{\mathcal{T}} |x(t) - F(A, t)| \, dt. \tag{6.1.2}$$

We are now able to formally state the approximation problem in the L_1-norm.

APPROXIMATION PROBLEM IN THE L_1-NORM. Let $x(t)$ be a given real-valued continuous function defined on a set $\mathcal{T}$, and let $F(A, t)$ be a real-valued approximating function depending continuously on $t \in \mathcal{T}$ and on m parameters A.

Determine the parameters $A^* \in E_m$ such that

$$L_1(x, A^*) \le L_1(x, A)$$

for all $A \in E_m$.

A solution to this problem is said to be the *best approximation* in the L_1-norm.

In this chapter the set $\mathcal{T}$ is usually standardized to be the interval $[0, 1]$. The results thus obtained are readily extended to any closed interval of the real line.

Given a function $x(t)$ to approximate, there are four steps in the solution of this problem:

- Existence of a solution.
- Uniqueness of a solution.
- Characteristic and other special properties of a solution.

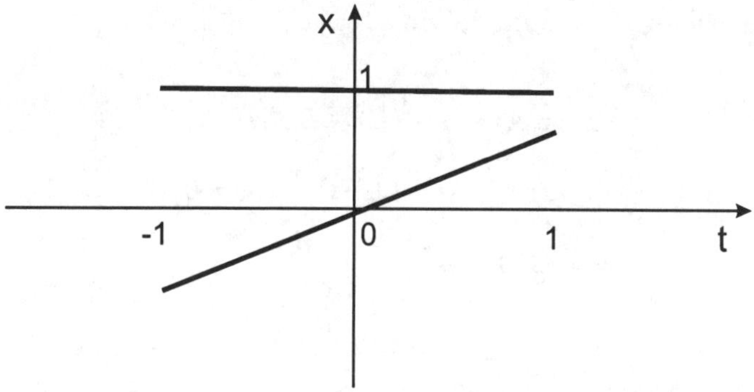

Figure 6.1. Non-uniqueness of L_1-approximations

- Computation of a solution.

Now we briefly describe all these phases.

Existence of best L_1-approximations. Existence of a solution is established using standard compactness arguments (see general results on the existence of best approximations in the L_p-norms ($1 \le p \le \infty$) (Akhiezer, 1958; Rice, 1964)).

Uniqueness of best L_1-approximations. In general, the uniqueness of a solution in the L_1-norm is not guaranteed, because the L_1-norm is not strictly convex. For instance, let $x(t) = 1$, $F(A,t) = L(A,t) = a_1 t$. In this case, any approximating linear function $L(A^*,t) = a_1^* t$ for which $|a_1^*| \le 1$ is the best L_1-approximation to $x(t)$ (see Fig. 6.1).

Now we consider the linear approximating polynomials

$$L(A,t) = \sum_{j=1}^{m} a_j \phi_j(t).$$

If the set $\{\phi_j(t)\}$ is a Chebyshev set, then the best L_1-approximation is unique (Akhiezer, 1958; Rice, 1964).

We recall that the set $\{\phi_j(t)\}_1^m$ is a Chebyshev set on $[0,1]$ if:

(1) $\{\phi_j(t)\}$ are continuous and linear independent on $[0,1]$;

(2) each nontrivial polynomial has on $[0,1]$ no more than m zeros.

For example, the set $\{1, t, t^2, \ldots, t^{m-1}\}$ is Chebyshev, but $\{t, t^2, \ldots, t^m\}$ is not.

Characterization of L_1-approximations. In order to characterize the best L_1-approximations, we should define

$$L_1(x, A) = \int_0^1 |x(t) - L(A, t)|\, dt,$$
$$Z(A) = \{t \mid L(A, t) - x(t) = 0\},$$
$$\operatorname{sgn} t = \begin{cases} +1, & t > 0, \\ -1, & t < 0, \end{cases}$$

and $\operatorname{sgn} 0 = 0$.

Now we are able to formulate the characterization lemma for the best L_1-approximations.

LEMMA 6.1.1 (Rice, 1964). *The relation*

$$L_1(x, A^*) \le L_1(x, A^* + sA) \tag{6.1.3}$$

holds for all s if and only if

$$\left| \int_0^1 L(A, t) \operatorname{sgn}[x(t) - L(A^*, t)]\, dt \right| \le \int_{Z(A^*)} |L(A, t)|\, dt. \tag{6.1.4}$$

Furthermore, if the strict inequality occurs in (6.1.4), *then the strict inequality occurs in* (6.1.3) *for all nonzero s.*

Lemma 6.1.1 is true as soon as one assumes that $x(t)$ and the $\phi_i(t)$ are merely integrable.

There is the following important corollary to this theorem.

COROLLARY 6.1.1. *If $L(A^*, t)$ is the best L_1-approximation to a continuous function $x(t)$, and if $\mu(Z) = 0$, then for all A from* (6.1.4) *it follows that*

$$\int_0^1 L(A, t) \operatorname{sgn}[x(t) - L(A^*, t)]\, dt = 0. \tag{6.1.5}$$

In the case of approximation by algebraic polynomials $L(A, t)$, it follows from (6.1.5) that the optimal coefficients $A^* = (a_1^*, \dots, a_m^*)$ satisfy the simultaneous equations

$$\int_0^1 t^{j-1} \operatorname{sgn}\left[x(t) - \sum_{j=1}^{m} a_j^* t^{j-1}\right] dt = 0, \qquad j = 1, \dots, m. \tag{6.1.6}$$

Computation of best L_1-approximations through interpolation. Here we consider only the important particular case where a simple explicit solution is possible (Rice, 1964, p. 105).

Immediately from (6.1.5) it follows that

$$\int_0^1 \phi_j(t)\,\mathrm{sgn}[x(t) - L(A^*,t)]\,dt = 0, \qquad j = 1,2,\dots,m. \tag{6.1.7}$$

Assume that there is a set of points $0 = t_0 < t_1 < \cdots < t_m < t_{m+1} = 1$, and a sign function

$$s(t) = \begin{cases} +1, & t_j < t < t_{j+1},\ j \text{ is even}, \\ -1, & t_j < t < t_{j+1},\ j \text{ is odd}, \end{cases} \tag{6.1.8}$$

such that

$$\int_0^1 \phi_j(t)s(t)\,dt = 0, \qquad j = 1,2,\dots,m. \tag{6.1.9}$$

Then any polynomial $L(A^*,t)$ defined by the relation

$$\mathrm{sgn}[x(t) - L(A^*,t)] = s(t) \tag{6.1.10}$$

is the best L_1-approximation. In this case, $L(A^*,t)$ can be determined from the solution of the simultaneous equations

$$L(A^*,t_j) = x(t_j), \qquad j = 1,2,\dots,m, \tag{6.1.11}$$

given the function $x(t) - L(A^*,t)$ changes its sign only at those m points $\{t_j\}$.

Thus the L_1-approximation problem is replaced by an *interpolation problem*, the interpolation taking place at points $\{t_j\}$ which are independent of $x(t)$.

This characteristic feature of best L_1-approximations leads to a simple and practical solution of a large number of L_1-approximation problems, and in our study, it is used further in order to establish the invariance of L_1-approximations to rare impulsive noises, and therefore their robustness to gross errors.

The following theorem states the exact result.

THEOREM 6.1.1 (Rice, 1964). *Assume that there exists a sign function $s(t)$ as in* (6.1.8) *for which* (6.1.9) *is valid. Let $L(A^*,t)$ interpolate $x(t)$ at $\{t_j\}$. If $x(t)$ is such that $x(t) - L(A^*,t)$ changes its sign at these t_j and at no other points, then $L(A^*,t)$ is the best L_1-approximation to $x(t)$.*

In the case of approximation by algebraic polynomials, the points $\{t_j\}$ can be determined for the best L_1-approximation to t^m. The function

$$t^m - \sum_{j=1}^{m} a_j t^{j-1}$$

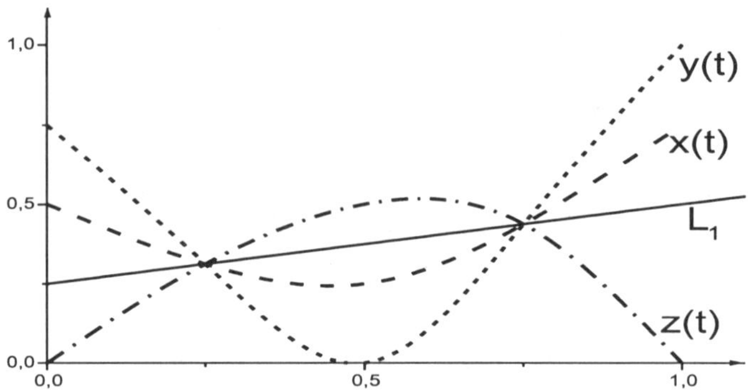

Figure 6.2. $L(A^*,t) = a_1^* + a_2^* t$ is the best L_1-approximation to $x(t)$, $y(t)$, and $z(t)$

has no more than m changes of sign, and by the definition of a Chebyshev system of functions, we have that if

$$\int_0^1 \phi_j(t)s(t)\,dt = 0, \qquad j = 1,2,\dots,m,$$

then $s(t)$ must have at least m changes of sign. Therefore, the best L_1-approximation to t^m is determined from the solution of the interpolation problem (6.1.11).

The corresponding points of interpolation are given by the following result.

LEMMA 6.1.2 (Bernstein, 1926). *In the case of interpolation by algebraic polynomials*

$$L(A,t) = \sum_{i=1}^{m} a_j t^{j-1},$$

the points of interpolation t_j *are of the form*

$$t_j = \sin^2 \frac{j\pi}{2(m+1)}, \qquad j = 1,2,\dots,m. \tag{6.1.12}$$

Finally, we observe that if

$$\operatorname{sgn}[y(t) - L(A^*,t)] = \pm \operatorname{sgn}[x(t) - L(A^*,t)],$$

then the best L_1-approximation $L(A^*,t)$ is the same to both functions $x(t)$ and $y(t)$. This is illustrated by Fig. 6.2: the straight line $L(A^*),t$ is the best L_1-approximation to $x(t)$, $y(t)$, and $z(t)$.

6.1.2. The examples of L_1-approximations

Further we show that high robustness of the sample median and L_1-norm estimators of regression parameters is determined by the specific property of best L_1-approximations expressed by their invariance to rare impulses of a high level (Shevlyakov, 1976; Guilbo, 1979; Shevlyakov, 1996).

Consider some simple examples illustrating the property of robustness of the best L_1-approximations.

L_1-approximation by a constant. Let the function

$$x(t) = \theta + e(t)$$

be defined on the interval $[0, 1]$, where θ is some constant to be estimated and $e(t)$ is a rectangular impulse of magnitude h and duration T^+. Set the problem of estimation of a constant parameter θ as the problem of approximation to $x(t)$ by a constant value, comparing the solutions of the following two approximation problems:

- in the L_1-norm

$$a_1^* = \arg\min_a \int_0^1 |x(t) - a|\, dt = \arg\min_a L_1(x, a), \tag{6.1.13}$$

- in the L_2-norm

$$a_2^* = \arg\min_a \int_0^1 [x(t) - a]^2\, dt = \arg\min_a L_2(x, a), \tag{6.1.14}$$

It is easy to show that the solution of problem (6.1.13) is of the form

$$a_1^* = \widehat{\theta}_1 = \begin{cases} \theta, & T^+ < 1/2, \\ \theta + h/2, & T^+ = 1/2, \\ \theta + h, & T^+ > 1/2, \end{cases} \tag{6.1.15}$$

and for problem (6.1.14),

$$a_2^* = \widehat{\theta}_2 = \theta + hT^+, \tag{6.1.16}$$

both solutions being independent of the location of the impulse $e(t)$ on $[0, 1]$.

These solutions essentially differ from each other:

- solution (6.1.15) is discontinuous having a threshold character and depending on the duration of an impulse; on the contrary, solution (6.1.16) is continuous;

- under certain conditions (when the duration is less than half of interval), the solution (6.1.15) of the L_1-approximation problem does not depend on the magnitude of an impulsive noise, and it is *exactly equal* to the estimated parameter;

- the mean square solution (6.1.16) does always depend both on the magnitude and duration of an impulse, the estimator error being proportional to its magnitude.

Now consider a simple stochastic model of the impulsive noise and compare the efficiency of the L_1-norm and L_2-norm estimators.

Let $e(t)$ be a sequence of positive impulses of magnitude h on the interval $[0, T]$, the duration of a single impulse and the pause being exponentially distributed with the parameters λ and μ, respectively. In this case the noise is a Markov process with two states: 0 and h. In this simple model, it is easy to find the errors of the L_1- and L_2-estimators:

$$e_1 = \begin{cases} 0, & T^+ < T/2, \\ h/2, & T^+ = T/2, \\ h, & T^+ > T/2, \end{cases} \tag{6.1.17}$$

$$e_2 = hT^+/T. \tag{6.1.18}$$

For sufficiently large T, T^+ is approximately normal with (Gnedenko *et al.*, 1969)

$$\mathsf{E}T^+ = \frac{T_1}{T_1 + T_2}T, \qquad \mathsf{Var}\,T^+ = \frac{2T_1^2T_2^2}{(T_1 + T_2)^3}T, \tag{6.1.19}$$

where $T_1 = 1/\lambda$, $T_2 = 1/\mu$. The relations for the mean squared errors are of the form

$$\mathsf{E}e_1^2 = h^2\mathsf{P}\{T^+ > T/2\}, \qquad \mathsf{E}e_2^2 = \frac{h^2}{T^2}\mathsf{E}\,(T^{+2}). \tag{6.1.20}$$

Setting $N = T/\sqrt{T_1T_2}$, $k = T_2/T_1$, for (6.1.20) we obtain

$$\mathsf{E}e_1^2 = h^2\{1 - \Phi[(k-1)(k+1)^{12}N^{1/2}/k^{3/4}]\},$$

$$\mathsf{E}e_2^2 = h^2\left[\frac{1}{(k+1)^2} + \frac{2k^{3/2}}{N(k+1)^3}\right],$$

$$\Phi(x) = \frac{1}{\sqrt{2\pi}}\int_{-\infty}^{x} e^{-t^2/2}\,dt.$$

Table 6.1 presents the values of relative mean squared errors for various k with $N = 10$.

It follows from Table 6.1 that L_1-estimators dominate over L_2-estimators in the case of sufficiently rare noises.

Table 6.1. The relative mean squared errors for L_1- and L_2-approximations

k	1	1.5	2	2.5	∞
$\mathsf{E}e_1^2/\mathsf{E}e_2^2$	1.98	0.18	5.3×10^{-4}	$\approx 10^{-6}$	0

L_1-approximation by a sine function. Consider another simple example where an approximated function is the sine function $\theta(t) = \theta \sin t$, and the noise is the impulse of magnitude h and duration T^+ symmetric about the center of the interval $[0, \pi]$. We estimate $\theta(t)$ by minimizing the L_1-functional

$$a^* = \arg\min_a \int_0^\pi |x(t) - a \sin t|\, dt = \arg\min_a L_1(x, a), \qquad (6.1.21)$$

where $x(t) = \theta \sin t + e(t)\theta > 0$,

$$e(t) = \begin{cases} h > 0, & t \in [\pi/2 - T^+/2, \pi/2 + T^+/2], \\ 0, & \text{otherwise.} \end{cases}$$

Obviously, it suffices to consider the value of the L_1-functional in the interval $a \in [\theta, \theta + h]$:

$$L_1(x, a) = 2(a - \theta)(1 - 2\sin(T^+/2)) + hT^+. \qquad (6.1.22)$$

From (6.1.22) it follows that for $T^+ < \pi/3$ the minimum is attained at $a^* = \theta$ irrespective of h.

Thus the error of the L_1-estimator does not depend on the magnitude of the impulsive noise, provided its duration is short enough.

L_1-approximation by a third-degree polynomial. Here we follow the example of (Barrodale, 1968). Compare the L_1- and L_2-approximations by the third-degree polynomial $L(A, t) = a_3t^3 + a_2t^2 + a_1t + a_0$. The approximated function $x(t)$ is defined on the discrete set $\mathscr{T} = \{t_i = 0, 1, \dots, 9\}$ as $x(t_i) = x_i = e_i$, i.e., we approximate the noise, or in other words, the estimated polynomial is zero (in his original work, Barrodale used $L(A, t) = t^3 - 10t^2 + 21t$). The values e_i are set in such a way that $e_i = 0$ at some points, and they are gross errors at other points.

We set $\widehat{e}_{1i} = x(t_i) - L(A^*, t_i)$, $\widehat{e}_{2i} = x(t_i) - L(B^*, t_i)$, $\quad i = 0, 1, \dots, 9$, where $L(A^*, t)$ is the best L_1-approximation, and $L(B^*, t)$ is the best L_2-approximation.

The quality of approximation is determined by the closeness of the rows for e_i and $\widehat{e}_i$.

The computation results presented in Table 6.2 clearly illustrate the stability of L_1-approximations in the case of the rare noise. The items (a), (b), (c),

Table 6.2. L_1- and L_2-approximations to the rare noise by a third-degree polynomial

	t_i	0	1	2	3	4	5	6	7	8	9
	e_i	0	0	20	0	0	0	0	0	0	0
(a)	$\widehat{e_1}$	0.0	0.0	20.0	0.0	0.0	0.0	0.0	0.0	0.0	0.0
	$\widehat{e_2}$	−0.4	−5.0	13.5	−5.9	−3.9	−1.3	1.0	2.2	1.6	−1.8
	e_i	0	0	0	0	30	15	0	0	0	0
(b)	$\widehat{e_1}$	0.0	0.0	0.0	0.0	30.0	15.0	0.0	0.0	0.0	0.0
	$\widehat{e_2}$	4.2	−2.4	−6.8	−9.4	19.7	5.3	−7.8	−4.9	−1.2	3.2
	e_i	0	15	0	0	0	20	0	0	20	0
(c)	$\widehat{e_1}$	0.0	15.0	0.0	0.0	0.0	20.0	0.0	0.0	20.0	0.0
	$\widehat{e_2}$	−5.5	12.1	−2.8	−4.3	−6.6	11.3	−9.9	10.8	−5.9	0.8
	e_i	0	15	20	20	0	0	0	0	0	0
(d)	$\widehat{e_1}$	0.0	0.7	0.0	0.7	−14.3	−7.1	0.0	5.0	5.7	0.0
	$\widehat{e_2}$	−3.0	2.1	3.9	5.4	−10.0	−4.1	1.3	4.4	3.4	−3.5
	e_i	0	0	0	0	−30	0	0	−30	0	0
(e)	$\widehat{e_1}$	0.0	0.0	0.0	0.0	−30.0	0.0	0.0	−30.0	0.0	0.0
	$\widehat{e_2}$	−0.7	0.1	2.5	5.8	−20.9	11.9	13.0	−18.2	7.4	−0.9
	e_i	0	10	0	−10	0	10	0	10	0	0
(f)	$\widehat{e_1}$	0.0	10.0	0.0	−10.0	0.0	10.0	0.0	10.0	0.0	0.0
	$\widehat{e_2}$	−4.6	9.5	0.9	−9.5	−1.0	7.0	−4.5	5.1	−3.5	0.6
	e_i	−20	0	20	0	0	0	0	0	0	0
(g)	$\widehat{e_1}$	0.0	9.9	3.5	0.0	−1.2	−1.0	0.0	1.0	1.2	0.0
	$\widehat{e_2}$	−3.5	6.3	0.4	−2.1	−2.3	−1.0	0.7	1.8	1.4	−1.6

(d), (e) and (f) obviously confirm this. The case (e) is of a particular interest: four observations are wrong, nevertheless, the L_1-approximation is absolutely precise. Here we may underline the property of L_1-approximations to reject the rare noise and to show its true location on the interval of processing. It follows from this table that the L_2-norm estimators also can approximately do this, but much less efficiently. In the case (f), the L_2-approximation has proved to be better than the L_1-norm estimator: L_1-approximations are highly sensitive to the values of noise near the boundaries of the interval of processing. Further in Section 6.3, we will explain this effect.

The above examples show that all L_1-approximations yield the estimators with zero errors independent on the magnitude of gross rare noises — this is just the manifestation of robustness of best L_1-approximations. This effect observed at the L_1-norm solutions is not occasional but it is caused by general properties of robustness of best L_1-approximations.

6.2. Stability of L_1-approximations

6.2.1. Stability: linear continuous case

Consider the best L_1-approximation to a continuous function $x(t)$ defined on $[0,1]$ by the linear polynomials

$$L(A^*,t) = \arg\min_{L(A,t)} \int_0^1 |x(t) - L(A,t)|\,dt, \tag{6.2.1}$$

$$L(A,t) = \sum_{j=1}^{m} a_j \phi_j(t), \qquad A = (a_1, a_2, \dots, a_m),$$

where $\{\phi_j(t)\}_1^m$ is a Chebyshev set of functions on $[0,1]$.

Under these assumptions, the best L_1-approximations exist and are unique (see Subsection 6.1.1).

In order to study the stability of L_1-approximations, we consider the functions

$$x(t) = \theta(t) + e(t), \tag{6.2.2}$$

where $\theta(t) = \sum \theta_j \phi_j(t)$ is the estimated component and $e(t)$ is a continuous impulsive noise function assuming positive, negative, and zero values on the sets

$$E^+ = \{t\colon e(t) > 0\}, \quad E^- = \{t\colon e(t) < 0\}, \quad E^0 = \{t\colon e(t) = 0\}.$$

The stability of L_1-approximations is expressed by the property of their invariance to the rare impulsive noise and is given by the following (Shevlyakov, 1976; Shevlyakov, 1996).

THEOREM 6.2.1. *The best L_1-approximation $L(A^*,t)$ to $x(t)$ (6.2.2) is exactly equal to $\theta(t)$*

$$L(A^*,t) = \theta(t), \quad a_j = \theta_j, \quad j = 1, \dots, m,$$

if and only if the inequality

$$\left| \int_{E^+} L(A,t)\,dt - \int_{E^-} L(A,t)\,dt \right| \le \int_{E^0} |L(A,t)|\,dt \tag{6.2.3}$$

holds.

Inequality (6.2.3) imposes certain restrictions on the sets E^+ and E^-.

EXAMPLE 6.2.1. For the approximation by a constant $L(A,t) = a_1$, $m = 1$, $\phi_1 = 1$, formula (6.2.3) yields $|T^+ - T^-| \le 1 - T^+ - T^-$ and therefore,

$$T^- \le 1/2, \qquad T^+ \le 1/2, \tag{6.2.4}$$

Table 6.3. The bound for the duration T^+

m	1	2	3	4	5	6
	0.5	0.25	0.146	0.095	0.067	0.051

where $T^+ = \mu(E^+)$, $T^- = \mu(E^-)$, and μ is the ordinary Lebesgue measure on the real line.

The boundary value 1/2 is the value of the breakdown point for the sample median.

In the case of approximation by algebraic polynomials

$$L(A,t) = \sum_{j=1}^{m} a_j t^{j-1}$$

under the one-sided impulsive noise ($\mu(E^-) = 0$), the following is true.

THEOREM 6.2.2. *Inequality* (6.2.3) *holds if and only if*

$$T^+ \le \sin^2 \frac{\pi}{2(m+1)}. \tag{6.2.5}$$

The right-hand side of (6.2.5) gives the upper bound for the duration of a single impulsive noise. Table 6.3 gives these values depending on the degree of an approximating polynomial.

EXAMPLE 6.2.2. If $m = 1$, then $T^+ \le 1/2$ (see Example 6.2.1). In the important case of the approximation by the straight line ($m = 2$), inequality (6.2.5) and Table 6.3 yield $T^+ \le 1/4$.

6.2.2. Stability: nonlinear continuous case

Consider the best L_1-approximation to a continuous function $x(t)$ defined on $[0,1]$ by approximating functions $F(A,t)$ that are non-linear with regard to the estimated parameters A^θ

$$F(A^*,t) = \arg\min_{F(A,t)} \int_0^1 |x(t) - F(A,t)|\,dt, \tag{6.2.6}$$
$$x(t) = F(A^\theta,t) + e(t).$$

Assume that the impulsive noise is of the same form as in the linear case. Then the stability property of the best L_1-approximations (6.2.6) looks as follows (Shevlyakov, 1982b; Shevlyakov, 1991).

THEOREM 6.2.3. *If the L_1-approximation $F(A^*, t)$ (6.2.6) equals $F(A^\theta, t)$ ($A^* = A^\theta$), then the inequality*

$$\left| \int_{E^+} (A, \operatorname{grad} F(A^\theta, t))\, dt - \int_{E^-} (A, \operatorname{grad} F(A^\theta, t))\, dt \right| \leq \int_{E^0} |(A, \operatorname{grad} F(A^\theta, t))|\, dt \quad (6.2.7)$$

holds for all A, where A^θ is the vector of the estimated parameters,

$$\operatorname{grad} F(A, t) = \left(\frac{\partial F(A,t)}{\partial a_1}, \ldots, \frac{\partial F(A,t)}{\partial a_m} \right)^T,$$

$(A, \operatorname{grad} F)$ is the scalar product.

The basic distinction of the non-linear case from the linear is that the restrictions on the sets E^+ and E^- depend on the true value of the estimated parameters A^θ.

COROLLARY 6.2.1. *Let*

$$F(A, t) = \sum c_j \exp(-\lambda_j t), \qquad A = (c_1, \ldots, c_k, \lambda_1, \ldots, \lambda_k),$$
$$m = 2k, \quad c_j \geq 0, \quad \lambda_j \geq 0, \quad j = 1, \ldots, k,$$

be the approximation by exponentials. Assume that $e(t)$ is a one-sided single impulse function defined on

$$E^+ = (\delta, \tau), \qquad 0 \leq \delta < \tau \leq 1.$$

Then inequality (6.2.7) implies the simultaneous inequalities

$$2[\exp(-\lambda_i^\theta \delta) - \exp(-\lambda_i^\theta \tau)] \leq 1 - \exp(-\lambda_i^\theta), \qquad i = 1, \ldots, k. \quad (6.2.8)$$

EXAMPLE 6.2.3. In the case $\delta = 0$, the solution of system (6.2.8) is of the form

$$\tau \leq \min_i \left[-\frac{1}{\lambda_i^\theta} \ln \frac{1 + \exp(-\lambda_i^\theta)}{2} \right],$$

and the boundary is determined by the exponential with the maximum value of λ_i^θ.

In the case $\tau = 1$, vice versa, the determining exponential has the minimum value of λ_i^θ:

$$\delta \geq \max_i \left[-\frac{1}{\lambda_i^\theta} \ln \frac{1 + \exp(-\lambda_i^\theta)}{2} \right].$$

6.2.3. Proofs

PROOF OF THEOREM 6.2.1. The proof is based on the characterization lemma 6.1.1 for the best L_1-approximations (see Subsection 6.1.1).

Theorem 6.2.1 is a direct corollary to Lemma 6.1.1: by substituting $L(A^*, t) = \theta(t)$ into (6.1.4) we arrive the required result. □

PROOF OF THEOREM 6.2.2. This proof is based on Theorem 6.1.1 and Lemma 6.2.1. In problem (6.2.1), assume that

$$x(t) = x_1(t) = \theta(t) + \alpha t^m, \qquad \alpha > 0,$$

where $\theta(t)$ is the algebraic polynomial

$$L(\Theta, t) = \sum_{j=1}^{m} \theta_j t^{j-1}.$$

The approximated function $x_1(t)$ is an algebraic polynomial of order m. Then the number of zeros of the function $x_1(t) - L(A, t)$ does not exceed m, hence the best L_1-algebraic polynomial $L(A^*, t)$ is determined from (6.1.11) with interpolation points (6.1.12)

$$\sum_{j=1}^{m} (a_j^* - \theta_j) t_i^{j-1} = \alpha t_i^m, \qquad i = 1, 2, \dots, m. \tag{6.2.9}$$

The determinant of (6.2.9) is the Vandermonde determinant, and therefore the solution of (6.2.9) is unique. The best polynomial $L(A^*, t)$ determined from (6.2.9) is also best for the functions

$$x(t) = x_1(t) + e(t), \tag{6.2.10}$$

where $e(t)$ is the continuous impulsive noise function from (6.2.2) with

$$E^+ = \bigcup_{i=0}^{m/2} (\delta_{2i}, \tau_{2i}), \qquad E^0 = [0, 1] \setminus E^+, \tag{6.2.11}$$

for even m;

$$E^+ = \bigcup_{i=0}^{(m-1)/2} (\delta_{2i+1}, \tau_{2i+1}), \qquad E^0 = [0, 1] \setminus E^+, \tag{6.2.12}$$

for odd m;

$$t_k \le \delta_k < \tau_k \le t_{k+1}, \quad t_0 = 0, \quad t_{m+1} = 1, \quad k = 0, 1, \dots, m.$$

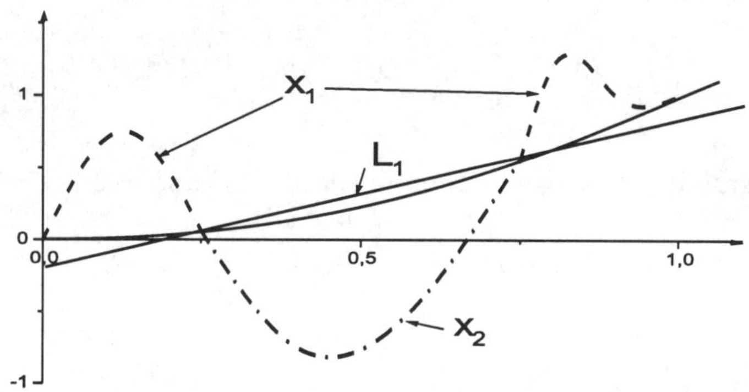

Figure 6.3. Invariance of L_1-approximations to impulsive noise functions

We immediately obtain this result from (6.1.6), because

$$\operatorname{sgn}[x(t) - L(A^*, t)] = \operatorname{sgn}[x_1(t) - L(A^*, t)].$$

In other words, the best L_1-algebraic polynomial $L(A^*, t)$ is invariant to the impulsive noise functions (6.2.11) and (6.2.12). Obviously, this assertion is also valid for negative impulsive noise functions. The case of the L_1-approximations by a linear polynomial $L(A, t) = a_1 + a_2 t$ is illustrated by Fig. 6.3 (see also Fig. 6.2).

Setting α tending to zero in (6.2.9), we obtain $a_j^* = \theta_j, j = 1, \dots, m$. Thus the solution of approximation problem (6.2.1) is equal exactly to the approximated function $\theta(t)$ independently of the impulsive noise values $e(t)$ (6.2.11) and (6.2.12).

The obtained stability property of L_1-approximations holds only under rather strict conditions on the location of impulsive noise (6.2.11) and (6.2.12) on $[0, 1]$. The admissible duration of impulses is bounded above by the distances between the interpolation points (6.1.12)

$$t_{j+1} - t_j, \quad j = 0, 1, \dots, m; \quad t_0 = 0, \quad t_{m+1} = 1.$$

It follows from (6.1.12) that the maximum distances are at the center and the minimum distances are at the boundaries of $[0, 1]$.

Now we demonstrate that the stability property of L_1-approximations by algebraic polynomials is provided for a single impulsive noise with duration bounded by the minimum distance between interpolation points

$$t_1 - t_0 = t_{m+1} - t_m = \sin^2 \frac{\pi}{2(m+1)}.$$

Consider a single impulse $e(t)$, $E^+ = (\delta, \tau)$, $0 \le \delta < \tau \le 1$, with duration $T^+ = \tau - \delta$ satisfying inequality (6.2.3). By Theorem 6.2.1, it suffices to show that the inequality

$$\left|\int_\delta^\tau L(A,t)\,dt\right| \le \int_0^\delta |L(A,t)|\,dt + \int_\tau^1 |L(A,)|\,dt \tag{6.2.13}$$

holds for all A.

By the stability property with impulses (6.2.11) and (6.2.12), we find that

$$\left|\int_{\delta_m}^1 L(A,t)\,dt\right| \le \int_0^{\delta_m} |L(A,t)|\,dt, \qquad \delta_m \ge t_m, \tag{6.2.14}$$

and

$$\left|\int_0^{\tau_0} L(A,t)\,dt\right| \le \int_{\tau_0}^1 |L(A,t)|\,dt, \qquad \tau_0 \le t_1, \tag{6.2.15}$$

hold for all A.

Consider an arbitrary interval (δ, τ) with duration satisfying inequality (6.2.5). Choose the point $t^* \in (\delta, \tau)$ from the condition

$$\frac{t^* - \delta}{\tau - t^*} = \frac{\delta}{1-\tau}.$$

In this case, the values $(t^* - \delta)/t^*$ and $(\tau - t^*)/(1 - t^*)$ also satisfy inequality (6.2.5). Then for the intervals $[0, t^*)$ and $(t^*, 1]$, the inequalities similar to (6.2.14) and (6.2.15)

$$\left|\int_\delta^{t^*} L(A,t)\,dt\right| \le \int_0^\delta |L(A,t)|\,dt,$$

$$\left|\int_{t^*}^\tau L(A,t)\,dt\right| \le \int_\tau^1 |L(A,t)|\,dt$$

hold for any A. Summing them, we arrive at (6.2.13), which completes the proof of Theorem 6.2.2. □

PROOF OF THEOREM 6.2.3. This proof is based on the result (Rice, 1965) similar to Lemma 6.1.1.

LEMMA 6.2.1. *A necessary condition for $F(A^*, t)$ to be the best approximation to $x(t)$ is that*

$$\left|\int_0^1 (A, \operatorname{grad} F(A^*, t)) \operatorname{sgn}[x(t) - F(A^*, t)]\,dt\right| \le \int_{Z(A^*)} |(A, \operatorname{grad} F(A^*, t))|\,dt \tag{6.2.16}$$

for all A, where

$$Z(A) = \{t \mid F(A,t) - x(t) = 0\}.$$

Theorem 6.2.3 now immediately follows from Lemma 6.2.1: it suffices to substitute $F(A^*,t) = F(A^\theta,t)$ into (6.2.16). □

PROOF OF COROLLARY 6.2.1. In this case, inequality (6.2.7) is of the form

$$\left|\int_\delta^\tau \sum_{j=1}^k (c_j - \lambda_j c_j^\theta t)\exp(-\lambda^\theta t)\,dt\right|$$
$$\le \int_0^\delta \left|\sum_{j=1}^k (c_j - \lambda_j c_j^\theta t)\right| \exp(-\lambda_j^\theta t)\,dt + \int_\tau^1 |c_j - \lambda_j c_j^\theta t| \exp(-\lambda_j^\theta t)\,dt \qquad (6.2.17)$$

for all $c_j, \lambda_j \ge 0, j = 1,2,\dots,k$.

We set $c_j^\theta = 0$, $j = 1,2,\dots,k$ in (6.2.17). Then, by integrating (6.2.17), we obtain

$$\sum_{j=1}^k (c_j/\lambda_j^\theta)[2(\exp(-\lambda_j^\theta\delta) - \exp(-\lambda_j^\theta\tau)) - (1 - \exp(-\lambda_j^\theta))] \le 0$$

for all $c_j, \lambda_j \ge 0, j = 1,2,\dots,k$, which yields (6.2.8). □

6.3. Robustness of the L_1-regression

In this section, we consider the stability (invariance) property of L_1-approximations to rare gross errors on finite point sets: most of the results derived above in continuous models are still valid in the discrete case.

6.3.1. The L_1-approximation on a finite point set

Let a function $x(t)$ be defined on a finite point set $\mathcal{T} = \{t_1, t_2, \dots, t_n\}$. Then the **best L_1-approximation** $L(A^*,t)$ to $x(t)$ satisfies the condition

$$\sum_{t\in\mathcal{T}} |x(t) - L(A^*,t)| \le \sum_{t\in\mathcal{T}} |x(t) - L(A,t)| \qquad (6.3.1)$$

for all A, where $L(A,t) = \sum_{j=1}^m a_j\phi_j(t)$.

The **existence** of the solution of problem (6.3.1) holds here as a particular case of the general result on the existence of L_1-approximations (see (Akhiezer, 1958)).

The **uniqueness** of the solution is not guaranteed even for Chebyshev sets $\{\phi_j(t)\}$. For instance, let $x(t)$ and $\mathcal{T}$ be defined by the points

$$\{(0,1),(1,-1),(2,-1),(3,1)\},$$

and $L(A,t) = a_1 + a_2 t$. Fig. 6.4 illustrates this situation. Any straight line going through the sides AB and CD is the best L_1-approximation.

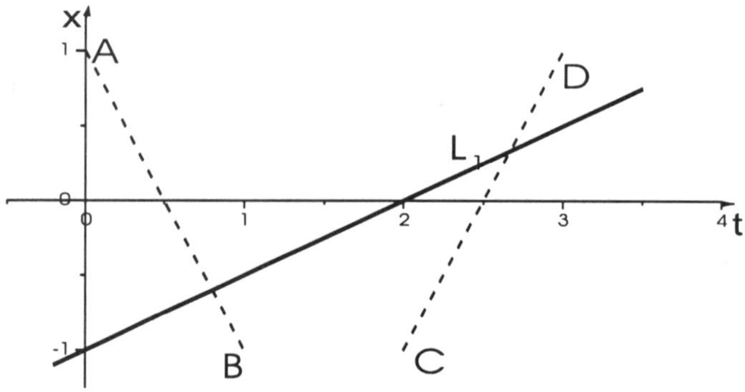

Figure 6.4. The non-uniqueness of the L_1-approximation by a straight line on a finite point set

On finite point sets, the following result is important: if $\{\phi_j(t)\}$ is a Chebyshev set then the best L_1-approximation $L(A^*, t)$ to $x(t)$ interpolates $x(t)$ in at least m points of $\mathcal{T}$ (Rice, 1964).

The **characterization** condition for the best L_1-approximations is given by the following analog of Lemma 6.1.1.

LEMMA 6.3.1. *The necessary and sufficient condition for $L(A^*, t)$ to be the best approximation to $x(t)$ on $\mathcal{T}$ is that*

$$\left| \sum_{t \in \mathcal{T}} L(A, t))\, \mathrm{sgn}[x(t) - L(A^*, t)] \right| \leq \sum_{Z(A^*)} |L(A, t)| \qquad (6.3.2)$$

for all A, where

$$Z(A) = \{t \mid t \in \mathcal{T},\ x(t) - L(A, t) = 0\}.$$

If (6.3.2) *holds with strict inequality for all A, then $L(A^*, t)$ is the unique best L_1-approximation to $x(t)$ on $\mathcal{T}$.*

6.3.2. Stability: linear discrete case

The least absolute values or L_1-norm estimators for linear regression model parameters

$$x_i = \theta(t_i) + e_i, \qquad \theta(t_i) = \sum_{j=1}^{m} \theta_j \phi_{ij},$$

$$x_i = x(t_i), \quad \phi_{ij} = \phi_j(t_i), \quad i = 1, \dots, n,$$

derived by the discretization of problem (6.2.1) are of the form

$$A^* = \arg\min_A \sum_{i=1}^{n} \left| x_i - \sum_{j=1}^{m} a_j \phi_{ij} \right|, \quad \widehat{\theta}(t_i) = L(A^*, t_i), \quad i = 1, 2, \dots, n, \tag{6.3.3}$$

where e_i are arbitrary variables assuming positive, negative, and zero values, respectively, on the sets:

$$I^+ = \{i : e_i > 0\}, \quad I^- = \{i : e_i < 0\}, \quad I^0 = \{i : e_i = 0\}.$$

We arrive at the discrete analog of Theorem 6.2.1.

THEOREM 6.3.1. *The L_1-norm estimator $L(A^*, t)$ (6.3.3) for linear regression is equal exactly to the true value $\theta(t)$*

$$L(A^*, t) = \theta(t) \quad a_j = \theta_j, \quad j = 1, \dots, m, \quad t = t_i, \quad i = 1, 2, \dots, n,$$

if and only if

$$\left| \sum_{i \in I^+} L(A, t_i) - \sum_{i \in I^-} L(A, t_i) \right| \le \sum_{i \in I^0} |L(A, t_i)| \tag{6.3.4}$$

holds for all A.

In the discrete case, it is not easy to derive constructive restrictions on the sets I^+ and I^- (on the admissible number of gross errors) as before in the continuous case. In the problem of estimation of the location parameter ($m = 1$) using the notations for the numbers of positive and negative errors n^+ and n^-, respectively, from (6.3.4) we obtain

$$|a_1 n^+ - a_1 n^-| \le |a_1|(n - n^+ - n^-);$$

hence it follows that

$$n^+ \le [n/2], \qquad n^- \le [n/2], \tag{6.3.5}$$

where $[\cdot]$ is the integer part of a number.

Condition (6.3.5) is a well-known property of the sample median.

In the general case, we propose the following approximate method to verify the condition of robustness (inequality (6.3.4)), which is reduced to the standard procedure of checking the positive definiteness of the corresponding quadratic form. By using the Cauchy inequality

$$\left(\sum_{i=1}^{n} u_i \right)^2 \le n \sum_{i=1}^{n} u_i^2,$$

we transform inequality (6.3.4) as

$$\left\{\left|\sum_{i\in I^+} L(A,t_i) - \sum_{i\in I^-} L(A,t_i)\right|\right\}^2 \le \left\{\sum_{i\in I^0} |L(A,t_i)|\right\}^2$$
$$\le n_0 \sum_{i\in I^0}^{n} \{L(A,t_i)\}^2, \qquad (6.3.6)$$

where $n_0 = \mu(I^0)$ is the number of observations with zero errors.

Inequality (6.3.6) is weaker than condition (6.3.5), and it yields the upper bound for the numbers n^+ and n^- of admissible positive and negative gross errors. In the particular case of estimation of the location parameter ($L(A,t) = a_1$), the use of (6.3.6) yields exact bounds of form (6.3.5).

Numerical calculations show that restriction (6.3.6) is realistic enough for higher dimensions of the approximating polynomial. For instance, in the case of approximation by a straight line ($L(A,t) = a_1 + a_2 t$) the restriction on the number of one-sided (positive) errors is of the form $n^+ < 0.26n$, which is close to the exact result $n^+ \le [0.25n]$. In the general case of approximation by algebraic polynomials with one-sided errors, it follows from (6.3.7) that the restriction can be written in the form

$$n^+ \le \left[n \sin^2 \frac{\pi}{2(m+1)}\right]. \qquad (6.3.7)$$

Now we are ready to explain the results of Section 6.1 on approximation by a polynomial of degree three (see Table 6.2). From (6.3.7) and Table 6.3, we obtain $n^+ \le [0.095n]$ for $m = 4$, and $n^+ = 0$ for $n = 10$, i.e., a single outlier located on the boundary of the interval of processing destroys the L_1-approximation.

6.3.3. L_1-regression breakdown points

We recall that the notion of the *breakdown point* is due to (Hampel, 1968) as the measure of the global robustness of an estimator in the model of gross errors (the contamination scheme), and in this case it gives the maximum contamination fraction ε^* possessed by an estimator remaining within the boundaries of the parametric space (see Section 1.3).

We define the breakdown points ε^* and ε_n^* of L_1-approximations with regard to impulsive noise functions $e(t)$ in the continuous case and to gross errors in the discrete case, respectively, as follows.

DEFINITION 6.3.1. $\varepsilon^* = \sup\{\mu(E^+),\ \mu(E^-)\colon \|A^*\| < \infty\}$.

DEFINITION 6.3.2. $\varepsilon_n^* = \frac{1}{n}\sup\{n^+, n^-\colon \|A^*\| < \infty\}$, where n^+ and n^- are the numbers of positive and negative values of e_i.

The following result gives the breakdown points of L_1-approximations with respect to gross errors in response variables (Shevlyakov, 1992).

THEOREM 6.3.2. *In the continuous case, the breakdown point* ε^* *is of the form*

$$\varepsilon^* = \sup\left\{\mu(E^+), \mu(E^-) : \left|\int_{E^+} L(A,t)\,dt - \int_{E^-} L(A,t)\,dt\right| \le \int_{E^0} |L(A,t)|\,dt \forall A\right\}; \tag{6.3.8}$$

in the discrete case the breakdown point is

$$\varepsilon_n^* = \frac{1}{n}\max\left\{n^+, n^- : \left|\sum_{i\in I^+} L(A,t_i) - \sum_{i\in I^-} L(A,t_i)\right| \le \sum_{i\in I^0} |L(A,t_i)|, \ \forall A\right\}. \tag{6.3.9}$$

REMARK 6.3.1. In the case of approximation by algebraic polynomials, the upper bound for the breakdown point is

$$\varepsilon^* \le \sin^2 \frac{\pi}{2(m+1)}.$$

For instance, $\varepsilon^* = 1/2$ for $m = 1$ and $\varepsilon^* = 1/4$ for $m = 2$ (the breakdown point of the L_1-regression by a straight line).

6.3.4. Proofs

PROOF OF THEOREM 6.3.1. The proof is based on the characterization lemma 6.3.1 for discrete L_1-approximations. It suffices to substitute $L(A^*,t) = \theta(t)$ into (6.3.2). □

PROOF OF THEOREM 6.3.2. We derive relation (6.3.8) from Theorem 6.2.1 and Definition 6.3.1, and relation (6.3.9), from Theorem 6.3.1 and Definition 6.3.2 respectively. □

6.4. Final remarks

On robustness of the shape of an approximating function. The stability (invariance) conditions for best L_1-approximations in the form of inequalities (6.2.3), (6.2.7), and (6.3.4) derived with regard to arbitrary impulsive noise functions impose certain restrictions on the duration of impulses. These depend not only on the structure of the sets E^+, E^-, and E^0 in the continuous case, or of the sets I^+, I^-, and I^0 in the discrete case, but also on the chosen system of polynomials $L(A,t)$.

In the case of approximation by algebraic polynomials, the upper bound for the duration of a single impulse is given by inequality (6.2.5), and it is determined by the minimal distance between the points of Bernstein alternance

(Bernstein, 1926)

$$t_i = \frac{1}{2} - \frac{1}{2}\cos\frac{i\pi}{m+1}, \qquad i = 0, 1, \dots, m+1.$$

These minimum interdistance points (see Subsection 6.2.3)

$$t_1 - t_0 = t_{m+1} - t_m = \sin^2\frac{\pi}{2(m+1)}$$

are located at the boundaries of the interval $[0, 1]$; in other words, this is the least favorable location of the noise. Therefore, the best algebraic polynomials are highly sensitive to the values of an approximated function near the boundaries of an approximation interval (Akhiezer, 1958).

Consider the right-hand part of inequality (6.2.5). With the increasing degree m of the approximating polynomial, the upper bound decreases as $O(1/m^2)$: the invariance property of L_1-approximations is manifested the more stronger, the lower is the degree of the approximating polynomial.

Consider the L_1-approximation to some continuous function $e(t)$ defined on the interval $[0, \pi]$ by the trigonometric sums

$$L(A, t) = \sum_{j=1}^{m} a_j \sin jt.$$

Recalling the elementary trigonometric identity

$$\sum_{j=1}^{m} a_j \sin jt = \sin t \sum_{j=1}^{m} b_j (\cos t)^{j-1},$$

we can pose the problem of L_1-approximation as follows:

$$\min_{b_1,\dots,b_m} \int_0^{\pi} \left| \widetilde{e}(t) - \sum_{j=1}^{m} b_j (\cos t)^{j-1} \right| \sin t\, dt = \min_{b_1,\dots,b_m} \int_{-1}^{1} \left| \widetilde{e}(y) - \sum_{j=1}^{m} b_j y^{j-1} \right| dy,$$

where $y = \cos t$, $\widetilde{e}(t) = e(t)/\sin t$.

Obviously, here the problem is reduced to the problem of approximation by algebraic polynomials. Now from this point of view we consider the example of the approximation to a single impulse by the sine function presented in Subsection 6.1.2. For this problem, the maximum weight is assigned to the values of the approximated function at the center of the interval $[0, \pi]$, namely, in the interval $t \in (\pi/3, 2\pi/3)$, the least favorable location of the impulse. The corresponding interval $y \in (-1/2, 1/2)$ yields the limiting location provided the invariance property of approximation by a constant ($m = 1$).

Thus it is always possible to check the invariance property using the transformation $y = \cos t$, which makes the original points of Bernstein alternance equidistant:

$$\frac{1}{\pi}\arccos(1-2t_i) = \frac{i}{m+1}, \qquad i = 0, 1, \dots, m+1.$$

So we have another order of decrease of the upper bound of the duration of impulsive noises, namely $O(1/m)$. Hence one may speak of the comparative stability or robustness of different systems of functions.

More on robustness of L_1-approximations by trigonometric sums. Now we present some results on the invariance property of L_1-approximations by trigonometric sums. Consider the impulsive noise function $e(t)$ defined on the interval $0, k\pi$, where k is an integer. Denote $e_i(t)$ the value of the noise on the interval $[(i-1)\pi, i\pi]$, $i = 1, 2, \dots, k$:

$$e_i(t) = \begin{cases} h_i(t) \ge 0, & t \in E_i^+, \\ 0, & \text{otherwise.} \end{cases}$$

Then

$$\min_{a_1,\dots,a_m} \int_0^{k\pi} \left| e(t) - \sum_{j=1}^m a_j \sin jt \right| dt = \min_{a_1,\dots,a_m} \left\{ \sum_{i=1}^k \int_{(i-1)}^{i\pi} \left| e(t) - \sum_{j=1}^m a_j \sin jt \right| dt \right\}$$
$$= \min_{b_1,\dots,b_m} \left\{ \sum_{i=1}^k \int_{-1}^{1} \left| \widetilde{e}_i(y) - \sum_{j=1}^m b_j y^{j-1} \right| dy, \right\}$$

where $y = \cos t$,

$$\widetilde{e}_i(t) = \begin{cases} |e_i(t)/\sin t| & \text{for even } i, \\ -|e_i(t)/\sin t| & \text{for odd } i. \end{cases}$$

Here we have the following analog of Lemma 6.1.1 for the problem of approximation to a group of functions $\widetilde{e}_i(t)$ by algebraic polynomials (Shevlyakov, 1976).

LEMMA 6.4.1. *A necessary and sufficient condition for the inequality*

$$\sum_{i=1}^k L_1(\widetilde{e}_i, B^*) \le \sum_{i=1}^k L_1(\widetilde{e}_i, B^* + sB) \tag{6.4.1}$$

to hold for all s is

$$\left| \sum_{i=1}^k \int_{-1}^{1} L(B,y)\, \mathrm{sgn}[\widetilde{e}_i(t) - L(B^*, y)]\, dy \right| \le \sum_{i=1}^k \int_{Z_i(A^*)} |L(B,y)| dy. \tag{6.4.2}$$

Furthermore, if the strict inequality occurs in (6.4.2), *then the strict inequality occurs in* (6.4.1) *for all nonzero s.*

Here

$$L_1(\widetilde{e}_i, B) = \int_{-1}^{1} |\widetilde{e}_i(y) - L(B,y)|\, dy,$$
$$Z_i = Z_i(B) = \{y \mid L(B,y) - \widetilde{e}_i(y) = 0\}.$$

The proof completely repeats that of Lemma 6.1.1 (see the latter in (Rice, 1964)).

The analog of Theorem 6.2.1 gives the invariance condition for best L_1-approximations $L(B^*,y) = 0$

$$\left|\sum_{i=1}^{k}(-1)^{i-1}\int_{E_i^+} L(B,y)\, dy\right| \le \sum_{i=1}^{k}\int_{E_i^0} |L(B,y)|\, dy. \tag{6.4.3}$$

for all B.

Moreover, we must guarantee the uniqueness of the best L_1-approximations to a family of functions.

THEOREM 6.4.1. *If* (6.4.2) *holds with strict inequality for all B, then $L(B^*,y)$ is a unique best L_1-approximation to the family of functions $\{\widetilde{e}_i\}$.*

To establish the uniqueness, assume that this family has two best approximations $L(B_1,y)$ and $L(B_2,y)$, and set $B = B_1 - B_2$. Then if

$$\left|\sum_{i=1}^{k}\int_{-1}^{1} L(B,y)\,\mathrm{sgn}[\widetilde{e}_i(t) - L(B_2,y)]\, dy\right| < \sum_{i=1}^{k}\int_{Z_i(B_2)} |L(B,y)|\, dy,$$

it follows from Lemma 6.4.1 with $s = 1$ that

$$\sum_{i=1}^{k} L_1(\widetilde{e}_i, B_2) < \sum_{i=1}^{k} L_1(\widetilde{e}_i, B_2 + sB) = \sum_{i=1}^{k} L_1(\widetilde{e}_i, B_1),$$

which contradicts the assumption that both $L(B_1,y)$ and $L(B_2,y)$ are the best L_1-approximations.

Now we consider invariance condition (6.4.3) for $k = 2$, $m = 1$ (the approximation by a sine function on $[0.\, 2\pi]$)

$$|b\mu(E_1^+) - b\mu(E_2^+)| \le |b|[2 - \mu(E_1^+) + |b|[2 - \mu(E_2^+)]. \tag{6.4.4}$$

Setting $\mu(E_1^+) = \mu_1$ and $\mu(E_2^+) = \mu_2$, we rewrite (6.4.4) as

$$|\mu_1 - \mu_2| \le 4 - \mu_1 - \mu_2, \quad 0 \le \mu_i \le 2, \quad i = 1, 2.$$

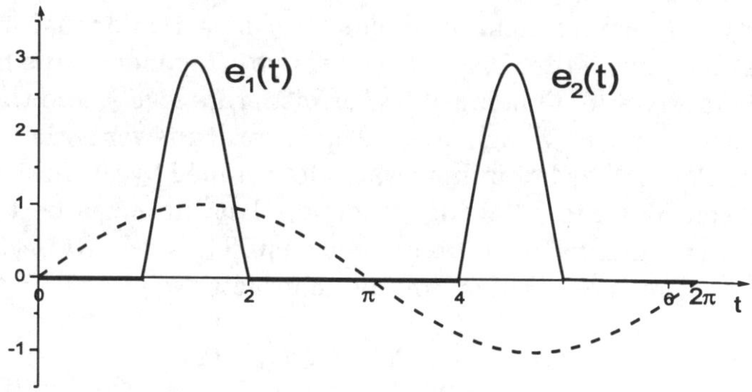

Figure 6.5. Approximation to the one-sided noise by a sine

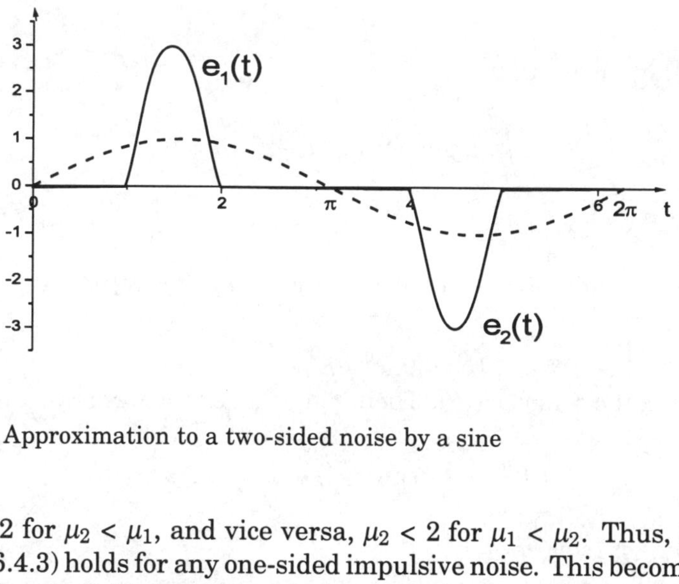

Figure 6.6. Approximation to a two-sided noise by a sine

Then $\mu_1 < 2$ for $\mu_2 < \mu_1$, and vice versa, $\mu_2 < 2$ for $\mu_1 < \mu_2$. Thus, invariance condition (6.4.3) holds for any one-sided impulsive noise. This becomes obvious if we consider the initial problem of approximation by a sine function. Fig. 6.5 illustrates this.

Another situation occurs when the noise has opposite signs on $[0, \pi]$ and $[\pi, 2\pi]$, for example, $e_1(t) > 0$ and $e_2(t) < 0$ as in Fig. 6.6.

In this case, condition (6.4.3) takes the form

$$|\mu_1 + \mu_2| \le 4 - \mu_1 - \mu_2, \qquad 0 \le \mu_i \le 2,$$
$$\mu_1 + \mu_2 \le 2. \tag{6.4.5}$$

Here invariance is observed only if (6.4.5) is valid.

On robustness of L_1-approximations to mixtures of noises. The invariance property of L_1-approximations holds for the particular case of impulsive noises. What happens if the noise consists of two components: the first defined on the whole interval $[0, T]$ being of low or moderate levels, and the second in the form of rare impulses of high level? The natural answer is that the error of L_1-approximation will be low or moderate, determined by the first component.

Now we consider the following example. Let the noise be the additive mixture of two continuous functions defined on $[0, T]$: $e(t) = e_1(t)+e_2(t)$. Assume that the latter component is of an impulse character with

$$e_2(t) \gg |e_1(t)|, \qquad t \in [0, T].$$

We have to show that if the condition $T^+ < T/2$ holds (T^+ is the total duration of $e_2(t)$), then the error of L_1-approximation to a constant is determined by the characteristics of the first component, i.e.,

$$|\widehat{e}(t)| \le \max_t |e_1(t)|.$$

From (6.1.6) it follows that

$$\int_0^T \operatorname{sgn}[e_1(t) + e_2(t) - \widehat{e}]\, dt = 0,$$

or

$$\int_{E^0} \operatorname{sgn}[e_1(t) - \widehat{e}]\, dt + \int_{E^+} \operatorname{sgn}[e_1(t) + e_2(t) - \widehat{e}]\, dt = 0, \tag{6.4.6}$$

where $E^+ = \{t \mid e_2(t) > 0\}$, $E^0 = [0, T] \setminus E^+$.

Assume that $\widehat{e} > \max |e_1(t)|$. Then (6.4.6) takes the form

$$-\mu(E^0) + \int_{E^+} \operatorname{sgn}[e_1(t) + e_2(t) - \widehat{e}]\, dt = 0. \tag{6.4.7}$$

Given $\mu(E^0) = T - T^+$, $T^+ < T/2$ and

$$\int_{E^+} \operatorname{sgn}[e_1(t) + e_2(t) - \widehat{e}]\, dt \le T^+,$$

it follows from (6.4.7) that

$$-T + T^+ + \int_{E^+} \operatorname{sgn}[e_1(t) + e_2(t) - \widehat{e}]\, dt \le -T + T^+ + T^+ < 0,$$

which contradicts (6.4.6). Therefore, $\widehat{e} < \max |e_1(t)|$.

The same reasoning can be used while considering the general case of L_1-approximations by polynomials.

On robustness of L_1-regression and breakdown points. In general, the results of Section 6.3 on L_1-approximation in the discrete case follow from the results of Section 6.2 for the continuous case.

The breakdown points 6.3.1 and 6.3.2 defined with respect to impulsive noise functions are similar to those defined with respect to gross errors. For example, in the case of the sample median $\varepsilon^* = 1/2$, and the same result we obtain from Remark 6.3.1 for $m = 1$; in the regression problem of approximation by a straight line ($m = 2$) the breakdown point $\varepsilon^* = 1/4$ (Hampel *et al.*, 1986), the same value is also given by the upper boundary for the breakdown points in Remark 6.3.1. Experimental study shows that this upper bound is really attainable .

On the computation of L_1-approximations. These questions are discussed in Chapter 8.

7

Robust estimation of correlation

Various groups of robust estimators of the correlation coefficient are studied in the case of contaminated bivariate normal distribution. Conventional and new robust estimators are considered in finite samples by Monte Carlo and in asymptotics by the influence functions technique.

Comparing the behavior of these estimators, we reveal the best in each group and show that some of them possess optimal robustness properties. In particular, an asymptotically minimax variance robust estimator of the correlation coefficient is designed for ε-contaminated bivariate normal distributions. For the estimator suggested, consistency and asymptotic normality is proved, and an explicit expression for its asymptotic variance is given. The limiting cases of this minimax variance estimator are the classical sample correlation coefficient with $\varepsilon = 0$ and the median correlation coefficient as $\varepsilon \to 1$.

We also show that two-stage algorithms based on preliminary rejection of outliers with subsequent application of the sample correlation coefficient to the rest of the data have quite high robustness.

The most advantageous approaches are applied to robust estimation of the correlation matrix, and a two-stage algorithm with rejection of outliers in each two-dimensional cut of a multivariate space manifests its high robustness.

7.1. Introductory remarks

Less attention is devoted in the literature to robust estimators of association and correlation as compared to robust estimators of location and scale. On the other hand, it is necessary to study these problems due to their widespread occurrence (estimation of correlation and covariance matrices in regression and multivariate analysis, estimation of correlation functions of stochastic processes, etc.), and also due to great instability of classical methods of estimation with outliers in the data (Devlin *et al.*, 1975; Gnanadesikan and Kettenring, 1972; Huber, 1981; Pasman and Shevlyakov, 1987; Rocke and

Woodruff, 1996; Rousseeuw and Leroy, 1987; Shevlyakov, 1997a; Shevlyakov and Khvatova, 1998b).

7.1.1. Non-robustness of the sample correlation coefficient

The simplest problem of correlation analysis is to estimate the correlation coefficient ρ in the case of observed values $(x_1,y_1),\dots,(x_n,y_n)$ of a bivariate random variable (X,Y). Its classical estimator is given by the sample correlation coefficient

$$\mathbf{r} = \frac{\sum_{i=1}^n (x_i - \bar{x})(y_i - \bar{y})}{\left(\sum_{i=1}^n (x_i - \bar{x})^2 \sum_{i=1}^n (y_i - \bar{y})^2\right)^{1/2}}, \tag{7.1.1}$$

where $\bar{x} = n^{-1}\sum x_i$, and $\bar{y} = n^{-1}\sum y_i$ are the sample means.

On the one hand, the sample correlation coefficient is a statistical counterpart of the correlation coefficient of a distribution

$$\rho = \frac{\mathrm{Cov}(X,Y)}{(\mathsf{Var}\,X\ \mathsf{Var}\,Y)^{1/2}}, \tag{7.1.2}$$

where $\mathsf{Var}\,X$, $\mathsf{Var}\,Y$, and $\mathrm{Cov}(X,Y)$ are the variances and the covariance of the random variables X and Y. On the other hand, it is an efficient maximum likelihood estimator of the correlation coefficient ρ for the bivariate normal distribution

$$\mathcal{N}(x,y;\mu_1,\mu_2,\sigma_1,\sigma_2,\rho) = \frac{1}{2\pi\sigma_1\sigma_2\sqrt{1-\rho^2}} \exp\left\{-\frac{1}{2(1-\rho^2)} \times \left[\frac{(x-\mu_1)^2}{\sigma_1^2} - 2\rho\frac{(x-\mu_1)(y-\mu_2)}{\sigma_1\sigma_2} + \frac{(y-\mu_2)^2}{\sigma_2^2}\right]\right\}, \tag{7.1.3}$$

where $\mu_1 = \mathsf{E}X$, $\mu_2 = \mathsf{E}Y$, $\sigma_1^2 = \mathsf{Var}\,X$, $\sigma_2^2 = \mathsf{Var}\,Y$.

In the contamination model described by a mixture of normal densities $(0 \le \varepsilon < 0.5)$

$$f(x,y) = (1-\varepsilon)\mathcal{N}(x,y;\mu_1,\mu_2,\sigma_1,\sigma_2,\rho) + \varepsilon\mathcal{N}(x,y;\mu_1',\mu_2',\sigma_1',\sigma_2',\rho'), \tag{7.1.4}$$

the sample correlation coefficient is strongly biased with regard to the estimated parameter ρ, i.e., for any positive $\varepsilon > 0$ there exists $k = \sigma_1'/\sigma_1 = \sigma_2'/\sigma_2 \gg 1$ such that $\mathsf{E}\mathbf{r} \approx \rho'$.

The presence of even one or two outliers in the data can completely destroy the sample correlation coefficient up to the change of its sign, as can be seen from Fig. 7.1.

Thus, we can see that the sample correlation coefficient is extremely sensitive to presence of outliers in the data, and hence it is necessary to use its robust counterparts.

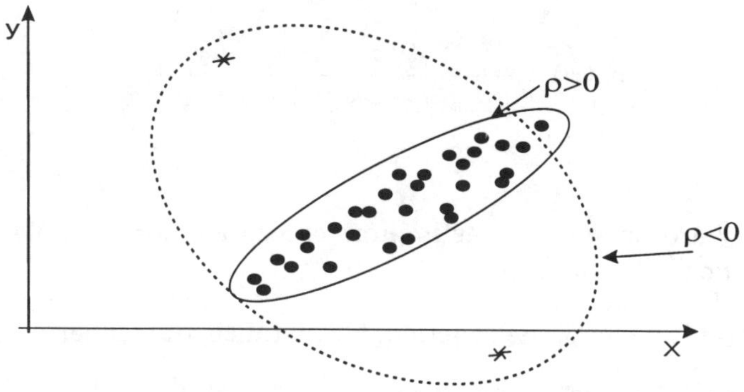

Figure 7.1. Point cloud and outliers with their impact on the sample correlation coefficient

Most of the robust estimators of the correlation coefficient described in the literature (Gnanadesikan and Kettenring, 1972; Devlin *et al.*, 1975; Huber, 1981; Pasman and Shevlyakov, 1987; Shevlyakov, 1997a) can be obtained from the following heuristic considerations:

- robust estimation of correlation via direct robust counterparts of the sample correlation coefficient;
- robust estimation of correlation via nonparametric measures;
- robust estimation of correlation via robust regression;
- robust estimation of correlation via robust estimation of the variances of the linear transformed data;
- robust estimation of correlation via two-stage robust estimators with preliminary rejection of outliers from the data and subsequent application of a classical estimator (for example, the sample correlation coefficient) to the rest of the observations.

Now we list these groups of estimators.

7.1.2. Robust correlation via direct robust counterparts of the sample correlation coefficient

A natural approach to robustifying the sample correlation coefficient is to replace the linear procedures of averaging by the corresponding nonlinear robust counterparts (Gnanadesikan and Kettenring, 1972; Devlin *et al.*, 1975;

Huber, 1981)

$$\mathbf{r}_\alpha(\psi) = \frac{\Sigma_\alpha \psi(x_i - \widehat{x})\psi(y_i - \widehat{y})}{\left(\Sigma_\alpha \psi^2(x_i - \widehat{x})\Sigma_\alpha \psi^2(y_i - \widehat{y})\right)^{1/2}}, \tag{7.1.5}$$

where

- $\widehat{x}$ and $\widehat{y}$ are some robust estimators of location, for example, the sample medians $\operatorname{med} x$ and $\operatorname{med} y$;
- $\psi = \psi(z)$ is a monotone function, for instance, the Huber ψ-function;
- Σ_α is a robust analog of a sum.

The latter transformation is based on trimming the outer terms of the variational series with subsequent summation of the remaining terms:

$$\Sigma_\alpha z_i = nT_\alpha(z) = n(n-2r)\sum_{i=r+1}^{n-r} z_i, \qquad r = [\alpha n],$$

where $[\cdot]$ stands for the integer part. For $\alpha = 0$, the operations of ordinary and of robust summation coincide: $\Sigma_0 = \Sigma$.

It is easy to see that estimator (7.1.5) has the following properties:

- it is invariant under translation and scale transformations of the observations x_i and y_i: $x_i \to a_1 x_i + b_1, \quad y_i \to a_2 y_i + b_2$;
- the normalization condition $|\mathbf{r}_\alpha| \le 1$ holds only for $\alpha = 0$;
- in the case of linearly dependent observations, $|\mathbf{r}_\alpha| = 1$.

Observe that in the experimental study of estimator (7.1.5), the condition of normalization was never violated under the mixture of normal distribution densities.

Further in Section 7.3, we use the following versions of estimator (7.1.5):

$$\mathbf{r}_\alpha = \frac{\Sigma_\alpha(x_i - \operatorname{med} x)(y_i - \operatorname{med} y)}{\left(\Sigma_\alpha(x_i - \operatorname{med} x)^2 \Sigma_\alpha(y_i - \operatorname{med} y)^2\right)^{1/2}},$$

where $\alpha = 0.1, 0.2$, and

$$\mathbf{r}_0(\psi_H) = \frac{\Sigma\psi_H(x_i - \operatorname{med} x)\psi_H(y_i - \operatorname{med} y)}{\left(\Sigma\psi_H^2(x_i - \operatorname{med} x)\Sigma\psi_H^2(y_i - \operatorname{med} y)\right)^{1/2}},$$

where

$$\psi_H(z) = \max(-c, \min(z, c)), \qquad c = 5 \operatorname{MAD} z.$$

7.1.3. Robust estimation of correlation via nonparametric measures

An estimation procedure can be endowed with robustness properties with the use of nonparametric rank statistics. The best known of them are the quadrant (sign) correlation coefficient (Blomqvist, 1950)

$$\mathbf{r}_Q = \frac{1}{n} \sum \operatorname{sgn}(x_i - \operatorname{med} x) \operatorname{sgn}(y_i - \operatorname{med} y), \tag{7.1.6}$$

that is the sample correlation coefficient between the signs of deviations from medians, and the rank correlation coefficient of Spearman (Spearman, 1904)

$$\mathbf{r}_S = \frac{\sum [R(x_i) - \overline{R}(x)] [R(y_i) - \overline{R}(y)]}{\left(\sum [R(x_i - \overline{R}(x)]^2 \sum [R(y_i) - \overline{R}(y)]^2\right)^{1/2}}, \tag{7.1.7}$$

that is the sample correlation coefficient between the observation ranks $R(x_i)$ and $R(y_i)$. For computing, it is more convenient to use the transformed version of (7.1.7) (Kendall and Stuart, 1963)

$$\mathbf{r}_S = 1 - \frac{S(d^2)}{6(n^3 - n)}, \qquad S(d^2) = \sum [R(x_i) - R(y_i)]^2.$$

Observe that formula (7.1.5) yields some of the above estimators:

- the sample correlation coefficient with $\alpha = 0, \widehat{x} = \bar{x}, \widehat{y} = \bar{y}, \psi(z) = z$;
- the quadrant correlation coefficient (7.1.6) with $\alpha = 0$, $\widehat{x} = \operatorname{med} x$, $\widehat{y} = \operatorname{med} y$, $\psi(z) = \operatorname{sgn} z$;
- the Spearman correlation coefficient (7.1.7) with $\alpha = 0, \widehat{x} = \overline{R}(x), \widehat{y} = \overline{R}(y)$, $\psi(z) = R(z)$.

For $\alpha = 0.5$, $\widehat{x} = \operatorname{med} x$, $\widehat{y} = \operatorname{med} y$, $\psi(z) = z$, formula (7.1.5) yields the median estimator

$$\mathbf{r}_{0.5} = \frac{\operatorname{med}(x_i - \operatorname{med} x)(y_i - \operatorname{med} y)}{\left(\operatorname{med}(x_i - \operatorname{med} x)^2 \operatorname{med}(y_i - \operatorname{med} y)^2\right)^{1/2}}.$$

7.1.4. Robust correlation via robust regression

The problem to estimate the correlation coefficient is directly related to the linear regression problem of fitting the straight line of the conditional expectation

$$\mathsf{E}(X \mid Y = y) = \mu_1 + \beta_1(y - \mu_2),$$
$$\mathsf{E}(Y \mid X = x) = \mu_2 + \beta_2(x - \mu_1).$$

For the normal distribution,

$$\beta_1 = \rho\frac{\sigma_1}{\sigma_2}, \qquad \beta_2 = \rho\frac{\sigma_2}{\sigma_1}. \tag{7.1.8}$$

If the coefficients β_1 and β_2 are estimated by the LS method

$$\widehat{\beta}_1 = \arg\min_{\alpha_1,\beta_1} \sum (x_i - \alpha_1 - \beta_1 y_i)^2,$$
$$\widehat{\beta}_2 = \arg\min_{\alpha_2,\beta_2} \sum (y_i - \alpha_2 - \beta_2 x_i)^2,$$

then the sample correlation coefficient can be expressed in terms of the estimators $\widehat{\beta}_1$ and $\widehat{\beta}_2$ as

$$r^2 = \widehat{\beta}_1\widehat{\beta}_2. \tag{7.1.9}$$

Using formula (7.1.9), we suggest the robust estimator for a correlation coefficient

$$\widehat{\rho} = \sqrt{\widehat{\beta}_1\widehat{\beta}_2}, \tag{7.1.10}$$

where $\widehat{\beta}_1$ and $\widehat{\beta}_2$ are some robust estimators of the slope, for example, the LAV estimators or the L_1-norm estimators of regression coefficients

$$\widehat{\beta}_1 = \arg\min_{\alpha_1,\beta_1} \sum |x_i - \alpha_1 - \beta_1 y_i|,$$
$$\widehat{\beta}_2 = \arg\min_{\alpha_2,\beta_2} \sum |y_i - \alpha_2 - \beta_2 x_i|.$$

In this case, we denote estimator (7.1.10) as $\mathbf{r}_{\mathrm{LAV}}$. It is easy to show that, in contrast to the LS formulas which yield the parameters of the straight line of the conditional mean, the LAV estimators yield the parameters of the straight line of the conditional median of the normal distribution

$$\operatorname{med}\{X \mid Y = y\} = \operatorname{med} X + \beta_1(y - \operatorname{med} Y),$$
$$\operatorname{med}\{Y \mid X = x\} = \operatorname{med} Y + \beta_2(x - \operatorname{med} X).$$

Another possibility is given by the least median squares regression

$$\widehat{\beta}_1 = \arg\min_{\alpha_1,\beta_1} \operatorname{med}(x_i - \alpha_1 - \beta_1 y_i)^2,$$
$$\widehat{\beta}_2 = \arg\min_{\alpha_2,\beta_2} \operatorname{med}(y_i - \alpha_2 - \beta_2 x_i)^2. \tag{7.1.11}$$

The corresponding estimator is referred to as $\mathbf{r}_{\mathrm{LMS}}$.

Using formula (7.1.8), we arrive at the robust estimators

$$\mathbf{r}_{m1} = \widehat{\beta}_{m1}\frac{\widehat{\sigma}_1}{\widehat{\sigma}_2} \tag{7.1.12}$$

and

$$\mathbf{r}_{m2} = \widehat{\beta}_{m2}\frac{\widehat{\sigma}_1}{\widehat{\sigma}_2}, \tag{7.1.13}$$

where

$$\widehat{\beta}_{m1} = \operatorname{med}\left\{\frac{y - \operatorname{med} y}{x - \operatorname{med} x}\right\}$$

and

$$\widehat{\beta}_{m2} = \operatorname{med}\left\{\frac{y_i - y_j}{x_i - x_j}\right\} \qquad i \neq j,$$

$\widehat{\sigma}_1$ and $\widehat{\sigma}_2$ are some robust estimators of scale, for example, the median absolute deviation MAD.

The structure of these estimators can be explained as follows: the distribution density of the ratio of centered normal random variables is given by the Cauchy formula

$$f(z) = \frac{\sqrt{1-\rho^2}}{\pi}\left[\frac{\sigma_1}{\sigma_2}\left(z - \frac{\sigma_2}{\sigma_1}\rho\right)^2 + \frac{\sigma_2}{\sigma_1}(1-\rho^2)\right]^{-1},$$

hence formulas (7.1.12) and (7.1.13) yield consistent estimators of the distribution center $\rho\sigma_2/\sigma_1$. With regard to variance, they are close to the optimal maximum likelihood estimator.

7.1.5. Robust correlation via robust variances

We can write the obvious relation for any bivariate random variables (X, Y)

$$\frac{\operatorname{Var}(X+Y) - \operatorname{Var}(X-Y)}{\operatorname{Var}(X+Y) + \operatorname{Var}(X-Y)} = \frac{2\operatorname{Cov}(X,Y)}{\operatorname{Var}(X) + \operatorname{Var}(Y)};$$

hence we obtain the correlation coefficient ρ provided $\operatorname{Var} X = \operatorname{Var} Y$.

It is convenient to use the standard variables $\widetilde{X}$ and $\widetilde{Y}$ such that $\operatorname{Var}\widetilde{X} = 1$ and $\operatorname{Var}\widetilde{Y} = 1$; thus

$$\rho = \frac{\operatorname{Var}(\widetilde{X}+\widetilde{Y}) - \operatorname{Var}(\widetilde{X}-\widetilde{Y})}{\operatorname{Var}(\widetilde{X}+\widetilde{Y}) + \operatorname{Var}(\widetilde{X}-\widetilde{Y})}. \tag{7.1.14}$$

By introducing the robust scale functional

$$S(X) = S(F_X)\colon\ S(aX + b) = |a|S(X),$$

we can write the robust analog of variance in the form $S^2(\cdot)$, and the robust analog of (7.1.14) in the form

$$\rho^*(X,Y) = \frac{S^2(\widetilde{X}+\widetilde{Y}) - S^2(\widetilde{X}-\widetilde{Y})}{S^2(\widetilde{X}+\widetilde{Y}) + S^2(\widetilde{X}-\widetilde{Y})}, \tag{7.1.15}$$

where $\widetilde{X}$ and $\widetilde{Y}$ are normalized in the same scale, $\widetilde{X} = X/S(X)$ and $\widetilde{Y} = Y/S(Y)$ (Gnanadesikan and Kettenring, 1972; Huber, 1981).

The robust 'correlation coefficient' $\rho^*(X,Y)$ defined by (7.1.15) satisfies the principal requirements on the correlation coefficient

- $|\rho^*(X,Y)| \le 1$;
- if the random variables X and Y are linearly dependent, then $|\rho^*(X,Y)| = 1$;
- in the case of independent random variables X and Y, we generally have $\rho^*(X,Y) \ne 0$.

However, for the mean and median absolute deviations functionals, the latter property holds for the distributions F_X and F_Y that are symmetric about the center.

Replacing the functionals by their robust estimators in (7.1.15), we arrive at robust estimators of the correlation coefficient in the form

$$\widehat{\rho}^*(X,Y) = \frac{\widehat{S}^2\,(\widetilde{X}+\widetilde{Y}) - \widehat{S}^2\,(\widetilde{X}-\widetilde{Y})}{\widehat{S}^2\,(\widetilde{X}+\widetilde{Y}) + \widehat{S}^2\,(\widetilde{X}-\widetilde{Y})}. \tag{7.1.16}$$

For the median absolute deviation functional, expression (7.1.16) takes the form of the *median correlation coefficient* (Pasman and Shevlyakov, 1987)

$$\mathbf{r}_{\text{med}\,1} = \frac{\mathrm{MAD}^2(\widetilde{X}+\widetilde{Y}) - \mathrm{MAD}^2(\widetilde{X}-\widetilde{Y})}{\mathrm{MAD}^2(\widetilde{X}+\widetilde{Y}) + \mathrm{MAD}^2(\widetilde{X}-\widetilde{Y})}, \tag{7.1.17}$$

where

$$\widetilde{X} = X/\,\mathrm{MAD}\,X, \qquad \widetilde{Y} = Y/\,\mathrm{MAD}\,Y.$$

We have another asymptotically equivalent version of the median correlation coefficient (Shevlyakov, 1988; Shevlyakov and Jae Won Lee, 1997)

$$\mathbf{r}_{\text{med}\,2} = \frac{\mathrm{med}^2\,|u| - \mathrm{med}^2\,|v|}{\mathrm{med}^2\,|u| + \mathrm{med}^2\,|v|}, \tag{7.1.18}$$

where u and v are called the *robust principal coordinates*

$$u = \frac{x - \mathrm{med}\,x}{\mathrm{MAD}\,x} + \frac{y - \mathrm{med}\,y}{\mathrm{MAD}\,y}, \qquad v = \frac{x - \mathrm{med}\,x}{\mathrm{MAD}\,x} - \frac{y - \mathrm{med}\,y}{\mathrm{MAD}\,y}. \tag{7.1.19}$$

Furthermore,

- for the mean absolute deviation functional,

$$\mathbf{r}_{L_1} = \frac{\left(\sum |u_i|\right)^2 - \left(\sum |v_i|\right)^2}{\left(\sum |u_i|\right)^2 + \left(\sum |v_i|\right)^2}, \tag{7.1.20}$$

- for the standard deviation functional,

$$\mathbf{r}_{L_2} = \frac{\sum u_i^2 - \sum v_i^2}{\sum u_i^2 + \sum v_i^2}, \tag{7.1.21}$$

- for the trimmed standard deviation functional,

$$\mathbf{r}_{tr}(n_1, n_2) = \frac{\sum_{i=n_1+1}^{n-n_2} u_i^2 - \sum_{i=n_1+1}^{n-n_2} v_i^2}{\sum_{i=n_1+1}^{n-n_2} u_i^2 + \sum_{i=n_1+1}^{n-n_2} v_i^2}. \tag{7.1.22}$$

The particular cases of the latter formula appear in (Gnanadesikan and Kettenring, 1972; Devlin *et al.*, 1975) with $n_1 = n_2 = [\alpha n]$ and $\alpha = 0.1, \dots, 0.2$.

Observe that the general construction (7.1.22) yields $\mathbf{r}_{L_2}$ with $n_1 = 0$ and $n_2 = 0$, and, in the case of odd sample sizes, the median correlation coefficient with $n_1 = n_2 = [0.5(n-1)]$. In addition, formula (7.1.21) yields the sample correlation coefficient if we use classical estimators in its inner structure: the sample means for location and the standard deviations for scale in (7.1.19).

7.1.6. Robust correlation via rejection of outliers

The preliminary rejection of outliers from the data with the consequent application of a classical estimator (for example, the sample correlation coefficient) to the rest of the observations defines the other group of estimators. Their variety mainly depends on the variety of the rules for rejection of outliers. In details, we consider this approach in Section 7.5.

7.1.7. Robust correlation via robust covariances

This approach is based on the preliminary robust estimation of the covariance matrix (Huber, 1981). Here the opposite way is used: evaluating robust covariances via the preliminary robust estimation of scale and correlation.

7.2. Analysis: Monte Carlo experiment

In this section we study the above-introduced groups of robust estimators of a correlation coefficient in normal and contaminated samples. As a result, we demonstrate the most perspective robust estimators within each group.

Table 7.1. $n = 20$: expectations and variances of estimators for normal distribution

	$\mathbf{r}$	$\mathbf{r}_Q$	$\mathbf{r}_S$	$\mathbf{r}_{m1}$	$\mathbf{r}_{m2}$	$\mathbf{r}_{L_1}$	$\mathbf{r}_{\text{med}}$	$\mathbf{r}_{\text{LMS}}$
$\rho = 0.0$	0.00	−0.01	−0.01	0.00	0.00	0.00	0.00	0.03
	0.05	0.05	0.05	0.09	0.07	0.05	0.10	0.27
$\rho = 0.5$	0.49	0.32	0.46	0.45	0.51	0.45	0.42	0.49
	0.03	0.05	0.03	0.11	0.06	0.03	0.08	0.21
$\rho = 0.9$	0.90	0.69	0.87	0.88	0.91	0.86	0.83	0.90
	0.00	0.03	0.01	0.05	0.05	0.01	0.02	0.04

The behavior of the estimators has been examined under the ε-contaminated bivariate normal distributions

$$f(x,y) = (1-\varepsilon)\mathcal{N}(x,y;0,0,1,1,\rho) + \varepsilon\mathcal{N}(x,y;0,0,k,k,\rho'), \quad 0 \le \varepsilon < 1, \tag{7.2.1}$$

in samples $n = 20, 30, 60$ using Monte Carlo techniques. As a rule, the number of trials is set to 1000, and in particular cases, it is increased up to 10000 for the sake of accuracy.

Nearly all estimators described in Section 7.1, namely $\mathbf{r}$, $\mathbf{r}_Q$, $\mathbf{r}_S$, $\mathbf{r}_{0.1}$, $\mathbf{r}_{0.2}$, $\mathbf{r}_{0.5}$, $\mathbf{r}_{\psi_H}$, $\mathbf{r}_{m1}$, $\mathbf{r}_{m2}$, $\mathbf{r}_{\text{LAV}}$, $\mathbf{r}_{\text{LMS}}$, $\mathbf{r}_{L_1}$, $\mathbf{r}_{L_2}$, $\mathbf{r}_{\text{med}\,1}$, $\mathbf{r}_{\text{med}\,2}$, and some others, have been tested in our study. Here we present only a part of our results concerned with the best and typical representatives of the above-introduced classes of estimators. More information about this topic can be found in (Gnanadesikan and Kettenring, 1972; Devlin *et al.*, 1975; Pasman and Shevlyakov, 1987).

7.2.1. Monte Carlo results for normal data

First we give some results for the bivariate normal density $\mathcal{N}(0,0,1,1,\rho)$ with small, medium, and large values of the correlation coefficient.

From Tables 7.1–7.3 we can see that

- in the normal case, as expected, the best is the sample correlation coefficient $\mathbf{r}$ both by its bias and variance;

- the classical nonparametric estimators such as the quadrant correlation coefficient $\mathbf{r}_Q$ and the rank correlation $\mathbf{r}_S$ have comparatively moderate variances, but their biases increase together with the estimated value of ρ, especially for $\mathbf{r}_Q$;

- the regression estimators $\mathbf{r}_{m1}$, $\mathbf{r}_{m2}$ and the estimators based on robust variances $\mathbf{r}_{L_1}$ and $\mathbf{r}_{\text{med}}$ behave similarly well except $\mathbf{r}_{\text{LMS}}$ that is somewhat worse by its variance.

Table 7.2. $n = 30$: expectations and variances of estimators for normal distribution

	$\mathbf{r}$	$\mathbf{r}_Q$	$\mathbf{r}_S$	$\mathbf{r}_{m1}$	$\mathbf{r}_{m2}$	$\mathbf{r}_{L_1}$	$\mathbf{r}_{\text{med}}$	$\mathbf{r}_{\text{LMS}}$
$\rho = 0.0$	0.00	−0.05	−0.03	−0.01	0.00	0.00	−0.01	0.00
	0.03	0.03	0.03	0.07	0.04	0.07	0.08	0.16
$\rho = 0.5$	0.49	0.32	0.47	0.46	0.50	0.47	0.45	0.48
	0.02	0.03	0.02	0.07	0.04	0.02	0.05	0.13
$\rho = 0.9$	0.90	0.70	0.87	0.88	0.90	0.87	0.85	0.90
	0.00	0.02	0.00	0.03	0.03	0.00	0.01	0.06

Table 7.3. $n = 60$: expectations and variances of estimators for normal distribution

	$\mathbf{r}$	$\mathbf{r}_Q$	$\mathbf{r}_S$	$\mathbf{r}_{m1}$	$\mathbf{r}_{m2}$	$\mathbf{r}_{L_1}$	$\mathbf{r}_{\text{med}}$	$\mathbf{r}_{\text{LMS}}$
$\rho = 0.0$	0.00	0.00	0.00	0.00	0.00	0.00	0.00	0.09
	0.02	0.02	0.02	0.04	0.02	0.02	0.04	0.09
$\rho = 0.5$	0.49	0.33	0.48	0.48	0.50	0.48	0.47	0.50
	0.01	0.02	0.01	0.03	0.02	0.01	0.03	0.07
$\rho = 0.9$	0.90	0.71	0.89	0.90	0.90	0.89	0.88	0.90
	0.00	0.01	0.00	0.01	0.01	0.00	0.00	0.03

REMARK 7.2.1. The median correlation coefficients $\mathbf{r}_{\text{med}\,1}$ and $\mathbf{r}_{\text{med}\,2}$ are similar in their structure and behavior, so we use the notation $\mathbf{r}_{\text{med}}$ for both of them.

7.2.2. Monte Carlo results under contamination

Here we give some results in small samples under heavy contamination for $\varepsilon = 0.1$, $k = 10$ and $\rho' = -0.9$ in formula (7.2.1).

Finally, in Figure 7.2 and Figure 7.3, we present the scatters of estimators in bias-standard error axes for the normal and contaminated normal small samples.

From the results listed in Tables 7.4–7.6 and in Figures 7.2–7.3, it follows that

- the sample correlation coefficient is catastrophically bad under contamination;
- the classical nonparametric estimators $\mathbf{r}_Q$ and $\mathbf{r}_S$ behave moderately ill together with the regression estimators $\mathbf{r}_{m1}$ and $\mathbf{r}_{m2}$;
- the best estimators are the regression estimator $\mathbf{r}_{\text{LMS}}$ and the median correlation coefficient $\mathbf{r}_{\text{med}}$.

Table 7.4. $n = 20$: expectations and variances of estimators under heavy contamination with ($\varepsilon = 0.1$, $k = 10$, and $\rho' = -0.9$

	$\mathbf{r}$	$\mathbf{r}_Q$	$\mathbf{r}_S$	$\mathbf{r}_{m1}$	$\mathbf{r}_{m2}$	$\mathbf{r}_{L_1}$	$\mathbf{r}_{\text{med}}$	$\mathbf{r}_{\text{LMS}}$
$\rho = 0.0$	−0.74	−0.11	−0.24	−0.19	−0.45	−0.16	0.00	−0.02
	0.14	0.05	0.06	0.15	0.34	0.02	0.10	0.26
$\rho = 0.5$	−0.66	0.21	0.17	0.30	0.22	0.02	0.41	0.48
	0.34	0.01	0.06	0.13	0.13	0.07	0.08	0.09
$\rho = 0.9$	−0.55	0.48	0.37	0.71	0.65	0.70	0.81	0.90
	0.37	0.04	0.09	0.06	0.13	0.09	0.02	0.04

Table 7.5. $n = 30$: expectations and variances of estimators under heavy contamination $\varepsilon = 0.1$, $k = 10$, and $\rho' = -0.9$

	$\mathbf{r}$	$\mathbf{r}_Q$	$\mathbf{r}_S$	$\mathbf{r}_{m1}$	$\mathbf{r}_{m2}$	$\mathbf{r}_{L_1}$	$\mathbf{r}_{\text{med}}$	$\mathbf{r}_{\text{LMS}}$
$\rho = 0.0$	−0.86	−0.10	−0.28	−0.21	−0.49	−0.19	−0.01	−0.02
	0.29	0.04	0.03	0.10	0.03	0.05	0.08	0.17
$\rho = 0.5$	−0.81	0.18	0.07	0.26	0.09	−0.09	0.44	0.48
	0.14	0.04	0.05	0.09	0.12	0.03	0.05	0.12
$\rho = 0.9$	−0.84	0.50	0.37	0.74	0.68	0.07	0.83	0.89
	0.08	0.03	0.06	0.04	0.05	0.05	0.01	0.05

Table 7.6. $n = 60$: expectations and variances of estimators under heavy contamination with $\varepsilon = 0.1$, $k = 10$, and $\rho' = -0.9$

	$\mathbf{r}$	$\mathbf{r}_Q$	$\mathbf{r}_S$	$\mathbf{r}_{m1}$	$\mathbf{r}_{m2}$	$\mathbf{r}_{L_1}$	$\mathbf{r}_{\text{med}}$	$\mathbf{r}_{\text{LMS}}$
$\rho = 0.0$	−0.86	−0.10	−0.28	−0.21	−0.49	−0.19	−0.01	−0.02
	0.29	0.04	0.03	0.10	0.03	0.05	0.08	0.17
$\rho = 0.5$	−0.81	0.18	0.07	0.26	0.09	−0.09	0.44	0.48
	0.14	0.04	0.05	0.09	0.12	0.03	0.05	0.12
$\rho = 0.9$	−0.84	0.50	0.37	0.74	0.68	0.07	0.83	0.89
	0.08	0.03	0.06	0.04	0.05	0.05	0.01	0.05

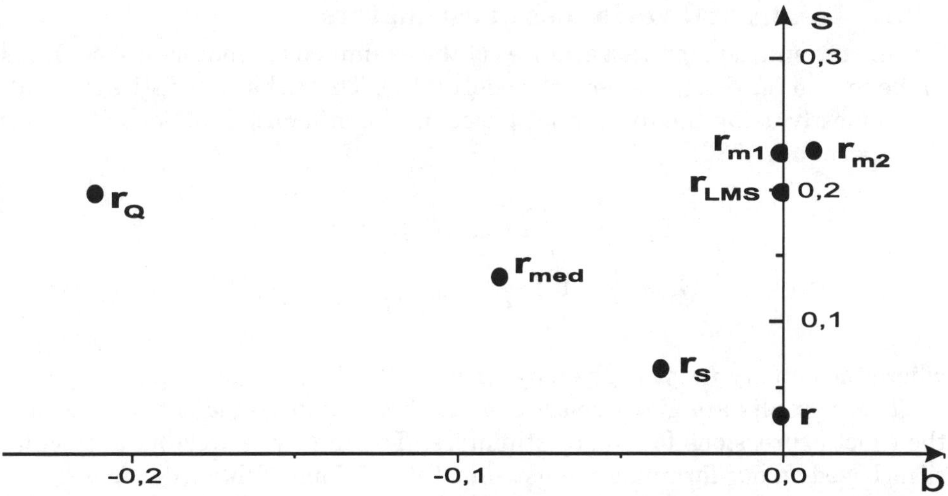

Figure 7.2. n = 20: biases and standard errors of estimators for normal distribution with $\rho = 0.5$

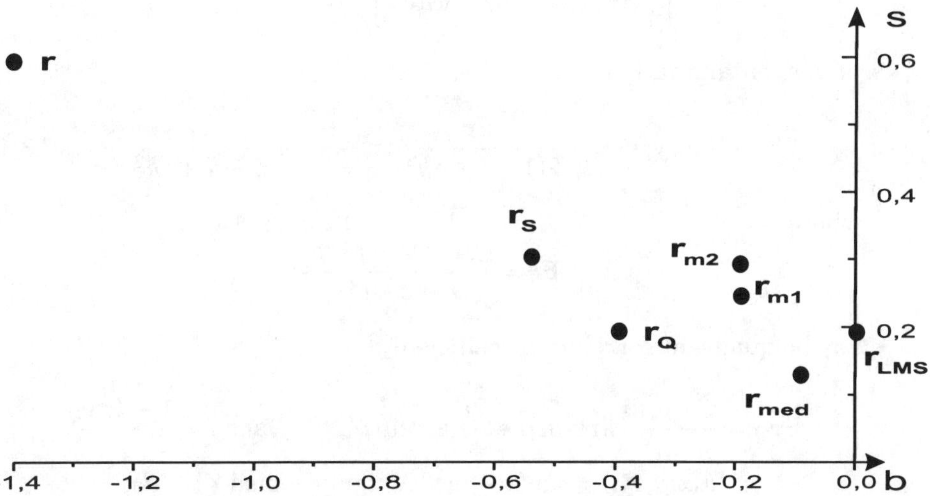

Figure 7.3. n = 20: biases and standard errors of estimators under heavy contamination with $\rho = 0.5$, $\varepsilon = 0.1$, $\rho' = -0.9$ and $k = 10$, (n = 20)

7.3. Analysis: asymptotic characteristics

7.3.1. Means and variances of estimators

The means and asymptotic variances of the examined estimators are obtained in the contamination (gross error) model (7.2.1). These characteristics are evaluated mostly using the techniques based on the influence functions $IF(x,y;\widehat{\rho})$ (Hampel *et al.*, 1986)

$$\mathsf{E}\widehat{\rho} \approx \rho + \int IF(x,y;\widehat{\rho})\, f(x,y)\, dx\, dy,$$

$$\mathsf{Var}\,\widehat{\rho} = n^{-1} \int IF^2(x,y;\widehat{\rho})\, f(x,y)\, dx\, dy,$$

where the density $f(x,y)$ is given by formula (7.1.4).

These results are given below. Due to their cumbersome nature, we omit the exact expressions for some estimators. The numerical results of calculations based on our formulas are listed in Table 7.7 and Table 7.8.

Direct robust analogs of the sample correlation coefficient, the means, asymptotic variances, and influence functions are of the form

- for the sample correlation coefficient of the bivariate normal distribution (Kendall and Stuart, 1962),

$$\mathsf{E}\mathbf{r} = \rho\left[1 - \frac{(1-\rho^2)}{2n} + O\left(\frac{1}{n^2}\right)\right], \qquad \mathsf{Var}\,\mathbf{r} = \frac{(1-\rho^2)^2}{n};$$

  under contamination,

$$IF(x,y;\mathbf{r}) = -\frac{\mathsf{E}\mathbf{r}}{2(1-\varepsilon+\varepsilon k^2)}(x^2+y^2) + \frac{xy}{1-\varepsilon+\varepsilon k^2}, \tag{7.3.1}$$

  where

$$\mathsf{E}\mathbf{r} = \frac{(1-\varepsilon)\rho + \varepsilon k^2\rho'}{1-\varepsilon+\varepsilon k^2};$$

- for the quadrant correlation coefficient,

$$\mathsf{E}\mathbf{r}_Q = \frac{2(1-\varepsilon)}{\pi}\arcsin\rho + \frac{2\varepsilon}{\pi}\arcsin\rho', \qquad \mathsf{Var}\,\mathbf{r}_Q = \frac{1-\mathsf{E}^2\mathbf{r}_Q}{n},$$
$$IF(x,y;\mathbf{r}_Q) = \operatorname{sgn}(x - \operatorname{med} X)\operatorname{sgn}(y - \operatorname{med} Y) - \rho_Q,$$

  where ρ_Q is the functional corresponding to the quadrant correlation coefficient

$$\rho_Q = \int \operatorname{sgn}(x - \operatorname{med} X)\operatorname{sgn}(y - \operatorname{med} Y)\, dF(x,y);$$

- for the Spearman rank correlation coefficient,

$$\mathsf{E}\mathbf{r}_S = \frac{6(1-\varepsilon)}{\pi}\arcsin\left(\frac{\rho}{2}\right) + \frac{6\varepsilon}{\pi}\arcsin\left(\frac{\rho'}{2}\right);$$

- for the median algorithm $\mathbf{r}_{0.5}$,

$$\mathsf{E}\mathbf{r}_{0.5} \approx \rho - 1.3\,\varepsilon\,C(\rho)\sqrt{1+\rho^2}, \qquad 0 \le C(\rho) \le 1,$$
$$\mathsf{Var}\,\mathbf{r}_{0.5} \approx 2/n, \quad |\rho| \ll 1, \qquad \mathsf{Var}\,\mathbf{r}_{0.5} \approx 1/n \quad |\rho| \approx 1.$$

For the regression group of estimators, we represent the results for the one of the best estimators from this group, namely that based on the median of slopes $\mathbf{r}_{m1}$ (7.1.12)

$$\mathsf{E}\mathbf{r}_{m1} = \rho + \varepsilon\arctan[(\rho' - \rho)\sqrt{1-\rho^2}/\sqrt{1-\rho'^2}] + o(\varepsilon),$$

$$\mathsf{Var}\,\mathbf{r}_{m1} = \frac{\pi^2(1-\rho^2)}{4n}\left\{1 + 2\varepsilon\sqrt{1-\rho^2}\left[\frac{1}{\sqrt{1-\rho^2}} - \frac{\sqrt{1-\rho'^2}}{(\rho-\rho')^2 + (1-\rho'^2)}\right]\right\}.$$

Another good estimator of this group, the $\mathbf{r}_{\mathrm{LMS}}$ based on the LMS regression, has the order of convergence $n^{-1/3}$ (Rousseeuw and Leroy, 1987).

As for the group based on robust variances, we are particularly interested in the median correlation coefficient, which proved its high robustness in the Monte Carlo study

$$\mathsf{E}\mathbf{r}_{\mathrm{med}} = \rho + 1.17\varepsilon(1-\rho^2)\,\mathrm{sgn}\,(\rho' - \rho) + o(\varepsilon).$$

The following results are concerned with the quantitative and qualitative robustness of this estimator also are of some interest.

THEOREM 7.3.1. *Under the bivariate normal distribution, the median correlation coefficient is a consistent and asymptotically normal estimator of the correlation coefficient* ρ *with the following asymptotic variance*

$$\mathsf{Var}\,\mathbf{r}_{\mathrm{med}} = \frac{(1-\rho^2)^2}{8n\phi^2(\zeta_{3/4})\zeta_{3/4}^2},$$

where $\zeta_{3/4} = \Phi^{-1}(3/4)$ *and* $\Phi(z)$ *is the standard normal distribution function*

$$\Phi(z) = \frac{1}{\sqrt{2\pi}}\int_{-\infty}^{z} e^{-t^2/2}\,dt, \qquad \phi(z) = \Phi'(z).$$

In the normal case, the asymptotic relative efficiency of the median correlation coefficient $\mathbf{r}_{\mathrm{med}}$ to the sample correlation coefficient $\mathbf{r}$ is 0.367.

Table 7.7. $\varepsilon = 0$, $\rho = 0.9$

	$\mathbf{r}$	$\mathbf{r}_Q$	$\mathbf{r}_S$	$\mathbf{r}_{\mathrm{LAV}}$	$\mathbf{r}_{m1}$	$\mathbf{r}_{L_1}$	$\mathbf{r}_{L_2}$	$\mathbf{r}_{\mathrm{med}}$
$\mathsf{E}\widehat{\rho}$	0.90	0.93	0.90	0.90	0.90	0.90	0.90	0.90
$n\,\mathsf{Var}(\widehat{\rho})$	0.04	0.13	0.05	0.09	0.07	0.06	0.04	0.10

Table 7.8. $\varepsilon = 0.1$, $\rho = 0.9$, $\rho' = -0.9$, $k = 10$.

	$\mathbf{r}$	$\mathbf{r}_Q$	$\mathbf{r}_S$	$\mathbf{r}_{\mathrm{LAV}}$	$\mathbf{r}_{m1}$	$\mathbf{r}_{L_1}$	$\mathbf{r}_{L_2}$	$\mathbf{r}_{\mathrm{med}}$
$\mathsf{E}\widehat{\rho}$	−0.75	0.57	0.71	0.74	0.84	0.79	0.72	0.88
$n\,\mathsf{Var}(\widehat{\rho})$	1.00	0.46	0.32	0.65	0.50	0.50	0.45	0.13

THEOREM 7.3.2. *The median correlation coefficient has the maximal breakdown point $\varepsilon^* = 1/2$, and its influence function is of the form*

$$IF(x,y;\mathbf{r}_{\mathrm{med}}) = \frac{1-\rho^2}{4\zeta_{3/4}^2\phi(\zeta_{3/4})}[\operatorname{sgn}(|x+y| - \zeta_{3/4}) - \operatorname{sgn}(|x-y| - \zeta_{3/4})].$$

From Table 7.7 and Table 7.8 we can see that the results of the asymptotic analysis confirm the preliminary conclusions of the Monte Carlo study:

- for the normal distribution, the best are the sample correlation coefficient $\mathbf{r}$ and the estimator $\mathbf{r}_{L_2}$, the latter being asymptotically equivalent to $\mathbf{r}$ in this case;

- for the normal distribution, the biases of estimators can be neglected, but not their variances;

- under contamination, the sample correlation coefficient is extremely poor both in bias and in variance, but robust estimators of location make $\mathbf{r}_{L_2}$ a little more robust;

- under heavy contamination, the best is obviously the median correlation coefficient;

- under heavy contamination, the bias of an estimator seems to be a more informative characteristic than its variance.

7.3.2. Proofs

PROOF OF THEOREM 7.3.1. Consider the asymptotic behavior of the median correlation coefficient

$$\mathbf{r}_{\mathrm{med}} = \frac{\mathrm{med}^2\,|u| - \mathrm{med}^2\,|v|}{\mathrm{med}^2\,|u| + \mathrm{med}^2\,|v|}, \tag{7.3.2}$$

where $u = x/\sqrt{2} + y/\sqrt{2}$, $v = x/\sqrt{2} - y/\sqrt{2}$ are the standardized variables.

For the sake of brevity, we use the following notations: $m_1 = \mathrm{med}\,|u|$ and $m_2 = \mathrm{med}\,|v|$ for medians, M_1 and M_2 for their asymptotic values. Observe that m_1 and m_2 converge in probability to M_1 and M_2 respectively.

First we demonstrate the consistency of the median correlation coefficient or, in other words, check that its asymptotic value coincides with the correlation coefficient ρ:

$$\rho = \frac{M_1^2 - M_2^2}{M_1^2 + M_2^2}. \tag{7.3.3}$$

For the distribution densities of the variables

$$|U| = |X/\sqrt{2} + Y/\sqrt{2}|, \qquad |V| = |X/\sqrt{2} - Y/\sqrt{2}|$$

the following is true:

$$f_{|U|}(z) = \begin{cases} \frac{2}{\sqrt{2\pi}\sqrt{1+\rho}} \exp\left(-\frac{z^2}{2(1+\rho)}\right), & z \geq 0, \\ 0, & z < 0; \end{cases}$$

$$f_{|V|}(z) = \begin{cases} \frac{2}{\sqrt{2\pi}\sqrt{1-\rho}} \exp\left(-\frac{z^2}{4(1-\rho)}\right), & z \geq 0, \\ 0, & z < 0. \end{cases}$$

The medians of these distributions are derived from the equations

$$\int_0^{M_1} f_{|U|}(z)\,dz = \frac{1}{2}, \qquad \int_0^{M_2} f_{|V|}(z)\,dz = \frac{1}{2},$$

and the explicit expressions of them are

$$M_1 = \sqrt{(1+\rho)}\,\Phi^{-1}(3/4), \qquad M_2 = \sqrt{(1-\rho)}\,\Phi^{-1}(3/4), \tag{7.3.4}$$

where $\Phi(z)$ is the standard normal distribution function.

Relation (7.3.3) and the consistency of the median correlation coefficient (7.3.2) immediately follow from (7.3.4).

Now we obtain the expression for the asymptotic variance of the median correlation coefficient. The difference between the estimator and its asymptotic value can be written as

$$\begin{aligned} \mathbf{r}_{\mathrm{med}} - \rho &= \frac{\mathrm{med}^2\,|u| - \mathrm{med}^2\,|v|}{\mathrm{med}^2\,|u| + \mathrm{med}^2\,|v|} - \frac{M_1^2 - M_2^2}{M_1^2 + M_2^2} \\ &= \frac{2}{M_1^2 + M_2^2} \frac{M_2^2 m_1^2 - M_1^2 m_2^2}{m_1^2 + m_2^2}, \end{aligned} \tag{7.3.5}$$

whereas the numerator of the latter fraction tends to zero as $n \to \infty$, therefore $m_1 \xrightarrow{\mathsf{P}} M_1$ and $m_2 \xrightarrow{\mathsf{P}} M_2$ like it should be due to the consistency just proved.

In view of the asymptotic normality of the sample medians,

$$m_1 = M_1 + \xi_1 + o(1/\sqrt{n}), \qquad m_2 = M_2 + \xi_2 + o(1/\sqrt{n}),$$

where

$$\xi_1 \sim \mathcal{N}\left(0, \frac{1}{2f_1(M_1)\sqrt{n}}\right), \qquad \xi_2 \sim \mathcal{N}\left(0, \frac{1}{2f_2(M_2)\sqrt{n}}\right),$$

$$f_1(z) = \frac{2}{\sqrt{2\pi}\sqrt{(1+\rho)}} \exp\left(-\frac{z^2}{2(1+\rho)}\right),$$

$$f_2(z) = \frac{2}{\sqrt{2\pi}\sqrt{(1-\rho)}} \exp\left(-\frac{z^2}{2(1-\rho)}\right);$$

M_1 and M_2 are given by (7.3.4).

Then it is easy to show that the asymptotic bias (7.3.5) can be rewritten as

$$\mathbf{r}_{\text{med}} - \rho = \frac{4M_1M_2}{(M_1^2 + M_2^2)^2}(M_2\xi_1 - M_1\xi_2) + o(1/\sqrt{n}). \tag{7.3.6}$$

Therefore, by virtue of the independence of ξ_1 and ξ_2, we arrive at the asymptotic variance of the median correlation coefficient

$$\mathsf{Var}\,\mathbf{r}_{\text{med}} = \frac{16M_1^2M_2^2}{(M_1^2 + M_2^2)^4}(M_2^2\sigma_1^2 + M_1^2\sigma_2^2), \tag{7.3.7}$$

where

$$\sigma_1^2 = \frac{1}{4f_1^2(M_1)n}, \qquad \sigma_2^2 = \frac{1}{4f_2^2(M_2)n}.$$

By substituting (7.3.4) into (7.3.7), we obtain the asymptotic variance as in Theorem 7.3.1, which completes the proof. □

REMARK 7.3.1. As concerns the asymptotic normality, the latter follows either directly from representation (7.3.6), or by reasoning due to (Huber, 1964, p. 78, Lemma 5): the numerator of the second fraction in (7.3.5) is asymptotically normal, the denominator tends in probability to the positive constant $M_1^2+M_2^2$, hence, $n^{1/2}\rho_n$ is asymptotically normal (Cramér, 1946, 20.6).

PROOF OF THEOREM 7.3.2. The median correlation coefficient $\mathbf{r}_{\text{med}}$ is constructed of the median absolute deviations MAD u and MAD v whose breakdown point is 1/2, hence $\varepsilon^*(\mathbf{r}_{\text{med}}) = 1/2$.

Differentiating the functional for the median correlation coefficient, we obtain

$$IF(x,y;\mathbf{r}_{\mathrm{med}}) = \frac{d}{ds}\,\rho_{\mathrm{med}}\left((1-s)F + s\Delta_{xy}\right)\Big|_{s=0} = \frac{d}{ds}\left(\frac{\mathrm{MAD}^2\,|u| - \mathrm{MAD}^2\,|v|}{\mathrm{MAD}^2\,|u| + \mathrm{MAD}^2\,|v|}\right)\Big|_{s=0}$$
$$= \frac{d}{ds}\left(\frac{M_1^2 - M_2^2}{M_1^2 + M_2^2}\right)\Big|_{s=0} = \frac{4M_1M_2}{(M_1^2+M_2^2)^2}(M_2M_1' - M_1M_2')\Big|_{s=0}$$
$$= \frac{4M_1M_2}{(M_1^2+M_2^2)^2}(M_2 IF(x,y;\mathrm{MAD}\,u) - M_1 IF(x,y;\mathrm{MAD}\,v)),$$

where $\Delta_{x_0y_0}$ is a bivariate analog of the Heaviside function

$$\Delta_{x_0y_0} = \begin{cases} 1, & x \ge x_0,\ y \ge y_0, \\ 0, & \text{otherwise}. \end{cases}$$

Since the influence function of the median absolute deviation is of the form (Hampel *et al.*, 1986)

$$IF(x,y;\mathrm{MAD}\,z) = \frac{1}{4\zeta_{3/4}\phi(\zeta_{3/4})}\,\mathrm{sgn}(|z| - \zeta_{3/4}),$$

we conclude that Theorem 7.3.2 is true. □

7.4. Synthesis: minimax variance correlation

In this section we use the Huber minimax approach to design a robust estimator of the correlation coefficient for ε-contaminated bivariate normal distributions. The Huber results on robust M-estimators of location and scale in ε-contamination models are extended to the problems of robust estimation of ρ. Consistency and asymptotic normality of the robust estimator obtained are proved, and an explicit expression for its asymptotic variance is obtained.

The problem of robust estimation of correlation is reduced to the problem of robust estimation of scale, therefore the structure of the minimax estimator of ρ is determined by the structure of the minimax estimator of scale in ε-contamination models that is similar to the trimmed standard deviation. The level of trimming depends on the value of the contamination parameter ε. The limiting cases of the obtained robust estimator are the sample correlation coefficient and the median correlation coefficient with $\varepsilon = 0$ and as $\varepsilon \to 1$ respectively.

7.4.1. Bivariate distributions allowing for principal factorization

Let the parameters of location and scale of the random variables X and Y be $\mu_1 = \mu_2 = 0$, $\sigma_1 = \sigma_2 = 1$. We introduce the class of bivariate distribution densities corresponding to the class of estimators based on robust variances (see Subsection 7.1.5)

$$f(x, y; \rho) = \frac{1}{\beta_u(\rho)} g\left(\frac{u}{\beta_u(\rho)}\right) \frac{1}{\beta_v(\rho)} g\left(\frac{v}{\beta_v(\rho)}\right), \tag{7.4.1}$$

where u and v are the principal variables

$$u = (x+y)/\sqrt{2}, \qquad v = (x-y)/\sqrt{2};$$

$g(x)$ is a symmetric density $g(-x) = g(x)$ belonging to a certain class $\mathcal{G}$.

If the variance of the density g exists ($\sigma_g^2 = \int x^2 g(x)\,dx < \infty$) then the straightforward calculation yields

$$\operatorname{Var} X = \operatorname{Var} Y = (\beta_u^2 + \beta_v^2)\sigma_g^2/2, \qquad \operatorname{Cov}(X, Y) = (\beta_u^2 - \beta_v^2)\sigma_g^2/2,$$

and hence the correlation coefficient of the class (7.4.1) depends on the scale parameters β_u and β_v as follows:

$$\rho = \frac{\beta_u^2 - \beta_v^2}{\beta_u^2 + \beta_v^2}. \tag{7.4.2}$$

Now we assume that the variances of the random variables X and Y do not depend on the unknown correlation coefficient ρ:

$$\operatorname{Var} X = \operatorname{Var} Y = \text{const}\,(\rho).$$

Setting for convenience $\sigma_g = 1$, for β_u and β_v we obtain

$$\beta_u = \sigma\sqrt{1+\rho}, \qquad \beta_v = \sigma\sqrt{1-\rho},$$

and for densities (7.4.1),

$$f(x, y; \rho) = \frac{1}{\sigma\sqrt{1+\rho}} g\left(\frac{u}{\sigma\sqrt{1+\rho}}\right) \frac{1}{\sigma\sqrt{1-\rho}} g\left(\frac{v}{\sigma\sqrt{1-\rho}}\right). \tag{7.4.3}$$

Observe that class (7.4.1) and its subclass (7.4.3) contain the standard bivariate normal distribution density

$$f(x, y) = \mathcal{N}(x, y \mid 0, 0, 1, 1, \rho)$$

with

$$\beta_u(\rho) = \sqrt{1+\rho}, \quad \beta_v(\rho) = \sqrt{1-\rho}, \quad g(x) = \phi(x) = (2\pi)^{-1/2}\exp(-x^2/2).$$

REMARK 7.4.1. Using other forms of univariate distribution densities, say the Laplace or even the heavy-tailed Cauchy (with the apparent modification of the definition for ρ), we can construct bivariate analogs for the corresponding univariate distributions.

In what follows, we deal with subclass (7.4.3).

REMARK 7.4.2. Class (7.4.1) represents a rather rich family of bivariate distributions, and hence it surely can be used in multivariate analysis for purposes independent of robustness. In this context, it is introduced as a construction corresponding entirely to the class of estimators based on robust variances (Subsection 7.1.4): it can be shown that the ML estimator of ρ in the class (7.4.1) is just the estimator of class (7.1.16).

Now we formulate the basic idea of introducing class (7.4.1): for any random pair (X, Y), the transformation $U = X + Y$, $V = X - Y$ yields uncorrelated random principal variables (U, V) (independent for the densities (7.4.1)), and estimation of their scale solves the problem of estimation of the correlation between (X, Y).

For distribution densities (7.4.3), the Fisher information is of the form

$$I(f) = \mathsf{E}_F \left(\frac{\partial \ln f}{\partial \rho} \right)^2 = \frac{1+\rho^2}{2(1-\rho^2)^2} I(g), \tag{7.4.4}$$

where $I(g)$ is the Fisher information for scale

$$I(g) = \int_{-\infty}^{\infty} \left[-x \frac{g'(x)}{g(x)} - 1 \right]^2 g(x)\, dx.$$

7.4.2. Estimation procedure

Given a sample $(x_1, y_1), (x_2, y_2), \dots, (x_n, y_n)$, we propose the following estimation procedure:

- transform the initial data as

$$u_i = (x_i + y_i)/\sqrt{2}, \quad v_i = (x_i - y_i)/\sqrt{2}, \quad i = 1, \dots, n;$$

- evaluate the M-estimators of scale $\widehat{\beta}_u$ and $\widehat{\beta}_v$ as the solutions of the equations

$$\sum \chi \left(\frac{u_i}{\widehat{\beta}_u} \right) = 0, \qquad \sum \chi \left(\frac{v_i}{\widehat{\beta}_v} \right) = 0, \tag{7.4.5}$$

  where $\chi(\cdot)$ is the score function;

- substitute these M-estimators of scale into formula (7.4.2) and evaluate the estimator of ρ in the form

$$\widehat{\rho}_n = \frac{\widehat{\beta}_u^2 - \widehat{\beta}_v^2}{\widehat{\beta}_u^2 + \widehat{\beta}_v^2}. \tag{7.4.6}$$

The optimal choice of the score function in (7.4.5) will be made later by applying the minimax approach.

7.4.3. Consistency and asymptotic normality

The asymptotic properties of the proposed estimator (7.4.6) are completely determined by the asymptotic properties of M-estimators of scale (7.4.5). Sufficient conditions of regularity providing the desired properties are imposed on the densities g and score functions χ. Assume that they satisfy conditions (F1)–(F2) and (χ1)–(χ4) (Section 4.3). Then the following is true.

THEOREM 7.4.1. *Under the above conditions of regularity, the estimator $\widehat{\rho}_n$ is consistent and asymptotically normal with variance*

$$\operatorname{Var}\widehat{\rho}_n = \frac{2(1-\rho^2)^2}{n} V(\chi, g), \tag{7.4.7}$$

where

$$V(\chi, g) = \frac{\int \chi^2(x) g(x)\, dx}{\left(\int x\chi'(x) g(x)\, dx\right)^2}.$$

is the asymptotic variance of M-estimators for scale.

PROOF. The consistency of (7.4.6) follows immediately from the consistency of M-estimators for scale: as $\widehat{\beta}_u$ and $\widehat{\beta}_u$ tend in probability to $\beta_u = \sigma\sqrt{1+\rho}$ and $\beta_v = \sigma\sqrt{1-\rho}$, $\widehat{\rho}_n$ tends in probability to ρ.

The asymptotic normality follows from the reasoning due to (Huber, 1964, p. 78, Lemma 5): the numerator of the fraction in (7.4.6) is asymptotically normal, the denominator tends in probability to the positive constant $c = \beta_u^2 + \beta_v^2$, hence $n^{1/2}\rho_n$ is asymptotically normal (Cramér, 1946, 20.6).

The exact structure of asymptotic variance is obtained by direct routine calculation using the asymptotic formula for the variance of the ratio of the random variables ξ and η (Kendall and Stuart, 1962)

$$\operatorname{Var}\frac{\xi}{\eta} = \left(\frac{\mathsf{E}\,\xi}{\mathsf{E}\,\eta}\right)^2 \left(\frac{\operatorname{Var}\xi}{\mathsf{E}^2\xi} + \frac{\operatorname{Var}\eta}{\mathsf{E}^2\eta} - \frac{2\operatorname{Cov}(\xi,\eta)}{\mathsf{E}\,\xi\,\mathsf{E}\,\eta}\right) + o\left(\frac{1}{n}\right), \tag{7.4.8}$$

where $\xi = \widehat{\beta}_u^2 - \widehat{\beta}_v^2$ and $\eta = \widehat{\beta}_u^2 + \widehat{\beta}_v^2$.

In view of independence of $\widehat{\beta}_u$ and $\widehat{\beta}_v$,

$$\mathsf{E}\xi = \beta_u^2 - \beta_v^2 + \sigma_u^2 - \sigma_v^2, \qquad \mathsf{E}\eta = \beta_u^2 + \beta_v^2 + \sigma_u^2 + \sigma_v^2,$$
$$\operatorname{Var}\xi = \operatorname{Var}\eta = 4(\beta_u^2\sigma_u^2 + \beta_v^2\sigma_v^2) + o(1/n),$$
$$\operatorname{Cov}(\xi,\eta) = 4(\beta_u^2\sigma_u^2 - \beta_v^2\sigma_v^2) + o(1/n),$$

where

$$\beta_u^2 = \sigma^2(1+\rho), \qquad \beta_v^2 = \sigma^2(1-\rho),$$
$$\sigma_u^2 = \beta_u^2 V(\chi, g)/n, \qquad \sigma_v^2 = \beta_v^2 V(\chi, g)/n.$$

By substituting these into (7.4.8) we arrive at (7.4.7), which completes the proof. □

EXAMPLE 7.4.1. From (7.4.7) we have the expression for the asymptotic variance of the sample correlation coefficient under the bivariate normal distribution with $\chi(x) = x^2 - 1$ and $g(x) = \phi(x)$: $\operatorname{Var} \mathbf{r} = (1-\rho^2)^2/n$.

EXAMPLE 7.4.2. The choice $\chi(x) = \operatorname{sgn}(|x|-1)$ and $g(x) = \phi(x)$ yields the asymptotic variance of the median correlation coefficient obtained in Section 7.3 (Theorem 7.3.1) by other approach.

Formula (7.4.7) for the asymptotic variance has two factors: the first depends only on ρ, the second $V(\chi, g)$ is the asymptotic variance of M-estimators for scale. Thus we can immediately apply the known minimax variance estimators of scale in the gross error model for minimax variance estimation of a correlation coefficient.

7.4.4. Minimax variance estimators

In (Huber, 1981) it was shown that, under rather general conditions of regularity, the M-estimators $\widehat{\beta}_n$ are consistent, asymptotically normal, and possess the minimax property with regard to the asymptotic variance $\operatorname{Var} \widehat{\beta}_n = V(\chi, g)$:

$$V(\chi^*, g) \le V(\chi^*, g^*). \tag{7.4.9}$$

Here g^* is the least informative (favorable) density minimizing the Fisher information $I(g)$ for scale in a certain class $\mathcal{G}$:

$$g^* = \arg\min_{g\in\mathcal{G}} I(g), \qquad I(g) = \int \left[-x\frac{g'(x)}{g(x)} - 1\right]^2 g(x)\,dx, \tag{7.4.10}$$

and the score function $\chi^*(x)$ is given by the ML method.

For the class of ε-contaminated normal distributions

$$\mathcal{G} = \{g\colon g(x) \ge (1-\varepsilon)\phi(x), 0 \le \varepsilon < 1\} \tag{7.4.11}$$

the minimax variance M-estimator of scale is defined by the score function (Huber, 1964; Huber, 1981)

$$\chi(x) = \begin{cases} x_0^2 - 1, & |x| < x_0, \\ x^2 - 1, & x_0 \le |x| \le x_1, \\ x_1^2 - 1, & |x| > x_1, \end{cases} \tag{7.4.12}$$

with $x_0 = x_0(\varepsilon)$ and $x_1 = x_1(\varepsilon)$. The exact relations for these parameters are given in Section 4.3, and their values are tabulated (Huber, 1981). This M-estimator is asymptotically equivalent to the trimmed standard deviation (Huber, 1981, p. 122).

The following result is immediately obtained from the above.

THEOREM 7.4.2. *In the class* (7.4.1) *of* γ*-contaminated bivariate normal distributions*

$$f(x,y) \geq (1-\gamma)\,\mathcal{N}(x,y \mid 0,0,1,1,\rho), \qquad 0 \leq \gamma < 1, \tag{7.4.13}$$

the minimax robust estimator of ρ *is given by the trimmed correlation coefficient* (7.1.22)

$$\mathbf{r}_{tr}(n_1,n_2) = \frac{\sum_{i=n_1+1}^{n-n_2} u_{(i)}^2 - \sum_{i=n_1+1}^{n-n_2} v_{(i)}^2}{\sum_{i=n_1+1}^{n-n_2} u_{(i)}^2 + \sum_{i=n_1+1}^{n-n_2} v_{(i)}^2}, \tag{7.4.14}$$

where the numbers n_1 *and* n_2 *of the trimmed smallest and greatest order statistics* $u_{(i)}$ *and* $v_{(i)}$ *depend on the value of the contamination parameter* $\varepsilon = 1-\sqrt{1-\gamma}$: $n_1 = n_1(\varepsilon)$ *and* $n_2 = n_2(\varepsilon)$. *The exact character of this dependence is given in* (Huber, 1981, 5.6).

PROOF. It suffices to check that densities (7.4.13) belong to class (7.4.1).

From (7.4.11), we obtain

$$\frac{1}{\sigma\sqrt{1+\rho}}\, g\left(\frac{u}{\sigma\sqrt{1+\rho}}\right) \geq (1-\varepsilon)\frac{1}{\sigma\sqrt{1+\rho}}\,\phi\left(\frac{u}{\sigma\sqrt{1+\rho}}\right),$$

$$\frac{1}{\sigma\sqrt{1-\rho}}\, g\left(\frac{v}{\sigma\sqrt{1-\rho}}\right) \geq (1-\varepsilon)\frac{1}{\sigma\sqrt{1-\rho}}\,\phi\left(\frac{v}{\sigma\sqrt{1-\rho}}\right).$$

By multiplying them, we obtain the restriction of the class of γ-contaminated bivariate normal distributions (7.4.13), where $\gamma = 2\varepsilon - \varepsilon^2$, which completes the proof. □

In the limiting case as $\gamma \to 1$, we observe that n_1 and n_2 tend to $[n/2]$, the estimators of scale $\widehat{\beta}_u$ and $\widehat{\beta}_v$ tend to the medians of absolute deviations $\operatorname{med}|u|$ and $\operatorname{med}|v|$, respectively, and hence $\widehat{\rho}$ tends to the median correlation coefficient

$$\mathbf{r}_{\text{med}} = \frac{\operatorname{med}^2|u| - \operatorname{med}^2|v|}{\operatorname{med}^2|u| + \operatorname{med}^2|v|}. \tag{7.4.15}$$

If $\gamma = 0$, then this estimator is asymptotically equivalent to the sample correlation coefficient $\mathbf{r}$.

REMARK 7.4.3. In applications, one should use robust estimators for unknown location and scale, namely the sample median and the median of absolute deviations, or the robust principal variables $(u_i, v_i)_1^n$

$$u = \frac{x - \operatorname{med} x}{\sqrt{2}\,\mathrm{MAD}\, x} + \frac{y - \operatorname{med} y}{\sqrt{2}\,\mathrm{MAD}\, y}, \qquad v = \frac{x - \operatorname{med} x}{\sqrt{2}\,\mathrm{MAD}\, x} - \frac{y - \operatorname{med} y}{\sqrt{2}\,\mathrm{MAD}\, y}.$$

REMARK 7.4.4. The asymptotic confidence intervals for $\widehat{\rho}$ can be constructed by using the Fisher transformation (Kendall and Stuart, 1962)

$$z = \frac{1}{2}\ln\frac{1+\widehat{\rho}}{1-\widehat{\rho}}.$$

In this case, the variance of z does not depend on ρ:

$$\operatorname{Var} z = \frac{2V(\chi,g)}{n-3},$$

and this fact is due to the structure of the multiplier $(1-\rho^2)^2$ in $V(\chi,g)$.

REMARK 7.4.5. The minimax approach can be also applied to the parametric class of exponential-power densities

$$g(x) = \frac{q}{2\Gamma(1/q)}\exp(-|x|^q), \qquad q \ge 1.$$

It follows from the results of Section 4.5 that the Laplace density minimizes the Fisher information for scale in this class, hence the minimax estimators for scale in the principal axes are given by the mean absolute deviations

$$\widehat{\beta}_u = n^{-1}\sum|u_i|, \qquad \widehat{\beta}_v = n^{-1}\sum|v_i|.$$

Therefore the minimax variance estimator of the correlation coefficient in this class is

$$\mathbf{r}_{L_1} = \frac{(\sum|u_i|)^2 - (\sum|v_i|)^2}{(\sum|u_i|)^2 - (\sum|v_i|)^2}.$$

In the literature, there is only one result on the minimax approach to robust estimation of ρ (Huber, 1981, p. 205): the quadrant correlation coefficient is asymptotically minimax with respect to bias over the mixture $F = (1-\varepsilon)G+\varepsilon H$ (G and H being symmetric distributions in $\mathbf{R}^2$). Although its bias is minimax, the quadrant correlation coefficient $\mathbf{r}_Q$ demonstrates moderate robustness in the Monte Carlo experiment. This can be explained by the properties of the chosen class of direct robust counterparts of the sample correlation coefficient ($\mathbf{r}_\psi$-estimators) for which the optimality of $\mathbf{r}_Q$ is established. It is more convenient to detect and eliminate the influence of outliers not in the initial axes x and y but in the principal axes u and v. Fig. 7.4 illustrates this effect: the outliers (marked by stars) in the principal axes should not necessarily be such in the initial axes, in other words, in these systems of coordinates, the extremes should not coincide.

The trimmed correlation coefficient (7.1.22) proved its high robustness in former experimental studies (Gnanadesikan and Kettenring, 1972; Devlin *et*

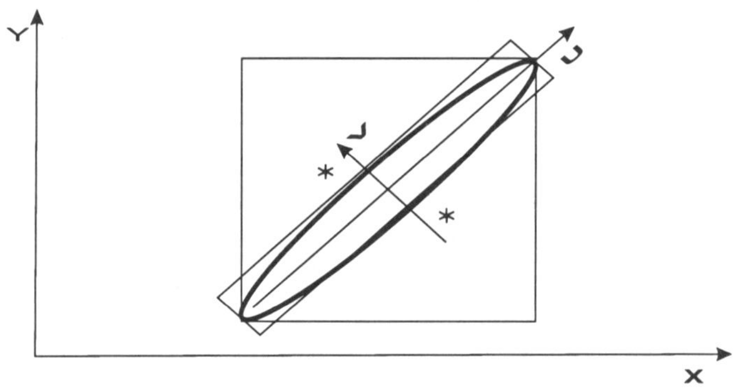

Figure 7.4. On outliers in the initial and principal axes

al., 1975), and its optimality for ε-contaminated models explains those results. There remains a general question for such models: how to choose the value of the contamination parameter ε? The practical recommendation is the following: if we assume that the value of γ does not exceed 0.2, and this choice is made with safety (in fact, in (Hampel, 1973; Huber, 1973) $\varepsilon \approx 0.1$ was suggested for robust location), then the corresponding value of ε for robust scale is approximately 0.1, and therefore the optimal estimator is the one-sided trimmed correlation coefficient (7.4.14) with $n_1 = 0$ and $n_2 \approx [0.1\,n]$.

REMARK 7.4.6. As concerns the quadrant correlation coefficient, it can serve as a moderate robust alternative to the sample correlation coefficient because of its minimaxity with regard to bias (Huber, 1981), its binomial sample distribution (Blomqvist, 1950), and its simple structure.

Summarizing the obtained results on the median correlation coefficient, namely that it possesses both the optimal qualitative (breakdown point $\varepsilon^* = 1/2$) and quantitative (minimax) robustness properties, we may regard the median correlation coefficient as the correlation analog of the sample median and the median of absolute deviations—the well-known robust estimators of location and scale having both quantitative minimax and highest qualitative robustness properties. Further in Section 8.4, we use these highly robust estimators along with their location and scale analogs for constructing a bivariate boxplot.

7.5. Two-stage estimators: rejection of outliers plus classics

Preliminary rejection of outliers from the data and subsequent application of a classical estimator (for example, the sample correlation coefficient) to the rest of the observations represents the next group of estimators. Their variety mainly depends on the rules for rejection of outliers.

7.5.1. Preliminaries on the rejection of outliers

In our setting, the problem of outlier rejection is subordinate to the problem of robust estimation of a correlation coefficient, though, certainly, that problem is of its own importance.

The concrete aims of detection of outlying observations from the bulk of the data may be very different. Here we consider the main two of them.

The first is in exposing the significant observations, which admit a specific interpretation. These observations may be quite 'good' showing new possibilities for unexpected improvements and discoveries of new effects. Darwin noticed that outliers indicate the vector of development.

The second important aim consists of eliminating gross errors away from the data for providing stability and efficiency of statistical inference.

In this study, we keep in view and try to pursue both aims, but the second is common within the robustness context.

First we recall how the problem of detection and/or rejection of outliers is solved in the univariate case.

Rejection of outliers in the univariate case. In this case, statistical methods mainly aim at the exposure of a single outlier in the data when the minimal or maximal order statistic is regarded as a candidate for an outlier (Dixon, 1950; Dixon, 1960; Grubbs, 1950). In classical statistics, this problem is usually set as a problem of testing hypotheses about an underlying distribution, say in the model of scale contamination

$$H_0\colon F(x) = \Phi(x),$$
$$H_1\colon F(x) = (1-\varepsilon)\Phi(x) + \varepsilon\Phi(x/k),$$

where $k > 1$.

The structure of a standard test for rejection of an extremal order statistic is as follows:

- form the difference between a candidate for an outlier and a chosen estimator of location (central tendency), for example, $x_{(n)} - \bar{x}$ or $\bar{x} - x_{(1)}$;

- standardize this difference by a chosen estimator of scale, say, by the standard deviation

$$B = \frac{x_{(n)} - \bar{x}}{s}, \quad \text{or} \quad B = \frac{\bar{x} - x_{(1)}}{s};$$

- compare the test statistic B_n with a certain bound: if

$$B = \frac{x_{(n)} - \bar{x}}{s} < \lambda_\alpha \tag{7.5.1}$$

  then the null hypothesis is accepted, and vice versa.

The threshold value λ_α is obtained from the chosen significance level

$$\mathsf{P}(B < \lambda_\alpha) = 1 - \alpha,$$

where the common choice is $\alpha = 0.01, 0.05, 0.1$.

Much is made for the development of this classical approach to rejection of outliers, various tests were proposed with the related tables of percentiles, but all of them have at least one shortcoming. Indeed, no more than one outlier can be detected with the use of such an approach: after rejection, the sample distribution is changed within the corresponding bounds.

More essential is another shortcoming of the classical approach: the classical estimators of location and scale usually used in their structures are sensitive to outliers, and this considerably reduces the power of tests. However, this situation can be improved by the use of robust statistics.

Here we give our old result on this subject (Shevlyakov, 1976; Guilbo and Shevlyakov, 1977).

For rejection of extremal order statistics, the following robust version of the test statistic B_n (7.5.1) is suggested:

$$B_m = \frac{x_{(n)} - \operatorname{med} x}{x_{(j)} - x_{(k)}} \quad \text{or} \quad B_m = \frac{\operatorname{med} x - x_{(1)}}{x_{(j)} - x_{(k)}}, \tag{7.5.2}$$

where the sample mean is replaced by the sample median, and the standard deviation, by the sample interquartile width of the form $x_{(j)} - x_{(k)}$, where $k = [n/4] + 1$ and $j = n - k + 1$.

Table 7.9 displays the 90% and 95% percentiles ($\alpha = 0.1$ and 0.05) for the test statistic B_m (7.5.2) under the normal null hypothesis for $n = 5, 7, 11, 15, 19$. These points are obtained by Monte Carlo modeling.

For the problem of rejection of a single outlier, the power of test is convenient to define as the ratio r of the number of contaminated samples with rejected outliers to the total number of contaminated samples. The following alternatives are considered:

Table 7.9. The percentiles of the test statistic B_m under the normal distribution

n	5	7	11	15	19
$\alpha = 0.05$	2.00	2.10	2.24	2.30	2.34
$\alpha = 0.1$	1.56	1.67	1.91	1.97	2.04

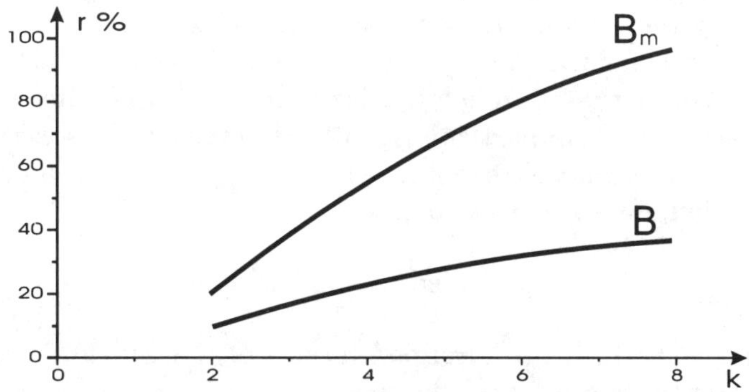

Figure 7.5. The power of the B and B_m tests under scale contamination

- the alternative of shift contamination

$$F(x) = (1 - \varepsilon)\mathcal{N}(0, 1) + \varepsilon\,\mathcal{N}(\mu, 1);$$

- and the alternative of scale contamination

$$F(x) = (1 - \varepsilon)\mathcal{N}(0, 1) + \varepsilon\,\mathcal{N}(0, k).$$

The power for the test statistics B_m and B has been examined by Monte Carlo techniques in samples $n = 5, 7, 11, 15$ for the both types of contamination. Fig. 7.5 shows the power of these tests under the scale contamination with $\varepsilon = 0.2$ and $\alpha = 0.1$. Similar dependencies also hold for the shift contamination.

The superiority of the robust test is obvious, and it is the larger contamination parameters ε, k, μ, and sample size n are, the higher this superiority is.

Certainly, it is possible to improve this statistic by replacing the interquartile width, say, by the median absolute deviation or another statistic, but other approaches have revealed their advantages for solution of these problems.

First of all, each robust procedure of estimation inherently possesses its own rule for rejection of outliers (Hampel *et al.*, 1986; Huber, 1981), and it may seem that then there is no need for any independent procedure for rejection, at least if to aim at estimation, and therefore no need for two-stage procedures. However, a rejection rule may be quite informal, for example, based on a priori knowledge about the nature of outliers, and, in this case, its use can improve the efficiency of estimation. Later, we will give some examples where two-stage procedures provide a reasonable level of efficiency as compared with optimal direct robust procedures.

Second, the above classical procedures of rejection of outliers have been moved aside by new technologies of data analysis created mainly by Tukey and other statisticians who used robust statistics (Mosteller and Tukey, 1977; Tukey, 1977). One of those technologies is the *boxplot* (its detailed construction is given in Section 8.4). In particular, it allows to regard the observation x_i as a candidate for an outlier if its distance from the sample median exceeds five times the median absolute deviation, i.e.,

$$|x_i - \operatorname{med} x| > 5 \operatorname{MAD} x.$$

This condition is nothing but a refined version of the '3σ-rule,' and this suggestion has proved its practical efficiency.

This approach based on data analysis technologies (much more refined than in the univariate case) can be also used in multivariate statistics.

Rejection of outliers in the multivariate case. Rejection of multiple outliers is much more complicated than in the univariate case for a number of related reasons:

- first, multivariate outliers can distort not only location and scale, but also the orientation and shape of the point-cloud in the space;
- second, it is difficult to figure out which type the outlier belongs to;
- third, these types are numerous (Rocke and Woodruff, 1996).

Thus, it might prove to be impossible to develop just one procedure which would be a reliable guard against outliers. There must be a variety of procedures for different types of outliers.

There are various rejection methods with the multivariate data. They are based on using discriminant, component, factor analysis, canonical correlation analysis, projection pursuit, etc. (Atkinson, 1985; Atkinson, 1994; Atkinson and Mulira, 1993; Atkinson and Riani, 2000; Barnett and Lewis, 1978; Davies and Gather, 1993; Hadi, 1992; Hawkins, 1980; Hawkins, 1993a; Rocke and Woodruff, 1996; Rousseeuw and Leroy, 1987; Rousseeuw and van Zomeren, 1990).

Now we describe the main approaches in this area.

In the multivariate space, the classical procedure of rejection is based on the use of the *Mahalanobis distances* d_i^2 between the points $\mathbf{x}_i$ in $\mathbf{R}^m$ and the sample mean $\overline{\mathbf{x}} = n^{-1}\sum \mathbf{x}_i$ for the data $\mathbf{x}_1, \dots, \mathbf{x}_n$:

$$d_i^2 = (\mathbf{x}_i - \overline{\mathbf{x}})^T \mathbf{S}^{-1} (\mathbf{x}_i - \overline{\mathbf{x}}), \qquad i = 1, 2, \dots, n, \tag{7.5.3}$$

where $\mathbf{x}_i = (x_{i1}, \dots, x_{im})^T$ and $\mathbf{S}$ is the sample covariance matrix

$$\mathbf{S} = \frac{1}{n}\sum_{i=1}^{n}(\mathbf{x}_i - \overline{\mathbf{x}})(\mathbf{x}_i - \overline{\mathbf{x}})^T.$$

The evaluated distances are ranked

$$d_{(1)}^2 \le d_{(2)}^2 \le \dots \le d_{(n)}^2,$$

and, obviously, those observations with greater Mahalanobis distances are the candidates for outliers. Like in the univariate case, one can construct the rejection test

$$d_{(n)}^2 < \lambda_\alpha,$$

where the bound λ_α is determined from the condition

$$\mathsf{P}(d_{(n)}^2 < \lambda_\alpha) = 1 - \alpha.$$

Recall that the distribution of the Mahalanobis statistic is χ^2.

In this case, the use of the classical sample mean and covariance matrix destroys the rejection procedure, if there are gross errors in the data because of the great sensitivity of these classical estimators to outliers.

This effect masking outliers is illustrated in Fig. 7.6 where the bulk of the data has a clearly expressed elliptical shape, but its classical estimator is strongly distorted and close to the circular due to the influence of four outliers marked by crosses and stars. This circular shape evidently masks two outliers (crosses) of four.

Thus the problem of rejecting outliers in the multivariate space obviously requires robust estimation of multivariate location and shape. The latter problem is one of the most difficult problems in robust statistics (Campbell, 1980; Campbell, 1982; Davies, 1987; Devlin *et al.*, 1981; Donoho, 1982; Hampel *et al.*, 1986; Huber, 1981; Lopuhaä, 1989; Maronna, 1976; Meshalkin, 1971; Rocke and Woodruff, 1993; Rousseeuw, 1985; Rousseeuw and Leroy, 1987; Shurygin, 1995; Shurygin, 2000; Stahel, 1981; Tyler, 1983; Tyler, 1991).

The multivariate location and shape problem is more difficult than, say, the problems of one-dimensional location and regression with error-free carriers. It is established that most known methods fail if the fraction of outliers is larger than $1/(m+1)$, where m is the dimension of the data (Donoho, 1982; Maronna, 1976; Stahel, 1981). This means that in high dimension, a very small fraction of outliers can completely destroy an estimation procedure.

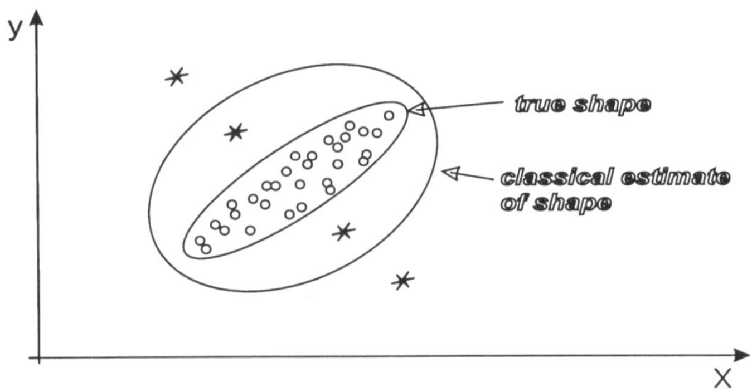

Figure 7.6. The masking effect of a classical estimator under contamination

It is very desirable to obtain location and shape estimators that are affine equivariant. A shape estimator $\widehat{\mathbf{C}}_n$ is said to be affine equivariant if and only if for any vector $\mathbf{b} \in \mathbf{R}^m$ and any nonsingular $m \times m$ matrix $\mathbf{A}$

$$\widehat{\mathbf{C}}_n(\mathbf{Ax} + \mathbf{b}) = \mathbf{A}\widehat{\mathbf{C}}_n(\mathbf{x})\mathbf{A}^T.$$

These requirements mean that measurement scales will change the estimators appropriately under translation and rotation.

REMARK 7.5.1. The Mahalanobis distance and its robust modifications being affine equivariant are widely used for constructing affine equivariant methods and algorithms of estimation of multivariate location and shape.

Now we just list the main approaches for finding robust estimators of multivariate location and shape, and thus for detection of outliers.

Rousseeuw proposes the minimum volume ellipsoid and minimum covariance determinant combinatorial estimators (Hampel *et al.*, 1986; Rousseeuw and Leroy, 1987), which are realized with the use of random search (Rousseeuw and Leroy, 1987), steepest descent (Hawkins, 1993a; Hawkins, 1993b), and heuristic optimization technique (Rocke and Woodruff, 1994).

Maximum likelihood and M-estimators (Campbell, 1980; Campbell, 1982; Huber, 1981; Lopuhaä, 1992; Maronna, 1976; Meshalkin, 1971; Shurygin, 1994a; Shurygin, 1995; Tyler, 1983; Tyler, 1991), S-estimators (Davies, 1987; Hampel *et al.*, 1986; Lopuhaä, 1989; Rousseeuw and Leroy, 1987) are iteratively computed from a good starting point (Rocke and Woodruff, 1993).

In (Gnanadesikan and Kettenring, 1972), the multivariate trimming iterative algorithm was suggested for elliptical rejection of outliers based on robustified versions of the Mahalanobis distance.

In (Maronna, 1976), iteratively computed weights were used depending on the Mahalanobis distances of observations from the robust location estimator for smooth eliminating of outliers.

Working independently, in (Atkinson, 1994; Hadi, 1992) they proposed the forward search (`FORWARD`) algorithm, which starts with a small randomly selected subset of observations intended to be outlier-free.

In (Rocke and Woodruff, 1996), the nature of multivariate outliers was analyzed; they found that outliers with the same shape as the main data are the hardest to find, and that the more compact the outliers, the harder they are to find. The proposed hybrid algorithm (Rocke and Woodruff, 1996) that uses search techniques from both (Hawkins, 1993a) and (Atkinson and Mulira, 1993), as well as from their own previous research (Rocke and Woodruff, 1993; Rocke and Woodruff, 1994), proves to be one of the best methods for multivariate outlier detection.

7.5.2. Algorithms for rejection of outliers in bivariate case

In the context of our study, we regard the problem of rejection of outliers to be subordinate to the problem of robust estimation of the correlation coefficient, so we are mainly interested in sufficiently simple and efficient procedures of rejection in the bivariate case, where, in general, there always exists a possibility of visual verification of the results. Moreover, this possibility might be used in the case with $m > 2$, since one can never rely entirely on formal methods in the multivariate space. We also suppose the elliptical shapes of the bulk of the data and outliers of the gross error nature.

All proposals can be separated into two groups: the algorithms for rejection when the expected fraction of outliers is assumed known, and when it is unknown. Obviously, the additional information about outliers makes rejection easier.

Rejection in the principal axes with the rectangular rule (RCT). Given a sample $(x_1, y_1), \dots, (x_n, y_n)$, use the standardized variables

$$\widetilde{x}_i = \frac{x_i - \operatorname{med} x}{\operatorname{MAD} x}, \quad \widetilde{y}_i = \frac{y_i - \operatorname{med} y}{\operatorname{MAD} y}, \quad i = 1, \dots, n. \tag{7.5.4}$$

Then transform (7.5.4) to the principal coordinates u and v

$$u_i = \frac{\widetilde{x}}{\sqrt{2}} + \frac{\widetilde{y}}{\sqrt{2}}, \quad v_i = \frac{\widetilde{x}}{\sqrt{2}} + \frac{\widetilde{y}}{\sqrt{2}}, \quad i = 1, \dots, n, \tag{7.5.5}$$

and trim all points (u_i, v_i) for which

$$|u_i| > 5 \operatorname{MAD} u \quad \text{or} \quad |v_i| > 5 \operatorname{MAD} v. \tag{7.5.6}$$

Rejection in the principal axes with the ellipse rule (ELL). Assume that the main data is normal or approximately normal, and the fraction of outliers in the data is known. Then the contours of the bivariate distribution density have the elliptical form, and we can use this for identifying outliers.

The steps of the ELL algorithm are as follows:

(i) trim the points (u_i, v_i) with the ellipse

$$\left(\frac{u_i - \operatorname{med} u}{k \operatorname{MAD} u}\right)^2 + \left(\frac{v_i - \operatorname{med} u}{k \operatorname{MAD} v}\right)^2 = 1,$$

where k is determined iteratively so that the given fraction of the data should lie inside the ellipse ('good' data);

(ii) begin with the initial estimators of location $m^* = (\overline{x}, \overline{y})$ and shape S^* for the 'good' data; calculate the Mahalanobis distances d_i^2 for all of the data $i = 1, \dots, n$;

(iii) find the sample median d^2_{med};

(iv) trim all the points with $d_i^2 > d^2_{\text{med}}$, i.e., $[n/2] - 1$ points;

(v) evaluate the sample covariance matrix S_{med} for the remained data;

(vi) re-calculate all the Mahalanobis distances with S_{med};

(vii) finally, trim the given fraction of points with the largest Mahalanobis distances.

Adaptive rejection with the maximal increment of the Mahalanobis distances (AD). In this case, we assume the fraction of outliers unknown. This algorithm may be referred as to the FORWARD algorithm. First, the outlier-free half of the data is determined, and then it is supplemented with the points from the other half until the maximal increment of the Mahalanobis distances is attained.

The steps of the AD algorithm are as follows:

(i) trim the points (u_i, v_i) lying outside of the ellipse

$$\left(\frac{u_i - \operatorname{med} u}{5 \operatorname{MAD} u}\right)^2 + \left(\frac{v_i - \operatorname{med} u}{5 \operatorname{MAD} v}\right)^2 = 1;$$

(ii) repeat steps (ii)–(vi) of the ELL algorithm;

(iii) find l^* such that

$$l^* = \arg\max_{l \le n}\{d^2_{(l)} : d^2_{(l)} < \operatorname{med} d^2 + 5 \operatorname{MAD} d^2\};$$

(iv) calculate the successive increments of the Mahalanobis distances starting from $d^2_{(k^*)}$:

$$d^2_{(k^*+1)} - d^2_{(k^*)}, \quad d^2_{(k^*+2)} - d^2_{(k^*+1)}, \quad \dots, \quad d^2_{(l^*)} - d^2_{(l^*-1)},$$

where $k^* = [n/2] + 1$;

(v) find the maximum of these increments, and let it be attained at $d^2_{(s^*+1)} - d^2_{(s^*)}$, where $k^* \le s^* \le l^*$;

(vi) trim $(n - s^*)$ points with the largest Mahalanobis distances.

The idea of this adaptive procedure seems natural: the rejection bound is constructed with the robustified Mahalanobis distances, and it is identified with the maximal jump of these distances among the plausible candidates for outliers.

7.5.3. Modeling two-stage robust estimators

In this subsection we represent the results of Monte Carlo studies of the above-introduced two-stage robust estimators in normal and contaminated samples.

The principle of a two-stage estimator of the correlation coefficient is the following: first, using some rule of rejection, trim the outliers, and, second, apply the classical sample correlation coefficient to the rest of the data.

Now we list all two-stage estimators, which were examined in our study:

$\mathbf{r}_{\mathrm{RCT}}$ based on rejection with the rectangle rule;

$\mathbf{r}_{\mathrm{ELL}}$ based on rejection with the ellipse rule;

$\mathbf{r}_{\mathrm{AD}}$ based on adaptive rejection;

$\mathbf{r}_{\mathrm{MVT}}$ based on rejection with the method of the ellipsoidal multivariate trimming (Gnanadesikan and Kettenring, 1972; Devlin *et al.*, 1975).

These estimators have been examined under the ε-contaminated bivariate normal distribution in samples $n = 20, 30, 60$. We give some results for the bivariate normal density $\mathcal{N}(0, 0, 1, 1, \rho)$ and ε-contaminated density with small, medium, and large values of the correlation coefficient.

From Tables 7.10 and 7.11 it follows that the best two-stage estimators $\mathbf{r}_{\mathrm{MVT}}$ and $\mathbf{r}_{\mathrm{AD}}$ are close in their performance to the best direct robust estimators $\mathbf{r}_{\mathrm{MAD}}$ and $\mathbf{r}_{\mathrm{med}}$ (see Tables 7.1–7.3). Similar results have been obtained in samples $n = 30$ and $n = 60$.

Table 7.10. $n = 20$: expectations and variances of estimators for normal distribution

	$\mathbf{r}_{\mathrm{RCT}}$	$\mathbf{r}_{\mathrm{ELL}}$	$\mathbf{r}_{\mathrm{MVT}}$	$\mathbf{r}_{\mathrm{AD}}$
$\rho = 0.0$	0.00	0.00	0.00	0.00
	0.08	0.05	0.11	0.09
$\rho = 0.5$	0.23	0.38	0.46	0.44
	0.07	0.05	0.07	0.07
$\rho = 0.9$	0.52	0.85	0.87	0.86
	0.08	0.01	0.01	0.01

Table 7.11. $n = 20$: expectations and variances of estimators under ε-contaminated normal distributions with $\varepsilon = 0.1$, $\rho' = -0.9$, $k = 10$

	$\mathbf{r}_{\mathrm{RCT}}$	$\mathbf{r}_{\mathrm{ELL}}$	$\mathbf{r}_{\mathrm{MVT}}$	$\mathbf{r}_{\mathrm{AD}}$
$\rho = 0.0$	−0.22	−0.01	−0.08	−0.02
	0.21	0.06	0.10	0.09
$\rho = 0.5$	0.11	0.38	0.41	0.43
	0.42	0.06	0.08	0.07
$\rho = 0.9$	0.70	0.83	0.84	0.87
	0.26	0.02	0.04	0.01

7.5.4. Some notes on robust estimation of correlation matrices

The estimation of correlation and covariance matrices, and also of their eigenvalues and eigenvectors, is one of the most important phases in the solution of various problems in multivariate statistical analysis. However, classical statistical estimators of correlation matrices and their characteristics, for instance, the sample correlation matrix, are very sensitive and unstable in presence of outliers in the data (Devlin *et al.*, 1981; Huber, 1981; Rousseeuw and Leroy, 1987).

The direct way to construct a robust estimator of the correlation matrix is to use good robust estimators of the matrix elements, especially those described in Section 7.4, for example, the median correlation coefficient $\mathbf{r}_{\mathrm{med}}$. However, this approach does not guarantee that a positive definite matrix is obtained. Special tricks ensuring this are required (Devlin *et al.*, 1981; Huber, 1981).

Another, affine equivariant, approach is based on estimation of the matrix as a whole, where the property of positive definiteness is satisfied primarily (Huber, 1981; Rousseeuw and Leroy, 1987). Within this approach, it is possible to select the group of perspective methods providing preliminary rejection of outliers with subsequent use of the classical sample correlation matrix.

In this subsection, we directly extend the bivariate rejection procedures RCT and ELL to the multivariate case successively applying them in each two-dimensional cut of the multivariate space. The adaptive procedure of rejection does not differ much from its bivariate analog; only at the first step, crude rejection is performed using the ELL rule with $k = 5$ (see Subsection 7.5.2), and as before, the adaptive phase of the procedure is determined by the maximal jump of the robustified Mahalanobis distances.

Let $\mathbf{x}_1, \dots, \mathbf{x}_n$ be a sample of the m-dimensional data

$$\mathbf{x}_i = (x_{i1}, \dots, x_{im})^T, \qquad i = 1, \dots, n.$$

The above-mentioned bivariate rejection procedures are performed in each two-dimensional cut:

$$(x_{1k}, x_{1l}), (x_{2k}, x_{2l}), \dots, (x_{nk}, x_{nl})), \quad k, l = 1, 2, \dots, m; \quad k < l.$$

The following estimators of a correlation matrix are studied:

$\mathbf{R}$ the sample correlation matrix;

$\mathbf{R}_{\mathrm{MVT}}$ the estimator based on the ellipsoidal multivariate trimming (Devlin *et al.*, 1981);

$\mathbf{R}_{\mathrm{AD}}$ the two-stage estimator with preliminary rejection of outliers by the adaptive algorithm and subsequent applying the sample correlation matrix to the rest of the data;

$\mathbf{R}_{\mathrm{med}}$ the estimator based on the element-wise median correlation coefficients.

These estimators are examined in samples $n = 30$ and $n = 50$ (the number of trials is 200) under the contamination model with various forms of mixture of normal distributions, symmetric and asymmetric, for instance, under the spherically symmetric contamination

$$\mathrm{SCN}(\mathbf{0}, \mathbf{P}) = (1 - \varepsilon)\mathrm{NOR}(\mathbf{0}, \mathbf{P}) + \varepsilon \mathrm{NOR}(\mathbf{0}, k^2\mathbf{E}), \tag{7.5.7}$$

where $0 \le \varepsilon < 0.5$, $k > 1$, $\mathbf{E}$ is a unity matrix, and the estimated 6×6 correlation matrix $\mathbf{P}$ is of the form

$$\mathbf{P} = \begin{pmatrix} \mathbf{P}_1 & \mathbf{0} \\ \mathbf{0} & \mathbf{P}_2 \end{pmatrix},$$

where

$$\mathbf{P}_1 = \begin{pmatrix} 1 & 0.95 & 0.3 \\ & 1 & 0.1 \\ & & 1 \end{pmatrix}, \qquad \mathbf{P}_2 = \begin{pmatrix} 1 & -0.499 & -0.499 \\ & 1 & -0.499 \\ & & 1 \end{pmatrix}.$$

Table 7.12. $n = 50$; expectations of eigenvalues: upper with normal; middle with spherical contamination ($\varepsilon = 0.1$, $k = 3$); lower with asymmetric contamination ($\varepsilon = 0.1$)

λ	2.029	1.499	1.499	0.943	0.028	0.002
	2.23	1.62	1.30	0.82	0.03	0.002
$\mathbf{R}$	1.99	1.48	1.13	0.80	0.43	0.17
	2.22	1.61	1.30	0.82	0.03	0.02
	2.29	1.63	1.26	0.79	0.03	0.002
$\mathbf{R}_{\text{MVT}}$	2.29	1.64	1.27	0.79	0.05	0.04
	2.28	1.63	1.25	0.80	0.03	0.01
	2.26	1.63	1.27	0.80	0.03	0.001
$\mathbf{R}_{\text{AD}}$	2.32	1.62	1.25	0.78	0.03	0.001
	2.26	1.62	1.27	0.80	0.03	0.01
	2.33	1.64	1.22	0.77	0.05	−0.01
$\mathbf{R}_{\text{med}}$	1.96	1.66	1.39	0.84	0.10	0.03
	2.32	1.64	1.21	0.77	0.12	−0.05

Its eigenvalues

$$\lambda_1 = 2.029, \quad \lambda_2 = \lambda_3 = 1.499, \quad \lambda_4 = 0.943, \quad \lambda_5 = 0.028, \quad \lambda_6 = 0.002$$

are the main targets for estimation. The particular structure of the estimated matrix $\mathbf{P}$ has some attractive features: a moderate dimension, a rather wide range of correlation coefficients and eigenvalues, for instance, it is nearly degenerate, and, finally, it is still simple (Devlin *et al.*, 1981).

Also, the asymmetric contamination model is used:

$$\text{ACN}(\mathbf{0}, \mathbf{P}) = (1 - \varepsilon)\text{NOR}(\mathbf{0}, \mathbf{P}) + \varepsilon\text{NOR}(\mathbf{m}, \mathbf{P}), \tag{7.5.8}$$

where $\mathbf{m} = 0.536\,\mathbf{a}_6$ and $\mathbf{a}_6 = (1/\sqrt{3})(0\ 0\ 0\ 1\ 1\ 1)^T$ is the eigenvector corresponding to the minimum (nearly zero) eigenvalue. The factor 0.536 is about 12 standard deviations along the direction $\mathbf{a}_6$, so the contamination is confined to the last principal component of $\mathbf{P}$. Such contamination is designed to mask the near singularity in $\mathbf{P}$ (Devlin *et al.*, 1981).

In these contamination models, the results of modeling for the above collection of estimators are given in Table 7.12.

The obtained results partly repeat those of the thorough study (Devlin *et al.*, 1981) of the latter and related problems (robust estimation of the elements of the correlation matrix, its eigenvalues and eigenvectors) with a rich collection of methods under variety of models. In particular, the MVT estimator proved to be one of the best among the examined estimators. In our study, it is found out that the proposed two-stage estimator $\mathbf{R}_{\mathrm{AD}}$ and the $\mathbf{R}_{\mathrm{MVT}}$ are rather close in their quality.

We confirm that the sample correlation matrix $\mathbf{R}$ is a very poor estimator under contamination. The spherical contamination of the data causes the 'regularization' effect on the estimated matrix $\mathbf{P}$ and, certainly, distorts its nearly degenerate true structure. The use of $\mathbf{R}$ cannot reveal this effect. The direct element-wise estimator $\mathbf{R}_{\mathrm{med}}$ is hardly better since it produces negative eigenvalues with estimated positive definite matrices. Thus, both of these estimators are unacceptable.

The MVT and two-stage algorithm $\mathbf{R}_{\mathrm{AD}}$ with preliminary rejection of outliers are the best among the considered but the biases of the estimators of eigenvalues are too large.

However, we observe that, at present, there are no reliable good robust estimators of the correlation matrix yet, and the problem of their design is still open.

8

Computation and data analysis technologies

In this chapter we describe computational algorithms for robust procedures of estimation of the data characteristics of means (central tendency), spread, association, extreme values, and distributions, which have been derived in preceding sections. The minimax estimators for location (univariate and multivariate) and regression are, naturally, included in the family of adaptive (in the Hogg sense) estimators which thereby become well-grounded.

Moreover, we propose some new data analysis algorithms independent of the previous contents but important for applications, in particular, for estimating data distribution and quantile functions.

Much attention is paid to the finite sample behavior of estimators, which is examined by Monte Carlo.

8.1. Introductory remarks on computation

8.1.1. Computation of M-estimators by the Newton method

Here we briefly describe the main computational methods for the M-estimators defined by the solution of the minimization problem

$$\widehat{\theta}_n = \arg\min_{\theta} \sum_{i=1}^{n} \rho(x_i - \theta), \tag{8.1.1}$$

or by the solution of the gradient equation

$$\sum_{i=1}^{n} \psi(x_i - \widehat{\theta}_n) = 0. \tag{8.1.2}$$

In the first case, direct methods of optimization may be applied, for example, the steepest descent procedure (Huber, 1964), and in the second case, the Newton iteration scheme with the sample median as the initial estimator (Huber, 1964; Dutter, 1977; Arthanari and Dodge, 1981; Huber, 1981).

In order to provide scale equivariancy for the Huber M-estimators, one should solve the equation

$$\sum_{i=1}^{n} \psi\left(\frac{x_i - \widehat{\theta}_n}{S_n}\right) = 0, \tag{8.1.3}$$

where S_n is some robust estimator of scale, for example, the median absolute deviation.

Newton method. It is based on the Lagrange expansion of the left-hand side of (8.1.2) in a neighborhood of the initial estimator $\widehat{\theta}_n^{(0)}$

$$\sum_{i=1}^{n} \psi\left(\frac{x_i - \widehat{\theta}_n}{S_n}\right) = \sum_{i=1}^{n} \psi\left(\frac{x_i - \widehat{\theta}_n^{(0)}}{S_n}\right) - (\widehat{\theta}_n - \widehat{\theta}_n^{(0)}) \times \frac{1}{S_n}\sum_{i=1}^{n} \psi'\left(\frac{x_i - \widehat{\theta}_n^{(0)} - \xi(\widehat{\theta}_n - \widehat{\theta}_n^{(0)})}{S_n}\right), \qquad 0 < \xi < 1. \tag{8.1.4}$$

Setting $\xi = 0$ in (8.1.4), we obtain the iterative procedure

$$\widehat{\theta}_n^{(k+1)} = \widehat{\theta}_n^{(k)} + S_n \frac{\sum_{i=1}^{n} \psi\left(\frac{x_i - \widehat{\theta}_n^{(k)}}{S_n}\right)}{\sum_{i=1}^{n} \psi'\left(\frac{x_i - \widehat{\theta}_n^{(k)}}{S_n}\right)}, \tag{8.1.5}$$

where $\widehat{\theta}_n^{(0)} = \operatorname{med} x$ and $S_n = \operatorname{med} |x - \operatorname{med} x|$. Here σ is supposed to be a nuisance parameter.

Often the one-step version of the Newton method is preferable as a simple and rather efficient variant. The denominator of (8.1.5) can be replaced by a constant value: for $0 \le \psi' \le 1$, any constant denominator greater than 1/2 provides convergence of iteration process (Huber, 1981). The following algorithm is an example of such a procedure:

$$\widehat{\theta}_n^{(k+1)} = \widehat{\theta}_n^{(k)} + \frac{S_n}{n} \sum_{i=1}^{n} \psi\left(\frac{x_i - \widehat{\theta}_n^{(k)}}{S_n}\right).$$

Given a piece-wise linear score function ψ, iterations (8.1.5) tend to the true solution of equation (8.1.3) in a finite number of steps.

In the case where the scale parameter σ is of independent interest, the simultaneous equations for M-estimators of location and scale can be used:

$$\sum_{i=1}^{n} \psi\left(\frac{x_i - \widehat{\theta}_n}{S_n}\right) = 0,$$
$$\sum_{i=1}^{n} \chi\left(\frac{x_i - \widehat{\theta}_n}{S_n}\right) = 0. \tag{8.1.6}$$

Numerical solution of (8.1.6) is associated with greater difficulties than that of (8.1.3) with preliminary estimation of scale. In general, these estimators do not possess minimax properties.

In this research, difficulties of simultaneous estimation of location and scale are overcome with the use of scale equivariant L_p-estimators, and in what follows we will describe numerical methods for their evaluation.

The above methods for computation of univariate estimators of location and scale can be naturally extended to the problems of calculation of regression parameters and multivariate location (Huber, 1981; Hampel *et al.*, 1986; Rousseeuw and Leroy, 1987).

Taking into account the character of minimax solutions obtained earlier, we now consider the method of re-weighted least squares (RWLS) (Holland and Welsch, 1977; Arthanari and Dodge, 1981; Mudrov and Kushko, 1983; Green, 1984), which allows to find the solution of an optimization problem using well elaborated methods of least squares.

8.1.2. Method of re-weighted least squares

In (Weiszfeld, 1937), the re-weighted least squares were first suggested for minimization of the sum of distances (for details, see (Green, 1984)).

We represent this method as applied to computing L_p-estimators of the location parameter

$$\widehat{\theta}_n = \arg\min_{\theta} \sum_{i=1}^{n} |x_i - \theta|^p, \qquad p \ge 1. \tag{8.1.7}$$

The goal function is transformed as

$$J_p(\theta) = \sum_{i=1}^{n} \frac{(x_i - \theta)^2}{|x_i - \theta|^{2-p}} = \sum_{i=1}^{n} w_i (x_i - \theta)^2, \tag{8.1.8}$$

where

$$w_i = w_i(\theta) = \frac{1}{|x_i - \theta|^{2-p}}. \tag{8.1.9}$$

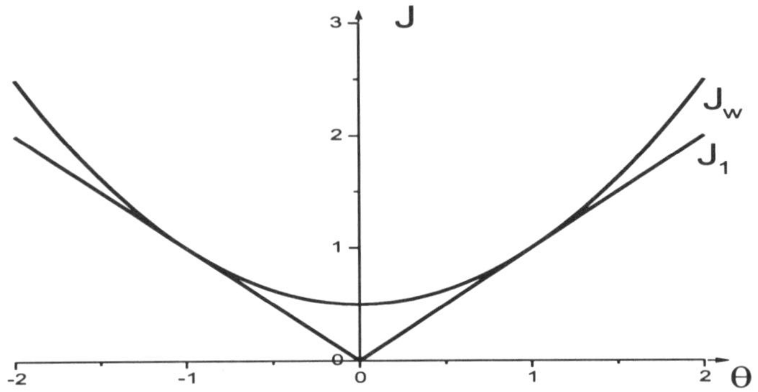

Figure 8.1. Approximation to $J_p(\theta)$ by $J_w(\theta)$

Formula (8.1.8) is the basis for the RWLS iteration scheme

$$\widehat{\theta}_n^{(k+1)} = \arg\min_\theta \sum_{i=1}^n w_i(\widehat{\theta}_n^{(k)})(x_i - \theta)^2, \tag{8.1.10}$$

where the LS estimator is chosen as the initial point: $\widehat{\theta}_n^{(0)} = \widehat{\theta}_{LS}$.

An explicit expression for the estimator of the location parameter is of the form

$$\widehat{\theta}_n^{(k+1)} = \frac{\sum w_i^{(k)} x_i}{\sum w_i^{(k)}}.$$

The sequence $\widehat{\theta}_n^{(0)}, \widehat{\theta}_n^{(1)}, \dots, \widehat{\theta}_n^{(k)}$, converges to the solution of problem (8.1.3) if $1 \le p \le 2$. This assertion is illustrated by Fig. 8.1: the graph of $J_w(\theta) = \sum w_i(x_i - \theta)^2$ is inscribed in the graph of $J_p(\theta)$ for $p \le 2$, and hence any θ reducing the value of $J_w(\theta)$ thus reduces the value of $J_p(\theta)$.

REMARK 8.1.1. The RWLS procedure can be directly rewritten for estimation of multivariate location and regression.

While realizing this method, it may occur that the denominator of (8.1.9) becomes zero for some i. To overcome this difficulty, we introduce the following goal function, instead of $J_p(\theta)$:

$$\widetilde{J}(\theta) = \sum_{i=1}^n \rho(x_i - \theta), \tag{8.1.11}$$

where

$$\rho(u) = \begin{cases} |u|^p, & |u| \ge \alpha, \\ \frac{\alpha}{2} + \frac{u^2}{2\alpha}, & |u| < \alpha. \end{cases}$$

For sufficiently small α, the minimum of $\widetilde{J}(\theta)$ can be made arbitrarily close to the minimum of $J_p(\theta)$.

Summarizing the above, we can say that the RWLS method makes evaluation of the LAV, LS, and intermediate between them L_p-estimators $(1 < p < 2)$ particular cases of a general computational scheme.

8.1.3. Computation of L_1-estimators

Besides the RWLS method, L_1-estimators can be evaluated by the methods of linear programming (Barrodale and Roberts, 1978; Armstrong *et al.*, 1979; Abdelmalek, 1980; Arthanari and Dodge, 1981; Gentle *et al.*, 1988; Fedorov, 1994).

Now we consider one of the most common schemes of reducing the L_1-norm minimization problem to a linear programming one (Dodge, 1987; Gentle, 1977; Gentle *et al.*, 1988).

The minimization problem in the L_1-norm

$$\text{minimize} \quad \sum_{i=1}^{n} \left| x_i - \sum_{j=1}^{m} \theta_j \phi_{ij} \right| \tag{8.1.12}$$

is equivalent to the problem of linear programming

$$\text{minimize} \quad \sum_{i=1}^{n} (u_i + v_i) \tag{8.1.13}$$

under the condition

$$\begin{aligned} & x_i - \sum_{j=1}^{m} \theta_j \phi_{ij} = u_i - v_i, && i = 1, 2, \dots, n; \\ & \theta_j = \theta_j^{(1)} - \theta_j^{(2)}, && j = 1, 2, \dots, m; \\ & u_i \ge 0, \quad v_i \ge 0, && i = 1, 2, \dots, n; \\ & \theta_j^{(1)} \ge 0, \quad \theta_j^{(2)} \ge 0, && j = 1, 2, \dots, m. \end{aligned} \tag{8.1.14}$$

As in any particular application of linear programming technique, it is always possible to simplify the general simplex-method procedure using peculiarities of problem (8.1.13) and (8.1.14) (see the above references).

8.2. Adaptive robust procedures

8.2.1. Preliminaries

Now we consider adaptive estimators which need simultaneous estimation of the data distribution function $F(x)$ or its characteristics (see also the surveys (Hogg, 1974; Ershov, 1979)).

Adaptive estimation of location and scale. In (Stone, 1975), adaptive estimators were suggested for the location parameter based on the method of maximum likelihood. Given a sample $x_1, x_2, \dots, x_n$, a nonparametric estimator $\widehat{f}(x)$ of the distribution density $f(x)$ is constructed and then this estimator is used in the ML equation, which is solved by the Newton method. The asymptotic normality and efficiency of this procedure is proved for a wide class of distributions $F(x)$. The Monte Carlo results of its comparison with the sample mean, sample median, and 0.25-trimmed mean are given for sample sizes $n = 40$ and the number of trials $M = 3000$. The weak point of the proposed procedure is its computational complexity, which is not compensated by its greater efficiency as compared with more simple estimators, for example, with the 0.25-trimmed mean.

Adaptive asymptotically efficient L- and R-estimators of location and scale are studied in (Bhattacharya, 1967; Beran, 1974; Sacks, 1975; van Eeden, 1970; Weiss and Wolfowitz, 1970; Wolfowitz, 1974).

Observe that estimation of a distribution function or its functions is a complicated problem, so the related algorithms are cumbersome in computation. Besides, the obtained estimators, as a rule, converge slowly. Thus it is natural to construct more crude but more simple and fast estimation procedures.

In (Jaeckel, 1971b), an adaptive procedure was introduced to find the parameter α in α-trimmed means and L-estimators with the best weight function over a finite collection of distributions.

In (Hogg, 1974), the following approach was applied to designing simple adaptive robust estimators. Let a distribution $F(x)$ belong to some finite collection of distributions $F_1, \dots, F_k$, and let $\widehat{\theta}_j$ be, in some sense, a 'good' estimator of the location parameter for the distribution F_j, $j = 1, \dots, k$. Consider the linear combination of such estimators

$$\widehat{\theta} = \sum_{j=1}^{k} a_j \widehat{\theta}_j, \quad \sum_{j=1}^{k} a_j = 1, \quad a_j \geq 0.$$

Hogg suggested to determine the shape of the underlying distribution $F(x)$ by the sample $x_1, x_2, \dots, x_n$ using some simple statistics, and then to assign a greater weight a_j to a more plausible distribution among F_j, $j = 1, \dots, k$.

As an example of the above described estimator, Hogg proposes the estimation procedure $\widehat{\theta}_n$ where the choice of a particular estimator in it depends on the comparative weight of distribution tails determined by the value of the sample kurtosis e_n, $e = \mathsf{E}(X - \mathsf{E}X)^4/\mathsf{E}^2(X - \mathsf{E}X)^2 - 3$:

$$\widehat{\theta}_n = \begin{cases} \underline{x}_{0.25}, & e_n < 2, \\ \bar{x}, & 2 \leq e_n < 4, \\ \bar{x}_{0.25}, & 4 \leq e_n \leq 5.5, \\ \operatorname{med} x, & e_n > 5.5, \end{cases}$$

where $\underline{x}_{0.25}$ stands for the mean of 25% of the minimum and 25% of the maximum sample order statistics, and $\bar{x}_\alpha$ is the trimmed mean (see Section 1.2). The properties of this adaptive estimator are studied by Monte Carlo, and these studies show its rather high efficiency over a wide class of distributions including the uniform and Cauchy.

Hogg and other authors (Hogg, 1972; Hogg, 1974; Randles and Hogg, 1973; Randles *et al.*, 1973) introduce more simple and convenient than kurtosis estimators for distribution tails, in particular, the measures based on distribution subranges, as $[F^{-1}(0.975) - F^{-1}(0.025)]/[F^{-1}(0.75) - F^{-1}(0.25)]$ (Crow and Siddiqui, 1967).

Estimators designed on the basis of similar statistics are simple and simultaneously can possess good characteristics over a wide class of distributions, in other words, they follow the 'spirit' of data analysis.

Adaptive estimation of regression. Under the conditions of the lack of information about the data distribution, in (Hogg, 1974) the following approach was proposed:

- Determine a preliminary robust estimator $\widehat{\boldsymbol{\theta}}$ of the vector $\boldsymbol{\theta}$ of regression parameters.
- Evaluate the residuals and determine the type of distribution tails for these residuals.
- Construct the final estimator using the information about distribution tails.

In what follows, we use some elements of the above approach for construction of adaptive robust estimators of location and regression based on the precise minimax solutions obtained in Chapter 3 and 5.

8.2.2. Adaptive robust estimation of the location parameter

Considering the problems of designing robust minimax estimators in Section 3.2, we have assumed that a priori information about the parameters of distribution classes is available. However, in practice of estimation these parameters are usually unknown and can be determined while processing the data.

To improve efficiency of estimation procedures, it seems useful to develop estimators that can adapt themselves to the changes in the data as the new data are being involved into processing. For small samples such an approach is heuristic and the simplest for the examination by Monte Carlo technique.

In the class $\widetilde{\mathscr{F}}_{12}$, the minimax robust estimator is based on the scale equivariant L_p-norm estimators (3.2.32) (see Subsection 3.2.5), where, instead of a priori limiting characteristics of the class $\bar{\sigma}^2$ and a^2, we use their estimators $\widehat{\bar{\sigma}}^2$ and $\widehat{a}^2$.

For estimating $\bar{\sigma}^2$ we can use, for example, the upper confidence limit for the estimator of variance, and for $1/(2a)$ the lower confidence limit for the nonparametric estimator of a distribution density at the center of symmetry:

$$\underline{\widehat{f}}(\mathrm{Med}) \le \widehat{f}(\mathrm{Med}).$$

Taking into account the form of the middle branch of the minimax estimator (3.2.32) ($1 < p < 2$), where both restrictions hold as equalities, we choose the estimators of variance and density as those of characteristics of the class $\widetilde{\mathscr{F}}_{12}$. For variance, this is the customary sample variance centered by the sample median instead of the sample mean

$$\widehat{\bar{\sigma}}^2 = \frac{1}{n} \sum_{i=1}^{n} (x_i - \mathrm{med}\, x)^2. \tag{8.2.1}$$

To avoid difficulties of nonparametric estimation of the distribution density, we estimate the density at its center of symmetry using the following obvious considerations.

Since the ith order statistic $x_{(i)}$ is a consistent estimator of the distribution quantile of level $i/(n+1)$, we have the following approximate relations for the central order statistics $x_{(k)}$ and $x_{(k+1)}$ ($n = 2k$ or $n = 2k+1$):

$$F(x_{(k)}) \cong \frac{k}{n+1}, \qquad F(x_{(k+1)}) \cong \frac{k+1}{n+1},$$

and

$$F(x_{(k+1)}) - F(x_{(k)}) = \frac{1}{n+1}. \tag{8.2.2}$$

Furthermore,

$$F(x_{(k+1)}) - F(x_{(k)}) = f(\xi)(x_{(k+1)} - x_{(k)}), \qquad x_{(k)} < \xi < x_{(k+1)}. \tag{8.2.3}$$

Then from (8.2.2) and (8.2.3) it follows that

$$\widehat{f}(\mathrm{Med}) = \frac{1}{(n+1)(x_{(k+1)} - x_{(k)})},$$

$$\widehat{a} = \frac{1}{2\widehat{f}(\mathrm{Med})} = \frac{(n+1)(x_{(k+1)} - x_{(k)})}{2}. \tag{8.2.4}$$

The behavior of the adaptive L_p-estimator $\widehat{\theta}_A$ is studied in samples $n = 20\,(10)\,100$ and $100\,(100)\,1000$ by Monte Carlo technique.

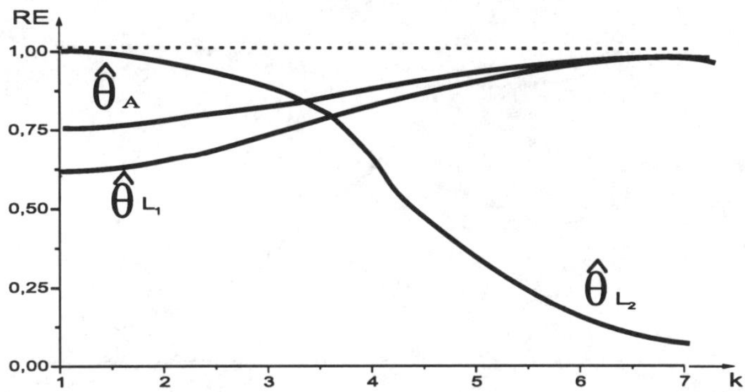

Figure 8.2. Relative efficiency under symmetric gross errors for $\varepsilon = 0.2$; $n = 20$

The robustness properties of $\widehat{\theta}_A$ are examined in the model of gross errors with symmetric contamination

$$f(x) = (1-\varepsilon)\mathcal{N}(x;0,1) + \varepsilon\mathcal{N}(x;0,k), \qquad 0 \le \varepsilon < 1, \qquad k > 1, \quad (8.2.5)$$

and with asymmetric contamination

$$f(x) = (1-\varepsilon)\mathcal{N}(x;0,1) + \varepsilon\mathcal{N}(x;\mu,1). \quad (8.2.6)$$

The adaptive properties of this algorithm are studied for the mixture of the normal and Laplace distributions

$$f(x) = (1-\varepsilon)\mathcal{N}(x;0,1) + \varepsilon L(x;0,1/\sqrt{2}), \qquad 0 \le \varepsilon < 1. \quad (8.2.7)$$

The parameters of model (8.2.7) are chosen to provide the constancy of expectation and variance: this allows us to examine the dependence of estimation efficiency on the distribution shape while the parameter ε varies from zero to one.

All the L_p-estimators are evaluated by the RWLS method with the initial LS estimator. The results of modeling are given in Figures 8.2–8.6.

From Figures 8.2–8.4 it follows that the adaptive estimator $\widehat{\theta}_A$ of the location parameter possesses high robustness properties: it practically coincides with the sample median for the high level of contamination (k exceeds 4–5) and considerably dominates over the sample mean in efficiency in this case. For smaller values of k, the adaptive estimator is more efficient than the sample median.

For small values of k, this algorithm is either a little inferior to the sample mean in efficiency (especially in small samples), or practically coincides with it. Starting from the values $k > 3$ and ε equal to 0.1–0.2, the adaptive estimator is superior to the sample mean in efficiency.

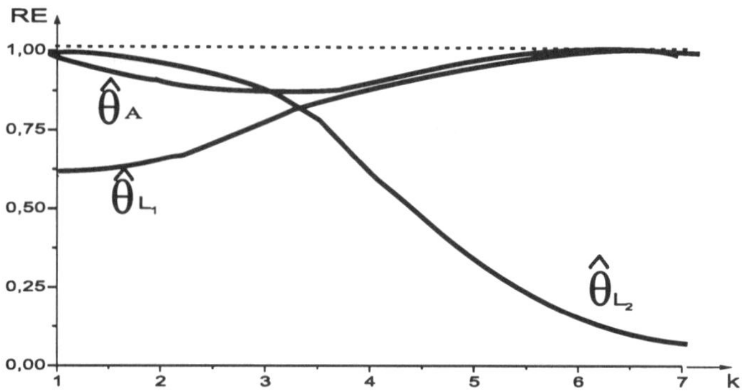

Figure 8.3. Relative efficiency under symmetric gross errors for $\varepsilon = 0.2$; $n = 1000$

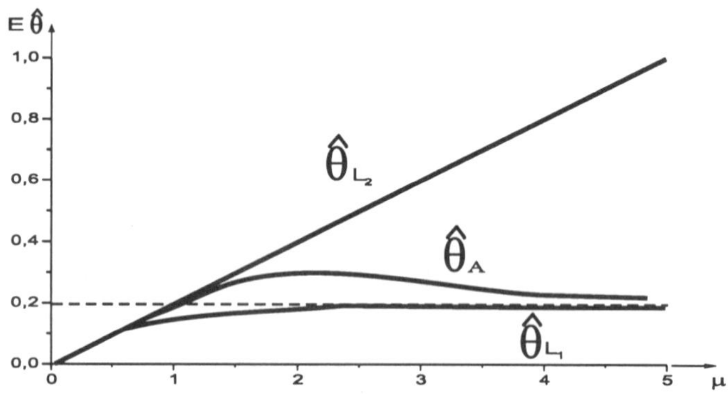

Figure 8.4. Expectations under asymmetric gross errors for $\varepsilon = 0.2$; $n = 20$

There are some peculiarities of the dependence of the adaptive estimator on the sample size n, this deserves a separate attention.

Since the estimators $\widehat{\widehat{\sigma}}^2$ and $\widehat{a}$ of the parameters of the class $\widetilde{\mathscr{F}}_{12}$ are consistent, the adaptive estimator $\widehat{\theta}_A$ tends in probability to the solution of the minimax problem in this class as $n \to \infty$. This is also confirmed by modeling: for $n > 200$ the behavior of $\widehat{\theta}_A$ practically coincides with asymptotics (see Fig. 8.5).

The adaptive properties of this algorithm begin to reveal themselves from $n > 100$ (see Fig. 8.6) when the estimator $\widehat{\theta}_A$ becomes 'tuned up' to the distribution shape: under the normal and Laplace distributions, it demonstrates nearly complete similarity in efficiency with the sample mean and sample median and preserves rather high efficiency in the intermediate zone between

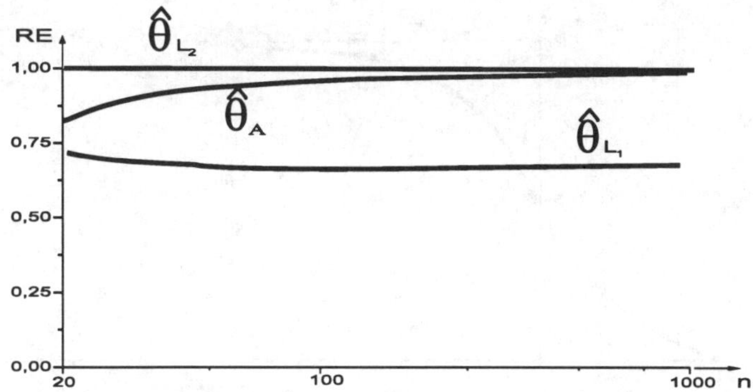

Figure 8.5. Dependence of the relative efficiency on the sample size n under the normal distribution

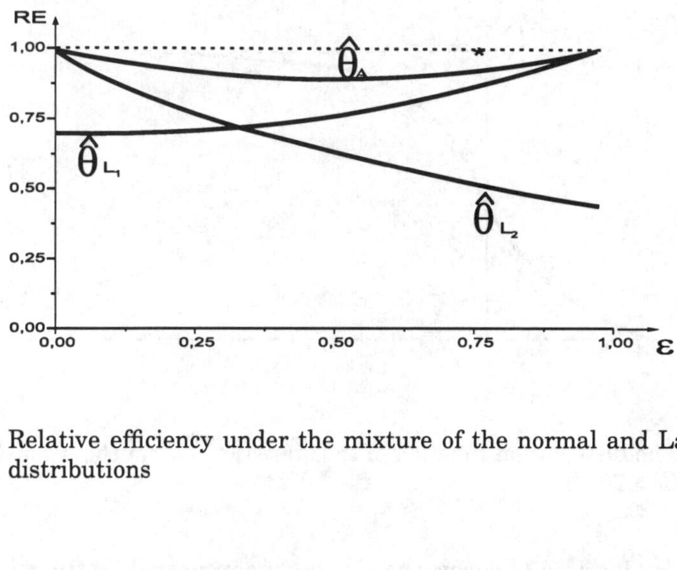

Figure 8.6. Relative efficiency under the mixture of the normal and Laplace distributions

them.

For small samples, the robustness properties dominate over the adaptive (see Fig. 8.2 and 8.4): this effect can be explained by the bias of the sample distribution of the switching statistic $\widehat{\sigma}^2/\widehat{a}^2$ with small n: its values determine the choice of the appropriate branch of the algorithm.

For normal samples $n = 20$, the distribution function of the statistic $\widehat{\sigma}^2/\widehat{a}^2$ is shown in Fig. 8.7; hence it follows that the sample mean occurs approximately for the 10% of cases ($\mathsf{P}(\widehat{\sigma}^2/\widehat{a}^2 < 2/\pi) \cong 0.1$), the sample median, for the 20% of cases ($\mathsf{P}(\widehat{\sigma}^2/\widehat{a}^2 > 2) \cong 0.2$), and the L_p-estimators, $1 < p < 2$, for the rest 70% of cases.

For samples of size $n = 100$, we have the opposite situation (see Fig. 8.8):

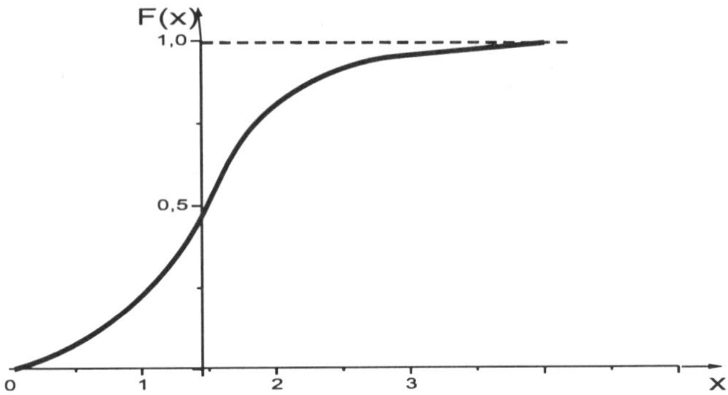

Figure 8.7. The distribution function of the statistic $\widehat{\sigma}^2/\widehat{a}^2$ in the normal case (n = 20)

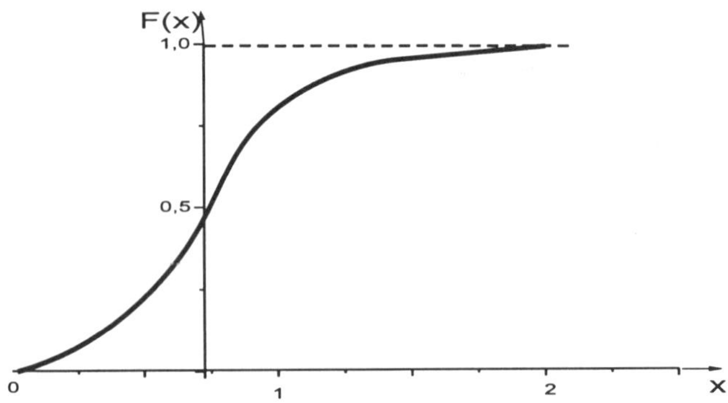

Figure 8.8. The distribution function of the statistic $\widehat{\sigma}^2/\widehat{a}^2$ in the normal case (n = 20)

approximately for 45% of cases, the sample mean branch of the adaptive estimator is realized, whereas the sample median occurs only for 5%.

Fig. 8.9 presents the dependence of the average value of the parameter p on the sample size n observed in our experiment (see Fig. 8.4). Its form also confirms the above mentioned peculiarity of the adaptive estimator: in small samples, it is close to the sample median, in large samples, it adaptively follows the true distribution law.

While processing small samples, we observe the high level of a priori uncertainty: with sufficiently high confidence, the data distribution law can be anyone from a reasonably chosen class of distributions. For instance, using classical goodness-of-fit tests, it is practically impossible to verify the true dis-

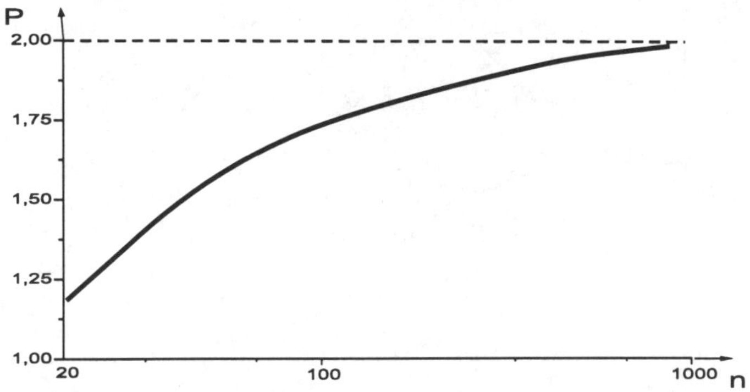

Figure 8.9. Dependence of the average value of the parameter p in the L_p-estimators on the sample size n in the normal case

tribution law in small samples, say to distinguish the normal distribution from the Laplace, or, what is much more unpleasant from the point of view of a user of statistics, to distinguish the Cauchy distribution from a normal with large variance.

We have examined the dynamics of applicability of the χ^2- and Kolmogorov tests to the problem of discrimination between the normal and Laplace distributions with equal expectations and variances. Fig. 8.10 shows dependence of the power of the χ^2-test on the sample size for this problem. It follows from this graph that, in small samples ($n < 60$) approximately in half of cases, the null hypothesis on the goodness-of-fit of the normal and Laplace distributions is accepted. Furthermore, with n increasing, the power decreases and from n greater than 200–300 the null hypothesis is surely rejected. The similar dependences have been modeled for the Kolmogorov test and for the normal and Cauchy distributions.

REMARK 8.2.1. The above results give an additional argument for the use of robust statistics, in particular for small samples.

Finally we consider the influence of asymmetric contamination (8.2.6) on the bias of the adaptive estimators. It can be seen from Fig. 8.4 that this estimator behaves differently under symmetric and asymmetric contamination. The bias of the $\widehat{\theta}_A$ is considerably smaller than that of the sample mean, it is a little more than the bias of the sample median under moderate values of the contamination parameter μ, and it has almost the same bias as the sample median under large values of μ. In this case, the adaptive estimator has a bounded bias as the bias of the sample median.

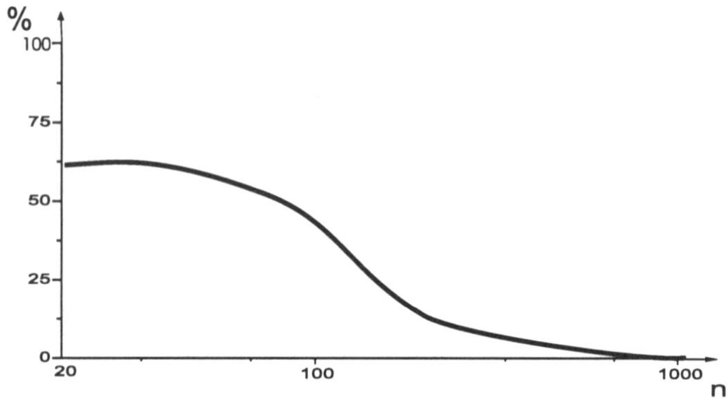

Figure 8.10. The power of the χ^2-test in the problem of discrimination between the normal and Laplace distributions

8.2.3. Adaptive robust estimation of the multivariate location parameter

Now we extend the above adaptive approach to the problem of minimax robust estimation of multivariate location in the class of spherically symmetric exponential-power distributions described in Section 3.4.4.

Consider the following adaptive algorithm called the ARML-estimator:

- Choose the initial L_1-norm estimator for $\boldsymbol{\theta}$

$$\widehat{\boldsymbol{\theta}}_{L_1} = \arg\min_{\boldsymbol{\theta}} \sum_{i=1}^{n} r_i, \quad p \geq 1, \quad r_i = \left(\sum_{j=1}^{m}(x_{ij} - \theta_j)^2\right)^{1/2}.$$

- Evaluate the residuals

$$\widehat{\mathbf{e}}_i = \mathbf{x}_i - \widehat{\boldsymbol{\theta}}_{L_1}, \qquad i = 1, \dots, n.$$

- Evaluate the estimators of the characteristics $\overline{\sigma}^2$ and a of the class $\mathscr{F}_{12q}$

$$\widehat{\overline{\sigma}}^2 = \frac{1}{nm}\sum_{i=1}^{n} \widehat{r}_i^2, \quad r_i = |\mathbf{e}_i|, \quad i = 1, \dots, n; \quad \hat{a} = \frac{\pi^{1/2}(n+1)^{1/m}\hat{r}_{(1)}}{\Gamma^{1/m}(m/2)},$$

where $r_{(1)}$ is the minimum order statistic of the sample $r_1, \dots, r_n$.

- Use the minimax L_p-norm estimator with $p = q^*$ of Section 3.2 with the estimators $\widehat{a}$ and $\widehat{\overline{\sigma}}^2$ as the characteristics of the class $\mathscr{F}_{12q}$.

Table 8.1. Relative efficiency of the ARML, L_2 and L_1-norm estimators under contamination: $\varepsilon = 0.1$, $n = 20$ (left), $n = 100$ (right)

k	1	2	3	4	5	k	1	2	3	4	5
$\widehat{\boldsymbol\theta}_A$	0.80	0.84	0.89	0.92	0.95	$\widehat{\boldsymbol\theta}_A$	0.96	0.94	0.90	0.90	0.93
L_2	1.00	0.92	0.67	0.40	0.30	L_2	1.00	0.94	0.70	0.42	0.31
L_1	0.70	0.76	0.83	0.88	0.93	L_1	0.73	0.78	0.86	0.89	0.93

We find $\widehat{a}$ from the relations

$$\mathsf{P}(r \le R) = F(R) = \frac{2\pi^{m/2}}{\Gamma(m/2)} \int_0^R f(t)t^{m-1}\,dt,$$

$$\widehat{F}(r_{(1)}) \cong \frac{2\pi^{m/2}}{\Gamma(m/2)} r_{(1)}^m \widehat{f}(0), \quad \widehat{F}(r_{(1)}) \cong \frac{1}{n+1}, \quad \widehat{f}(0) = \frac{1}{2\widehat{a}^m}.$$

The behavior of the ARML-algorithm $\widehat{\boldsymbol\theta}_A$ is studied by Monte Carlo in samples $n = 20$ and $n = 100$ under the ε-contaminated spherically symmetric bivariate normal distributions

$$f(x,y) = (1-\varepsilon)\mathcal{N}(x,y;0,0,1,1,0) + \varepsilon\mathcal{N}(x,y;0,0,k,k,0).$$

The number of trials is 1000. The L_1, L_2, and the ML estimators are also evaluated by the RWLS method using the initial LS estimator. The relative efficiency of estimators is defined as the ratio of the absolute values of the determinants of their sample covariance matrices.

The results of modeling are given in Table 8.1.

In general, the conclusion in the multivariate case coincides with that for the univariate case: the ARML-estimator has proved to be better than the L_1 and L_2-norm estimators both in small and large samples, especially under heavy contamination, and in small samples the ARML-estimator is close to the L_1-norm estimator.

8.2.4. Adaptive robust estimation of regression parameters

Consider now the realization of the adaptive approach to the problem of robust estimation of regression parameters (see Section 5.1 and Subsection 8.2.2) called the ARLI regression (Vilchevski and Shevlyakov, 1990a):

- Choose the initial L_1-estimator for $\boldsymbol\theta$

$$\widehat{\boldsymbol\theta}_{L_1} = \arg\min_{\boldsymbol\theta} \sum_{i=1}^n \Big|x_i - \sum_{j=1}^m \theta_j\phi_{ij}\Big|.$$

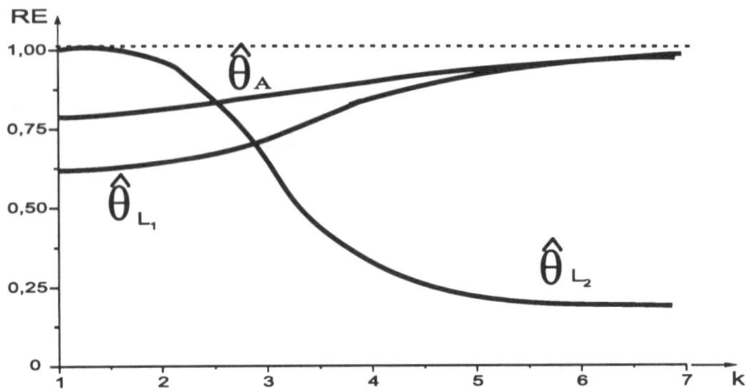

Figure 8.11. Relative efficiency of estimators for quadratic regression under symmetric gross errors: $\varepsilon = 0.2$ ($n = 20$)

- Evaluate the errors estimators

$$\widehat{e}_i = x_i - \sum_{j=1}^{m} \widehat{\theta}_{jL_1} \phi_{ij} \qquad i = 1, \dots, n.$$

- Evaluate the estimators of the characteristics of the class $\mathscr{F}_{12}$

$$\widehat{\sigma}^2 = \frac{1}{n-m} \sum_{i=1}^{n} \widehat{e}_i^2, \quad \widehat{a} = (n+1)[\widehat{e}_{(k+1)} - \widehat{e}_{(k)}], \quad n = 2k,\ n = 2k+1.$$

- Use the robust minimax M-estimator (3.2.11) with the score function ψ_{12}^*.

Here we directly apply the approach of Subsection 8.2.2 to the estimators of residuals. In this case the conclusions are qualitatively the same as above for location.

Fig. 8.11 illustrates the abovesaid for estimation of quadratic regression dependence under heavy contamination. The relative efficiency of the regression estimator is defined as $RE(\widehat{\boldsymbol{\theta}}) = \det \mathbf{V}(\widehat{\boldsymbol{\theta}}_{ML}) / \det \mathbf{V}(\widehat{\boldsymbol{\theta}})$.

8.3. Smoothing quantile functions by the Bernstein polynomials

The aim of this section is to demonstrate how smoothed variants of a sample quantile function can be used for the adaptive estimation of the underlying distribution and its characteristics. The idea of using precisely the quantile function, the inverse to the distribution function, is clearly formulated in

(Parzen, 1979b). The application of the Bernstein polynomials to this problem is proposed in (Vilchevski and Shevlyakov, 1985b).

8.3.1. Preliminaries

We begin with some definitions. Let X be a random variable with the distribution function $F(x) = \mathsf{P}(X \le x)$. The *quantile function* of X is

$$Q(t) = F^{-1}(t) = \inf\{x \mid F(x) \ge t\}, \qquad 0 < t < 1. \tag{8.3.1}$$

Given a sample $x_1, x_2, \dots, x_n$ from F, a natural estimator of the quantile function is the sample quantile function

$$\widetilde{Q}_n(t) = \widetilde{F}_n^{-1}(t) = \inf\{x \mid \widetilde{F}_n(x) \ge t\}, \qquad 0 < t < 1, \tag{8.3.2}$$

where

$$\widetilde{F}_n(x) = \frac{1}{n}\sum_{i=1}^{n} I(x_i \le x)$$

is the sample distribution function.

The sample quantile function can be computed explicitly in terms of the order statistics $x_{(1)} < x_{(2)} < \dots < x_{(n)}$

$$\widetilde{Q}_n(t) = x_{(i)}, \qquad \frac{i-1}{n} < t < \frac{i}{n}.$$

Several smooth quantile function estimators have been proposed, the Parzen difference kernel estimators (Parzen, 1979a)

$$\widetilde{Q}_n^K(t) = \int_0^1 \frac{1}{h_n} k\left(\frac{u-t}{h_n}\right) \widetilde{Q}_n(u)\, du$$

among them, where $k(\cdot)$ is a symmetric distribution density and h_n is the bandwidth parameter. These estimators are investigated in (Falk, 1985; Yang, 1985; Sheather and Marron, 1990).

In (Vilchevski and Shevlyakov, 1985b), the Bernstein polynomial estimator for the quantile function was introduced:

$$\widehat{Q}_n^B(t) = \sum_{i=1}^{n} \left[\binom{n-1}{i} t^{i-1}(1-t)^{n-i}\right] x_{(i)}, \qquad 0 \le t \le 1. \tag{8.3.3}$$

Now we dwell on the general properties of the Bernstein polynomials.

8.3.2. General properties of the Bernstein polynomials

The Bernstein polynomials (Bernstein, 1911; Lorentz, 1986) are widely used in the theoretical studies on convergence processes of approximations to functions continuous on $[0,1]$. The reason is that, first, these polynomials have a rather simple structure, and, second, they provide uniform convergence in the Chebyshev norm

$$\|f\| = \max_{t\in[0,1]} |f(t)|.$$

DEFINITION 8.3.1. The Bernstein polynomial being an uniform approximation to a continuous function $f(t)$ on a closed interval $[0,1]$ (the Weierstrass approximation theorem), is defined by

$$B_n(f,t) = \sum_{i=0}^{n} f\left(\frac{i}{n}\right)\binom{n}{i} t^i(1-t)^{n-i}. \tag{8.3.4}$$

The Bernstein polynomials possess the following basic properties:

(BP1) if a function $f(t)$ has k continuous derivatives, then $B_n^{(k)}(f,t) \to f^{(k)}(t)$ uniformly on $[0,1]$;

(BP2) if a bounded function $f(t)$ has a jump discontinuity at $t=c$, then

$$B_n(f,c) \to \frac{f(c_-)+f(c_+)}{2};$$

(BP3) for a twice continuously differentiable function $f(t)$, the asymptotic representation

$$B_n(f,t) = f(t) + \frac{f''(t)t(1-t)}{2n} + o\left(\frac{1}{n}\right) \tag{8.3.5}$$

holds true;

(BP4) if $f(t)$ is a monotonic function on $[0,1]$, then $B_n(f,t)$ is also monotonic on $[0,1]$.

REMARK 8.3.1. The Bernstein polynomial has an obvious probabilistic interpretation. Assume that t is the probability of success in a single trial of a random event. Hence, if the 'payoff' for exactly i successes is expressed as the value of f at i/n, then $B_n(f,t)$ gives the expected payoff from the n independent trials.

The above-mentioned properties make it possible to use the Bernstein polynomials for estimation of distribution laws and their characteristics (Vilchevski and Shevlyakov, 1985b; Brewer, 1986; Perez and Palacin, 1987; Sheather and Marron, 1990; Kaigh and Cheng, 1991; Cheng, 1995).

Now we generalize the above classical result on the asymptotic representation for the Bernstein polynomials. First we obtain the uniform convergence of approximations by the Bernstein polynomials to analytical functions.

THEOREM 8.3.1. *Let a function $f(t)$ defined on $[0,1]$ have a convergent Taylor expansion at any point on this interval:*

$$f(u) = \sum_{i=0}^{\infty} \frac{f^{(i)}(t)}{i!}(u-t)^i, \qquad u,t \in [0,1].$$

Then for any $\varepsilon > 0$, there exists n^* such that for all $n \geq n^*$ the inequality

$$\max_{t\in[0,1]} |f(t) - B_n(f,t)| \leq \varepsilon$$

is satisfied. Moreover, if $\max_{0\leq t\leq 1} |f^{(4)}(t)| \leq C$, then a more accurate inequality

$$\left| f(t) + \frac{t(1-t)}{2n} f''(t) + \frac{t(1-t)(1-2t)}{6n^2} f'''(t) - B_n(f,t) \right| \leq \frac{t^2(1-t)^2 3(n-2) + t(1-t)}{4!n^3} C \quad (8.3.6)$$

holds true.

PROOF. We represent $f(k/n)$ as the Taylor series with the Lagrange remainder

$$f\left(\frac{k}{n}\right) = f(t) + f'(t)\left(\frac{k}{n} - t\right) + \frac{f''(t)}{2!}\left(\frac{k}{n} - t\right)^2 + \frac{f'''(t)}{3!}\left(\frac{k}{n} - t\right)^3 + \frac{f''''(\theta_k)}{4!}\left(\frac{k}{n} - t\right)^4, \quad (8.3.7)$$

where $0 < \theta_k < 1$. From (8.3.7) it follows that the Bernstein polynomial is of the form

$$\begin{aligned} B_n(f,t) = {} & f(t)\sum_{i=0}^{n} \binom{n}{i} t^i(1-t)^{n-i} \\ & + f'(t)\sum_{i=0}^{n} \left(\frac{k}{n} - t\right)\binom{n}{i} t^i(1-t)^{n-i} \\ & + \frac{f''(t)}{2!}\sum_{i=0}^{n} \left(\frac{k}{n} - t\right)^2\binom{n}{i} t^i(1-t)^{n-i} \\ & + \frac{f'''(t)}{3!}\sum_{i=0}^{n} \left(\frac{k}{n} - t\right)^3\binom{n}{i} t^i(1-t)^{n-i} \\ & + \sum_{i=0}^{n} \frac{f''''(\theta_k)}{4!}\left(\frac{k}{n} - t\right)^4\binom{n}{i} t^i(1-t)^{n-i}. \end{aligned}$$

Since the fourth derivative is bounded, from the latter relation it follows that

$$\Bigg| B_n(f,t) - f(t)\sum_{i=0}^{n}\binom{n}{i}t^i(1-t)^{n-i} - f'(t)\sum_{i=0}^{n}\left(\frac{k}{n}-t\right)\binom{n}{i}t^i(1-t)^{n-i}$$
$$- \frac{f''(t)}{2!}\sum_{i=0}^{n}\left(\frac{k}{n}-t\right)^2\binom{n}{i}t^i(1-t)^{n-i} - \frac{f'''(t)}{3!}\sum_{i=0}^{n}\left(\frac{k}{n}-t\right)^3\binom{n}{i}t^i(1-t)^{n-i}\Bigg|$$
$$\le C\sum_{i=0}^{n}\left(\frac{k}{n}-t\right)^4\binom{n}{i}t^i(1-t)^{n-i}. \qquad (8.3.8)$$

Using rather cumbersome but obvious transformations, we obtain

$$\sum_{i=0}^{n}\binom{n}{i}t^i(1-t)^{n-i} = 1,$$
$$\sum_{i=0}^{n}\left(\frac{k}{n}-t\right)\binom{n}{i}t^i(1-t)^{n-i} = 0,$$
$$\sum_{i=0}^{n}\left(\frac{k}{n}-t\right)^2\binom{n}{i}t^i(1-t)^{n-i} = \frac{t(1-t)}{n},$$
$$\sum_{i=0}^{n}\left(\frac{k}{n}-t\right)^3\binom{n}{i}t^i(1-t)^{n-i} = \frac{t(1-t)(1-2t)}{n^2},$$
$$\sum_{i=0}^{n}\left(\frac{k}{n}-t\right)^4\binom{n}{i}t^i(1-t)^{n-i} = \frac{3(n-2)t^2(1-t)^2+t(1-t)}{n^3}. \qquad (8.3.9)$$

It remains to substitute (8.3.9) into (8.3.8). □

REMARK 8.3.2. The classical asymptotic representation for the Bernstein polynomials, namely (BP3), follows immediately from (8.3.6).

The proof of (BP1) can be obtained by a slight modification of the proof of Theorem 8.3.1.

For (BP4),

$$B_n'(f,t) = \sum_{i=0}^{n} f\left(\frac{i}{n}\right)\binom{n}{i}\left(it^{i-1}(1-t)^{n-i} - (n-i)t^i(1-t)^{n-i-1}\right)$$
$$= n\sum_{i=0}^{n-1}\left(f\left(\frac{i+1}{n}\right) - f\left(\frac{i}{n}\right)\right)\binom{n-1}{i}t^i(1-t)^{n-i-1}.$$

Therefore, if $f\left(\frac{i+1}{n}\right) - f\left(\frac{i}{n}\right)$ retains its sign, so does $B_n'(f,t)$.

8.3.3. The Bernstein polynomials for order statistics

Consider a sample $x_1, x_2, \dots, x_n$ from the distribution $F(x)$ defined on the interval $[a,b]$ and the corresponding set of order statistics $x_{(1)}, x_{(2)}, \dots, x_{(n)}$. Introduce the polynomial of the Bernstein type

$$\mathscr{B}_n^{[a,b]}(t) = a(1-t)^{n+1} + \sum_{i=1}^{n} x_{(i)} \binom{n+1}{i} t^i (1-t)^{n+1-i} + bt^{n+1}. \quad (8.3.10)$$

Let $Q(t)$ be the quantile function (8.3.1), the inverse to $F(x)$. Set

$$q_{n+1}(t) = \begin{cases} a, & 0 \le t < \frac{1}{n+1}, \\ x_{(i)}, & \frac{i}{n+1} \le t < \frac{i+1}{n+1}, \ 1 \le i \le n, \\ b, & t \ge 1. \end{cases}$$

Hence

$$\mathscr{B}_n^{[a,b]}(t) = B_{n+1}(q_{n+1}, t).$$

Obviously, the introduced function $q_{n+1}(t)$ is the inverse to the sample distribution function, and $q_{n+1}(t) \to Q(t)$ as $n \to \infty$, that is, for finite n, $q_{n+1}(t)$ is an estimator for the quantile function $Q(t)$.

It is known that the expectation of the sample distribution function coincides with its true value. Approximating the inverse to the sample distribution function, we may foresee the analogous assertion.

Now we introduce a more general construction than (8.3.10). We should distinguish the situations with finite and infinite bounds for $[a,b]$. Taking this remark into account, we define the Bernstein-type polynomials

$$\mathscr{B}_n^{[a,b]}(\Phi, t) = \Phi(a)(1-t)^{n+1} + \sum_{i=1}^{n} \Phi(x_{(i)}) \binom{n+1}{i} t^i (1-t)^{n+1-i} + \Phi(b)t^{n+1}, \quad (8.3.11)$$

$$\mathscr{B}_n^{[a,\infty]}(\Phi, t) = \Phi(a)(1-t)^{n} + \sum_{i=1}^{n} \Phi(x_{(i)}) \binom{n}{i} t^i (1-t)^{n-i}, \quad (8.3.12)$$

$$\mathscr{B}_n^{[-\infty,b]}(\Phi, t) = \sum_{i=0}^{n-1} \Phi(x_{(i+1)}) \binom{n}{i} t^i (1-t)^{n-i} + \Phi(b)t^{n}, \quad (8.3.13)$$

$$\mathscr{B}_n^{[-\infty,\infty]}(\Phi, t) = \sum_{i=0}^{n-1} \Phi(x_{(i+1)}) \binom{n-1}{i} t^i (1-t)^{n-1-i}. \quad (8.3.14)$$

Consider now the asymptotic behavior of the expectations for these polynomials. The expectation $\mathsf{E}_F\{\mathscr{B}_n^{[a,b]}(\Phi,t)\} \equiv I(\Phi,Q,t)$ can be written as

$$\begin{aligned}
I(\Phi,Q,t) &= \Phi(a)(1-t)^{n+1} + \Phi(b)t^{n+1} + \sum_{i=1}^{n} \mathsf{E}_F\{\Phi(x_{(i)})\}\binom{n+1}{i}t^i(1-t)^{n+1-i} \\
&= \Phi(a)(1-t)^{n+1} + \Phi(b)t^{n+1} \\
&\quad + \sum_{i=1}^{n}\binom{n+1}{i}t^i(1-t)^{n+1-i}\int_a^b n\Phi(x)\binom{n-1}{i-1}(F(x))^{i-1}(1-F(x))^{n-i}\,dF \\
&= \Phi(a)(1-t)^{n+1} + \Phi(b)t^{n+1} \\
&\quad + n\sum_{i=1}^{n}\binom{n+1}{i}t^i(1-t)^{n+1-i}\int_0^1 \Phi(Q(u))\binom{n-1}{i-1}u^{i-1}(1-u)^{n-i}\,du.
\end{aligned}$$

The behavior of $\mathsf{E}_F\{\mathscr{B}_n^{[a,b]}(\Phi,t)\}$ as $n \to \infty$ is as follows.

THEOREM 8.3.2. *Let* $\Psi(u) \equiv \Phi(Q(u))$ *be an analytical function on* $[0,1]$. *Then the expectation of the Bernstein-type polynomial* (8.3.11) *is*

$$I^{[a,b]}(\Phi,Q,t) = \mathsf{E}_F\{\mathscr{B}_n^{[a,b]}(\Phi,t)\} = \Phi(Q(t)) + \frac{t(1-t)}{n+2}\Phi''(Q(t)) + o\left(\frac{1}{n}\right).$$

PROOF. We write out the Taylor series for $\Psi \equiv \Phi(Q(u))$ at $u = \frac{i}{n+1}$. In this case, $I(\Phi,Q,t)$ is of the form

$$\begin{aligned}
I^{[a,b]}(\Phi,Q,t) &= \Phi(a)(1-t)^{n+1} + \Phi(b)t^{n+1} + n\sum_{s=0}^{\infty}\sum_{i=1}^{n}\binom{n+1}{i}t^i(1-t)^{n+1-i} \\
&\quad \times \frac{1}{s!}\Psi^{(s)}\left(\frac{i}{n+1}\right)\int_0^1\left(u-\frac{i}{n+1}\right)^s\binom{n-1}{i-1}u^{i-1}(1-u)^{n-i}\,du; \qquad (8.3.15)
\end{aligned}$$

furthermore, we introduce

$$\begin{aligned}
A_0 &= \Phi(a)(1-t)^{n+1} + \Phi(b)t^{n+1} \\
&\quad + n\sum_{i=1}^{n}\binom{n+1}{i}t^i(1-t)^{n+1-i}\Psi\left(\frac{i}{n+1}\right)\int_0^1\binom{n-1}{i-1}u^{i-1}(1-u)^{n-i}\,du, \\
A_1 &= n\sum_{i=1}^{n}\binom{n+1}{i}t^i(1-t)^{n+1-i}\Psi'\left(\frac{i}{n+1}\right) \\
&\quad \times \int_0^1\left(u-\frac{i}{n+1}\right)\binom{n-1}{i-1}u^{i-1}(1-u)^{n-i}\,du,
\end{aligned}$$

$$A_2 = n \sum_{i=1}^{n} \binom{n+1}{i} t^i (1-t)^{n+1-i} \frac{1}{2} \Psi'' \left(\frac{i}{n+1} \right)$$
$$\times \int_0^1 \left(u - \frac{i}{n+1} \right)^2 \binom{n-1}{i-1} u^{i-1} (1-u)^{n-i} \, du,$$
$$A_3 = n \sum_{s=3}^{\infty} \sum_{i=1}^{n} \binom{n+1}{i} t^i (1-t)^{n+1-i} \frac{1}{s!} \Psi^{(s)} \left(\frac{i}{n+1} \right)$$
$$\times \int_0^1 \left(u - \frac{i}{n+1} \right)^s \binom{n-1}{i-1} u^{i-1} (1-u)^{n-i} \, du.$$

In order to evaluate A_0, A_1, A_2, and A_3, we proceed as follows:

- We make use of the formulas $\Phi(a) = \Phi(Q(0))$, $\Phi(b) = \Phi(Q(1))$, where $a = Q(0)$, $b = Q(1)$.
- The integrals in the formulas for A_0, A_1, A_2, and A_3 are expressed in terms of the B-function

$$\int_0^1 u^k u^{i-1} (1-u)^{n-i} \, du = \frac{(k+i-1)!(n-i)!}{(n+k)!}.$$

- By the Laplace method, we arrive at the asymptotic relation

$$\int_0^1 \left(u - \frac{i}{n+1} \right)^s \binom{n-1}{i-1} u^{i-1} (1-u)^{n-i} \, du = \frac{1}{n^{s+1}} + o\left(\frac{1}{n^{s+1}} \right).$$

Taking the above relations into account, we obtain

$$A_0 = \Phi(a)(1-t)^{n+1} + \Phi(b) t^{n+1} + \sum_{i=1}^{n} \binom{n+1}{i} t^i (1-t)^{n+1-i} \Psi \left(\frac{i}{n+1} \right)$$
$$= \sum_{i=0}^{n+1} \binom{n+1}{i} t^i (1-t)^{n+1-i} \Phi \left(Q \left(\frac{i}{n+1} \right) \right) = B_{n+1}(\Phi(Q), t),$$
$$A_1 = 0,$$
$$A_2 = \frac{nt(1-t)}{2(n+1)(n+2)} \sum_{0}^{n-1} \Phi'' \left(Q \left(\frac{i+1}{n+1} \right) \right) \binom{n-1}{i} t^i (1-t)^{n-1-i};$$

furthermore, in view of the expansion

$$\Phi'' \left(Q \left(\frac{i+1}{n+1} \right) \right) = \Phi'' \left(Q \left(\frac{i-1}{n-1} - \frac{2i-1}{n^2-1} \right) \right)$$
$$= \Phi'' \left(Q \left(\frac{i-1}{n-1} \right) \right) + o\left(\frac{1}{n} \right),$$

we arrive at

$$A_2 = \frac{nt(1-t)}{2(n+1)(n+2)} \sum_0^{n-1} \Phi''\left(Q\left(\frac{i-1}{n-1}\right)\right) \binom{n-1}{i} t^i(1-t)^{n-1-i} + o\left(\frac{1}{n}\right)$$
$$= \frac{nt(1-t)}{2(n+1)(n+2)} B_{n-1}\left(\Phi''\left(Q\left(\frac{i-1}{n-1}\right)\right), t\right) + o\left(\frac{1}{n}\right),$$
$$A_3 = o\left(\frac{1}{n}\right).$$

By substituting the relations for A_0, A_1, and A_2 into (8.3.14) and taking into account the asymptotic representation (BP3) for the Bernstein polynomials, we obtain

$$\mathsf{E}_F\{\mathscr{B}_{n+1}(\Phi,t)\} = \Phi(Q(t)) + \frac{t(1-t)}{2(n+1)}\Phi''(Q(t)) + \frac{nt(1-t)}{2(n+1)(n+2)}\Phi''(Q(t)) + o\left(\frac{1}{n}\right)$$
$$= \Phi(Q(t)) + \frac{t(1-t)}{(n+2)}\Phi''(Q(t)) + o\left(\frac{1}{n}\right),$$

which completes the proof. □

EXAMPLE 8.3.1. We set

$$\Phi(u) = u, \qquad F(x) = \begin{cases} 0, & x < a, \\ \frac{x-a}{b-a}, & a \le x < b, \\ 1, & x \ge b. \end{cases}$$

Then the direct evaluation of (8.3.11)) yields

$$\mathsf{E}_F\{\mathscr{B}_{n+1}(u,t)\} = a + (b-a)t,$$

which is the exact expression for the inverse function to the uniform distribution on $[a,b]$. If we take $\Phi(u) = u^3$, then the corresponding calculation yields

$$\mathsf{E}_F\{\mathscr{B}_{n+1}(u^3,t)\} = (a+(b-a)t)^3 + \frac{t(1-t)}{n+2}6(a+(b-a)t) + \frac{6t(1-t)(1-2t)(b-a)^3}{(n+2)(n+3)},$$

which exactly corresponds to Theorem 8.3.2.

The proofs of the assertions describing the asymptotic behavior of the Bernstein-type polynomials (8.3.12), (8.3.13), and (8.3.14) practically coincide with the proof of the latter theorem. All differences refer to the choice of the points which the function $\Phi(Q(u))$ is expanded at.

We formulate the corresponding theorems.

THEOREM 8.3.3. *Let $\Psi(u) \equiv \Phi(Q(u))$ be an analytical function on* $[0,1]$. *Then the expectation of the Bernstein-type polynomial* (8.3.12) *is of the form*

$$I^{[a,\infty]}(\Phi, Q, t) = \mathsf{E}_F\{\mathscr{B}_n^{[a,\infty]}(\Phi, t)\}$$
$$= \Phi(Q(t)) - \frac{t}{n+1}\Phi'(Q(t)) + \frac{t(1-t)}{n+2}\Phi''(Q(t)) + o\left(\frac{1}{n}\right).$$

THEOREM 8.3.4. *Let $\Psi(u) \equiv \Phi(Q(u))$ be an analytical function on* $[0,1]$. *Then the expectation of the Bernstein-type polynomial* (8.3.13) *is of the form*

$$I^{[-\infty,b]}(\Phi, Q, t) = \mathsf{E}_F\{\mathscr{B}_n^{[-\infty,b]}(\Phi, t)\}$$
$$= \Phi(Q(t)) + \frac{1-t}{n+1}\Phi'(Q(t)) + \frac{t(1-t)}{n+2}\Phi''(Q(t)) + o\left(\frac{1}{n}\right).$$

THEOREM 8.3.5. *Let $\Psi(u) \equiv \Phi(Q(u))$ be an analytical function on* $[0,1]$. *Then the expectation of the Bernstein-type polynomial* (8.3.14) *is of the form*

$$I^{[-\infty,\infty]}(\Phi, Q, t) = \mathsf{E}_F\{\mathscr{B}_n^{[-\infty,\infty]}(\Phi, t)\}$$
$$= \Phi(Q(t)) + \frac{1-2t}{n+1}\Phi'(Q(t)) + \frac{t(1-t)}{n+2}\Phi''(Q(t)) + o\left(\frac{1}{n}\right).$$

EXAMPLE 8.3.2. Let the Bernstein-type polynomial (8.3.14) be used under the conditions of Example 8.3.1. In this case, for the expectation $\mathsf{E}_F\{\mathscr{B}_n^{[-\infty,\infty]}(\Phi, t)\}$ at $\Phi(u) = u$ we obtain

$$\mathsf{E}_F\{\mathscr{B}_n^{[-\infty,\infty]}(u, t)\} = a + (b-a)t + \frac{(b-a)(1-2t)}{n+1},$$

and if $\Phi(u) = u^3$, then

$$\mathsf{E}_F\{\mathscr{B}_n^{[-\infty,\infty]}(u^3, t)\} = (a + (b-a)t)^3 + 3\frac{(1-2t)}{n+1}(a + (b-a)t)^2(b-a)$$
$$+ 6\frac{t(1-t)}{n+2}(a + (b-a)t)(b-a)^2 + 6\frac{(5t^2 - 5t + 1)a + (9t^2 - 11t + 3)t(b-a)}{(n+1)(n+2)}$$
$$\times (b-a)^2 + 6\frac{(1-2t)(10t^2 - 10t + 1)}{(n+1)(n+2)(n+3)}(b-a)^3.$$

It is easy to see that this result completely agrees with Theorem 8.3.5.

REMARK 8.3.3. It is rather convenient to use the Bernstein approximations to the quantile function, as through them, it is easy to derive the moments characteristics of random variables. In particular, for the estimator of expectation we obtain

$$\widehat{\mu} = \int_a^b x\, d\widehat{F}(x) = \int_0^1 \widehat{Q}(t)\, dt.$$

Hence, using the Bernstein-type polynomials (8.3.11), we obtain

$$\hat{\mu} = \int_0^1 \mathscr{B}_n^{[a,b]}(u,t)\,dt = \frac{a + \sum_1^n x_{(i)} + b}{n+2}.$$

For approximation (8.3.14)), we obtain

$$\hat{\mu} = \int_0^1 \mathscr{B}_n^{[-\infty,\infty]}(u,t)\,dt = \frac{\sum_1^n x_{(i)}}{n}.$$

The estimator for the distribution density is expressed in terms of the quantile function in the parametric form, namely regarding t as a parameter:

$$f(x) = F'(x) \cong 1/B_n'(t), \qquad x = B_n(t).$$

8.3.4. The applications of the Bernstein polynomials and related constructions

One of the principal problems in data analysis and applied statistics is concerned with designing algorithms allowing to reduce the volumes of statistical information without significant loss of its specific properties and indications. The problems of determination of mean values, the methods of parametric approximation, etc. can be related to this class of problems. As a result of applying such methods, the initial data samples can be replaced by the corresponding mean values or by some functional dependences. It is obvious that the maximum of available information is contained in the complete initial data sample, and any its reduction necessarily leads to some loss of information. Thus, with small samples, it is reasonable to save the complete data. Otherwise, with large samples, it is necessary to compress the data arrays.

Now we introduce one of the possible methods of data compression based on the Bernstein polynomials.

Let $x_1, \ldots, x_n$ be a given sample, and the problem is to find a function $\phi(i; a_1, \ldots, a_m)$ reconstructing the initial data collection with a given accuracy so that $m \ll n$. There exist various algorithms of parametric approximation for solution of this problem, but as the location of the nodes is given (these nodes are equidistant), conventional approximations are usually unstable because they are, as a rule, solutions of ill-posed problems.

Consider a two-stage approach to the problems of compressing information which includes

(i) a preliminary approximation to the initial data by a continuous function;

(ii) an appropriate choice of the nodes and a final approximation to the obtained continuous function.

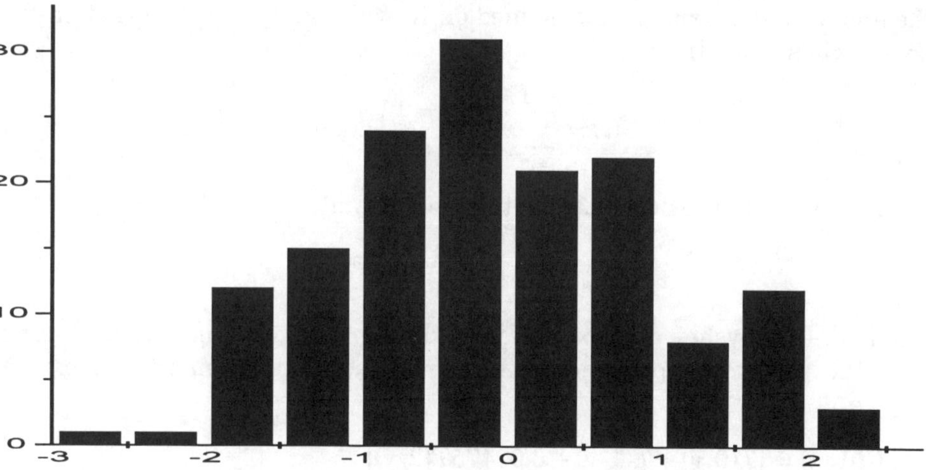

Figure 8.12. The histogram for the normal data

The preliminary approximation is made by the Bernstein polynomials after which we make use of use the Padé–Chebyshev approximations to the obtained function (Baker, 1975; Gilewicz, 1978).

The examples below are related with the following situations. Let a sample $x_1, \ldots, x_n$ be from a distribution $F(x)$. We assume that the sample elements are independent, that is, we can change the ordering of these elements. By this reason, we deal with the order statistics corresponding to a given sample.

It is desirable to

(i) find a dependence containing, as far as possible, the minimum number of parameters and allowing for reconstruction of the initial data;

(ii) construct an estimator (an approximation) of the quantile function $Q(y)$ corresponding to the distribution function $F(x)$;

(iii) construct an estimator (an approximation) of the distribution density $f(x) = F'(x)$.

REMARK 8.3.4. Generation of random numbers and all further calculations are made by the system of symbolic computations in `Maple V`.

EXAMPLE 8.3.3. Here we deal with the normal sample containing 150 random numbers with mean 0 and variance 1.

The histogram for this initial data is shown in Fig. 8.12.

Now construct the Bernstein polynomial for the set of order statistics. As the underlying distribution is defined on **R**, we should use the Bernstein-type polynomials (8.3.14)

$$\mathscr{B}_n^{[-\infty,\infty]}(t) = \sum_{i=0}^{n-1} x_{(i)} \binom{n-1}{i} t^i (1-t)^{n-1-i}.$$

We also evaluate the derivative of this polynomial

$$b_n^{[-\infty,\infty]}(t) = \frac{d}{dt} \mathscr{B}_n^{[-\infty,\infty]}(t).$$

The Padé–Chebyshev approximation to the Bernstein polynomial is obtained with the use of the package for numerical approximation `numapprox` of `Maple V`:

$$\begin{aligned}\text{PChB}(x) &= \left[T(0,y) + T(1,y) + 0.00475T(2,y)\right]^{-1} \\ &\times \left[0.898T(0,y) + 1.875T(1,y) + 1.14T(2,y) + 0.288T(3,y) + 0.122T(4,y)\right],\end{aligned}$$

where $T(n,y) = \cos(n \arccos y)$ is the Chebyshev polynomial and $y = 2x - 1$.

The error of the Bernstein approximation to this data does not exceed 0.088, and the error of the approximation to the same sample by the Padé–Chebyshev ratio is less than 0.13, whereas the maximum error of the Padé–Chebyshev approximation to the Bernstein polynomial is 0.061. The maximum error of approximation to the derivative of this polynomial is much greater (1.51), and it is attained on the boundary of the interval [0, 1].

From Fig. 8.13 it can be seen that these approximations fit the data rather accurately (except the boundaries of the interval).

Thus it is possible to reconstruct the data set to within accuracy 0.13 using the Padé–Chebyshev rational approximations with only five coefficients.

Fig. 8.14 presents the derivatives of the Bernstein polynomial and Padé–Chebyshev approximation, whence the smoothing properties of the latter approximation are obvious.

By Theorem 8.3.5 and the relation $\Phi(u) = u$, we conclude that the Bernstein polynomial and its Padé–Chebyshev approximation can be used as sufficiently good estimators for the quantile function $Q(y)$ corresponding to the normal distribution function.

To construct the distribution density, we use its parametric representation excluding the parameter t from $x = \mathscr{B}_n^{[-\infty,\infty]}(t)$ and substituting it into the estimator for this density $\widehat{f}(x) = 1/b_n^{[-\infty,\infty]}(t)$.

Fig. 8.15 presents this graph and the density estimator based on the Padé–Chebyshev approximation. It is seen from these graphs that the first graph fits well the histogram, whereas the second is close to the underlying normal density (the slashed line).

If to take the approximation $\mathscr{B}_n^{[-5,5]}(t)$ instead of the $\mathscr{B}_n^{[-\infty,\infty]}(t)$, the results will be practically the same.

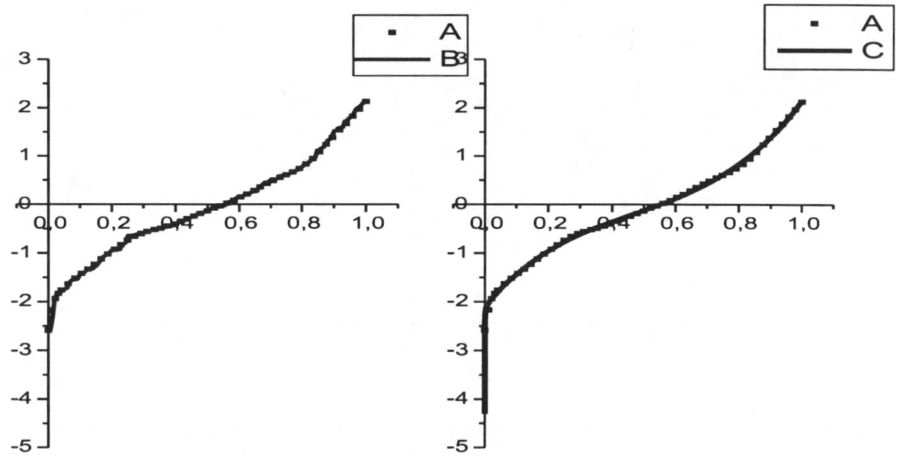

Figure 8.13. The graphs of the ordered data set (A), its Bernstein (B) and Padé-Chebyshev (C) approximations for the normal data

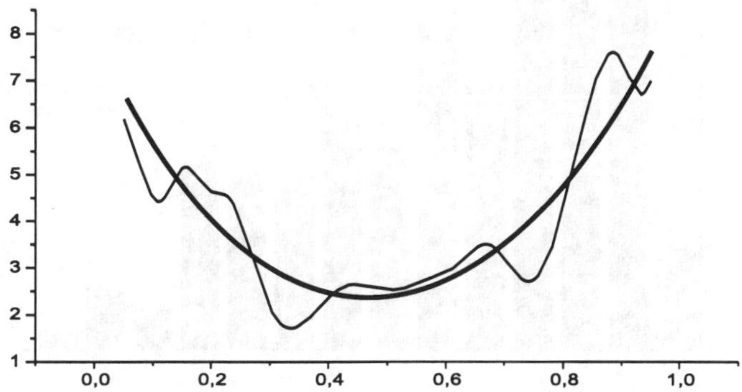

Figure 8.14. The derivatives of the Bernstein and Padé–Chebyshev approximations for the normal data

REMARK 8.3.5. Generation of random numbers in `Maple` is not accurate enough. In the next example, generation is performed with the use of `Statistica`, while the subsequent calculations are made in `Maple`.

EXAMPLE 8.3.4. Consider now the sample from the Beta(2, 6)-distribution and repeat all the stages of data processing as in Example 8.3.3.

The histogram for 150 sample elements from the Beta distribution is given in Fig. 8.16.

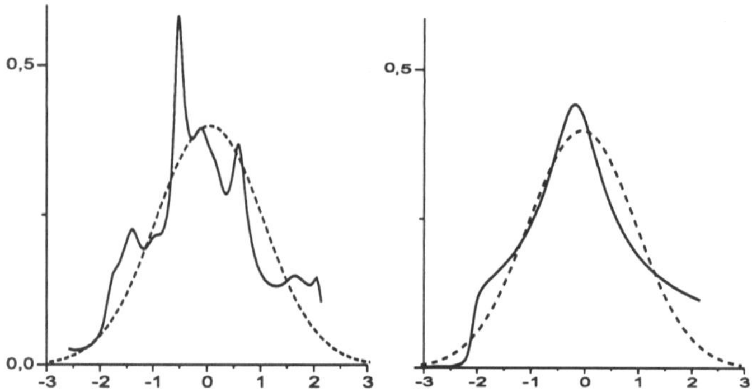

Figure 8.15. The normal density and its estimators based on the Bernstein and Padé–Chebyshev approximations for the normal data

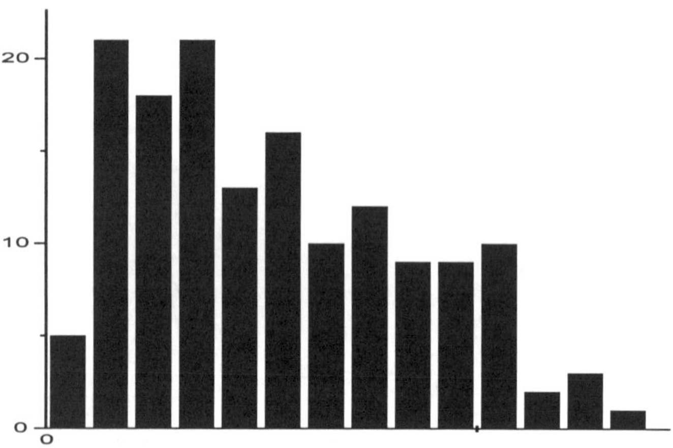

Figure 8.16. The histogram for the Beta-distributed data

Construct the Bernstein-type polynomial (8.3.11) corresponding to the finite Beta-distribution that is defined on $[0,1]$

$$\mathscr{B}_n^{[0,1]}(\Phi,t) = \Phi(0)(1-t)^{n+1} + \sum_{i=1}^{n} \Phi(x_{(i)}) \binom{n+1}{i} t^i (1-t)^{n+1-i} + \Phi(1)t^{n+1}$$

for the case $\Phi(u) = u$.

The Padé–Chebyshev approximation to the Bernstein polynomial is of the

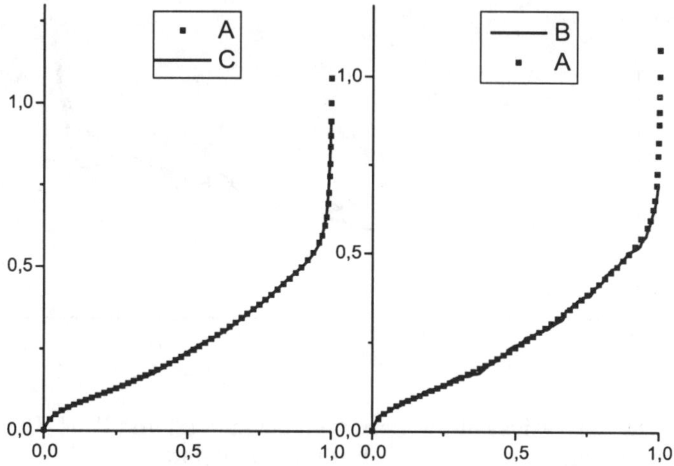

Figure 8.17. The graphs of the ordered data (A), its Bernstein (B) and Padé–Chebyshev (C) approximations for the Beta-distributed data

form

$$\begin{aligned}\mathrm{PChB}(x) &= \left[T(0,y) - 0.121T(1,y) - 0.85T(2,y)\right]^{-1}\\ &\times \left[0.244T(0,y) + 0.12T(1,y) - 0.21T(2,y) - 0.105T(3,y) - 0.0186T(4,y)\right],\end{aligned}$$

where $T(n,y) = \cos(n \arccos y)$ is the Chebyshev polynomial and $y = 2x - 1$.

The error of the Bernstein approximation to the initial sample does not exceed 0.1, and the error of the Padé–Chebyshev approximation to the same sample is less than 0.0664, whereas the maximum error of the Padé–Chebyshev approximation to the Bernstein polynomial is 0.074. The maximum error of approximation to the derivative of this polynomial does not exceed 0.23, and it is attained on the boundary of the interval $[0, 1]$.

From Fig. 8.17 it is seen that these approximations practically coincide and both fit well the ordered data with exceptions at the boundaries of the interval.

Thus it is possible to reconstruct the data set to within the accuracy 0.064 using the Padé–Chebyshev approximations with only five coefficients.

Fig. 8.18 presents the derivatives of the Bernstein and Padé–Chebyshev approximations, and the smoothing properties of the latter approximation are again confirmed.

REMARK 8.3.6. In Example 8.3.4, all approximations possess greater accuracy than under the normal distribution. This can be explained, first, by the shape of the underlying Beta distribution which is well approximated by polynomials,

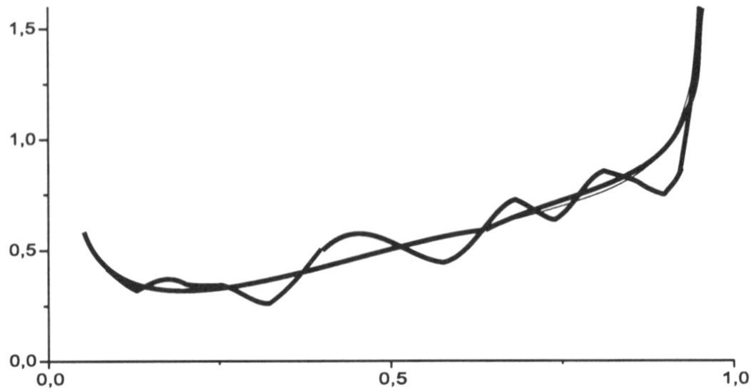

Figure 8.18. The derivatives of the Bernstein polynomial and Padé–Chebyshev approximations for the Beta-distributed data

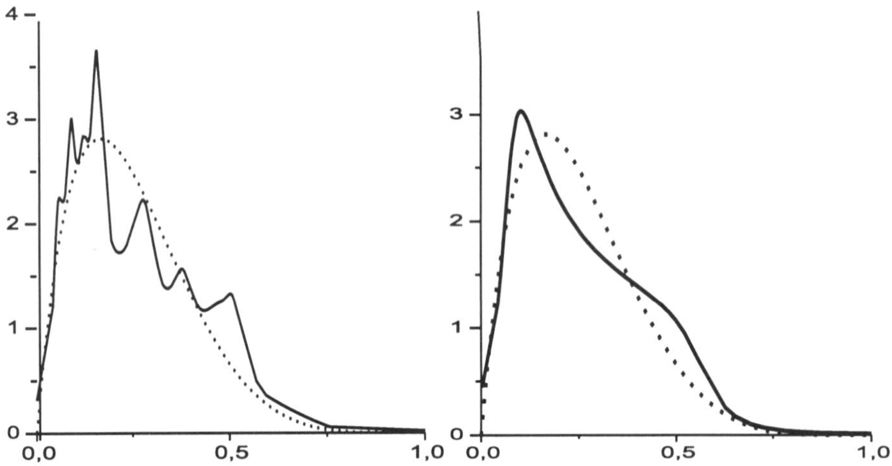

Figure 8.19. The Beta(2, 6)-density and its estimators based on the Bernstein and Padé–Chebyshev approximations for the Beta-distributed data

and, second, by the specific type (8.3.12) of the Bernstein approximation aimed at finite bounds of the data range.

As in Example 8.3.3, taking into account Theorem 8.3.5 and the relation $\Phi(u) = u$, we conclude that the Bernstein polynomial and its Padé–Chebyshev approximation can be used as reasonable estimators for the quantile function $Q(y)$ corresponding to the Beta distribution function.

To construct the distribution density, we use the procedure described in

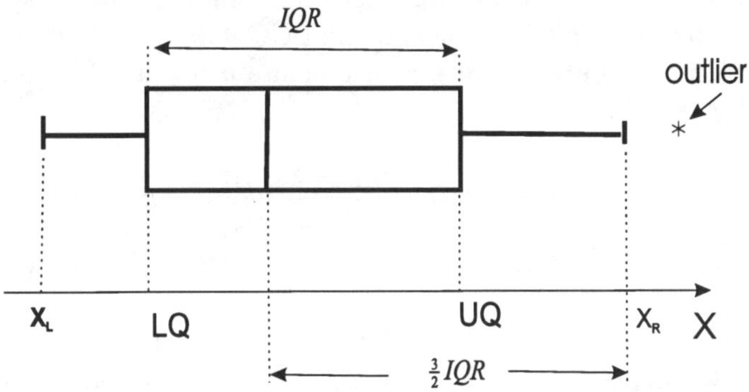

Figure 8.20. The boxplot based on the sample interquartile range

Example 8.3.3. Fig. 8.19 presents these density estimators. It is seen from these graphs that the first graph adapts to the behavior of the histogram (Fig. 8.16), whereas the second is close to the underlying Beta distribution density (the slashed line in both figures).

8.4. Robust bivariate boxplots

The univariate boxplot (Tukey, 1977) is a graphical tool for summarizing the distribution of a single random variable and for the fast visual comparison of different batches of data. Being a simple data analysis technique, it yields information about the location, scale, asymmetry, tails, and outliers of a data distribution, and moreover, it is available in many statistical packages.

In this section we recall the definition of the univariate boxplot, describe some its extensions to the bivariate case, and propose a simple robust bivariate boxplot technique aimed at detection of outliers.

Boxplot for the univariate data. A boxplot is the rectangle with the base equal to the sample interquartile range *IQR*, separated into two parts by the sample median (see Fig. 8.20).

From each side of 'the box', the two straight line segments ('mustaches') are drawn describing the distribution 'tails', and finally, the observations lying aside these domains are marked and plotted being the candidates for outliers. The left and right boundaries for the distribution 'tails' are given by

$$x_L = \max\left\{x_{(1)}, LQ - \frac{3}{2}IQR\right\}, \qquad x_R = \min\left\{x_{(n)}, UQ + \frac{3}{2}IQR\right\};$$

here LQ and UQ are the lower and upper sample quartiles, respectively. In general, they can be defined by $LQ = x_{(j)}$ and $UQ = x_{(n-j+1)}$ with $j = [0.25\,n]$, or, more accurately for small samples, by the medians of the left and right halves ($[0.5\,n]$) of the sample

$$
\begin{aligned}
&LQ = \frac{x_{(k)} + x_{(k+1)}}{2}, && UQ = \frac{x_{(3k)} + x_{(3k+1)}}{2}, && n = 4k,\\
&LQ = \frac{x_{(k)} + x_{(k+1)}}{2}, && UQ = \frac{x_{(3k+1)} + x_{(3k+2)}}{2}, && n = 4k + 1,\\
&LQ = x_{(k+1)}, && UQ = x_{(3k+2)}, && n = 4k + 2,\\
&LQ = x_{(k+1)}, && UQ = x_{(3k+3)}, && n = 4k + 3.
\end{aligned}
$$

This rule for the rejection of outliers is weaker than that proposed in Section 7.5, because the sample interquartile range is less resistant to outliers than the median absolute deviation. Nevertheless, one can use the other rule for constructing the boxplot 'mustache' as

$$x_L = \max\{x_{(1)}, \text{med} - 5\,\text{MAD}\}, \qquad x_R = \min\{x_{(n)}, \text{med} + 5\,\text{MAD}\}.$$

Observe that the boxplot summarizes information about a batch of data and it represents the following data characteristics:

- the sample median for location;
- the sample interquartile range for spread;
- the relation $(UQ - \text{med})/(\text{med} - LQ)$ for measuring asymmetry;
- $LQ - x_L$ and $x_R - UQ$ for the length of 'tails';
- the candidates for outliers.

Bivariate boxplots. As a rule, most of the proposed robust bivariate procedures are aimed at the data from a bivariate distribution of the elliptical shape. In the normal case, the contours of the joint density are elliptical. If the shape of the underlying distribution is far from elliptical, then normality may often be provided by rejection of the outliers and by transformation of the data.

However, the boxplot is a rather rough construction; it should allow departures from normality which preserve separation of the data into two groups: the central part consisting of the 'good' observations (its unimodal structure is strongly desirable) and the others, aside and sufficiently rare, 'bad'.

In general, if one can construct robust elliptical contours containing a given fraction of the data then it remains to inscribe a rectangle into the chosen contour, and so to define a boxplot.

In (Zani *et al.*, 1998), a computationally intensive form of the bivariate boxplot was proposed, successively peeling convex hulls to find the shape of the central part of the data. Those convex hulls are peeled until the first one is obtained which includes less than 50% of the data, and then this convex hull called the 50% hull is smoothed using a B-spline.

In (Goldberg and Iglewicz, 1992), a computationally less intensive version of boxplot was suggested by fitting ellipses to the data with the use of robust estimators of the parameters. The component-wise medians are used as location estimators. Then the required covariance matrix of the observations is estimated by sums of squares and products in these medians.

In (Atkinson and Riani, 2000), a fast very robust (mainly based on the LMS procedure) 'forward' search was applied to a large body of multivariate problems, constructing the bivariate boxplots included.

All the above described approaches to creating the bivariate boxplots require more or less intensive calculations. Now we propose a computationally simple technique that even allows to use calculations 'by hand' while constructing robust bivariate boxplots.

The robust bivariate boxplot based on the median correlation coefficient. The algorithm suggests the elliptical structures of the central part of the data, that is, we assume approximate normality of the data.

The main idea consists of transition to the principal axes of the ellipse of equal probability for the bivariate normal distribution (Cramér, 1946)

$$\begin{aligned} x' &= (x - \mu_1)\cos\phi + (y - \mu_2)\sin\phi, \\ y' &= -(x - \mu_1)\sin\phi + (y - \mu_2)\cos\phi, \end{aligned} \tag{8.4.1}$$

where

$$\tan 2\phi = \frac{2\rho\sigma_1\sigma_2}{\sigma_1^2 - \sigma_2^2}.$$

Fig. 8.21 illustrates the above.

We choose

- the component-wise medians $\widehat{\mu}_1 = \operatorname{med} x$ and $\widehat{\mu}_2 = \operatorname{med} y$ as estimators for location;
- the median absolute deviations $\widehat{\sigma}_1 = \operatorname{MAD} x$ and $\widehat{\sigma}_2 = \operatorname{MAD} y$ as estimators for scale;
- the median correlation coefficient $\mathbf{r}_{\text{med}}$ as an estimator for correlation;

and then use them in (8.4.1)

$$\begin{aligned} x' &= (x - \operatorname{med} x)\cos\phi + (y - \operatorname{med} y)\sin\phi, \\ y' &= -(x - \operatorname{med} x)\sin\phi + (y - \operatorname{med} y)\cos\phi, \end{aligned} \tag{8.4.2}$$

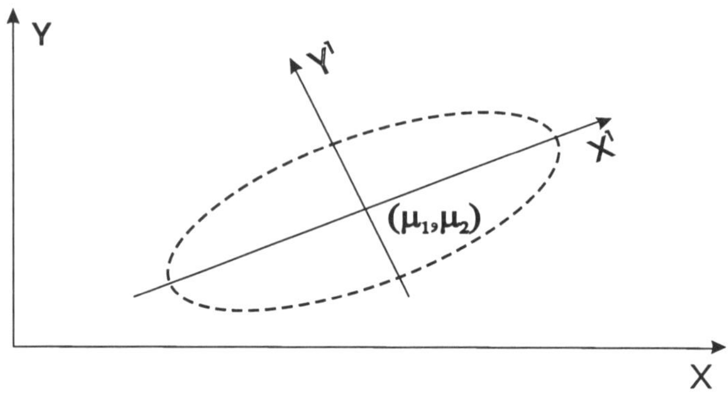

Figure 8.21. The principal axes for the normal density contours

where

$$\tan 2\phi = \frac{2\mathbf{r}_{\text{med}}\,\text{MAD}\,x\,\text{MAD}\,y}{(\text{MAD}\,x)^2 - (\text{MAD}\,y)^2}.$$

The boxplot itself consists of two rectangles: one into another with the sides parallel to the axes x' and y'. The lengths of these sides are defined similarly to the univariate case:

(i) the sides of the inner rectangle are equal to the sample interquartile ranges $IQR_{x'}$ and $IQR_{y'}$ evaluated by the samples $\{x'_i\}$ and $\{y'_i\}$;

(ii) the boundaries of the outer rectangle are defined by the median absolute deviations

$$x'_L = \max\{x'_{(1)}, \text{med}\,x' - 5\,\text{MAD}\,x'\}, \quad x'_R = \min\{x'_{(n)}, \text{med}\,x' + 5\,\text{MAD}\,x'\},$$
$$y'_L = \max\{y'_{(1)}, \text{med}\,y' - 5\,\text{MAD}\,y'\}, \quad y'_R = \min\{y'_{(n)}, \text{med}\,y' + 5\,\text{MAD}\,y'\}.$$

The inner rectangle describes the modal zone of a distribution, the outer rectangle is defined by the boundaries for the distribution tails (certainly, the indexes L and R for the left and right boundaries are used here as conventional), and the data values lying aside the outer rectangle are marked being the candidates for outliers. The relative location of these rectangles may indicate the departures from symmetry.

Summarizing the abovesaid, we state that the boxplot gives the following characteristics of the batch of observations:

LOCATION as the component-wise median and inner rectangle;

SPREAD/SCALE as the boundaries of the inner and outer rectangles;

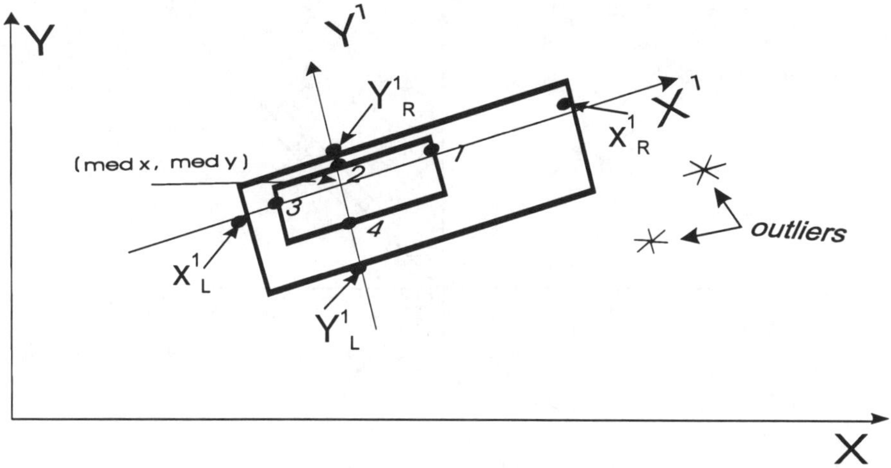

Figure 8.22. The robust bivariate boxplot

CORRELATION as the median correlation coefficient $\mathbf{r}_{\text{med}}$;

ASYMMETRY as the relative location of the rectangles;

'TAIL' AREAS as the boundaries of the outer rectangle;

ORIENTATION ON THE PLANE as the angle ϕ between the axes x' and x;

OUTLIERS as the marked observations.

Fig. 8.22 presents the bivariate boxplot defined so.

In Fig. 8.23, this boxplot technique is applied to the artificial data consisting of the following 15 points: the eight of them $(-10,-10)$, $(-3,-3)$, $(-2,-2)$, $(-1,-1)$, $(1,1)$, $(2,2)$, $(3,3)$, $(10,10)$ lie on the straight line $y = x$; other six $(-10,10)$, $(-5,5)$, $(-1,1)$, $(1,-1)$, $(5,-5)$, $(10,-10)$ lie on the line $y = -x$, and one point $(0,0)$ belongs to these both lines.

In this case,

(i) $\operatorname{med} x = 0$, $\operatorname{med} y = 0$; hence, $\operatorname{med} x' = 0$, $\operatorname{med} y' = 0$;

(ii) $\operatorname{MAD} x = \operatorname{MAD} y = 3$ and $\mathbf{r}_{\text{med}} = 1$; hence, $\phi = \pi/4$;

(iii) $IQR_{x'} = 2\sqrt{2}$, $IQR_{y'} = 0$; $\operatorname{MAD} x' = \sqrt{2}$, $\operatorname{MAD} y' = 0$.

It follows from the above results and from Fig. 8.23 that the obtained boxplot is degenerate: all the points lying on the line $y = -x$ are regarded as the candidates for outliers. Observe that the points $(-10,-10)$ and $(10,10)$ lying

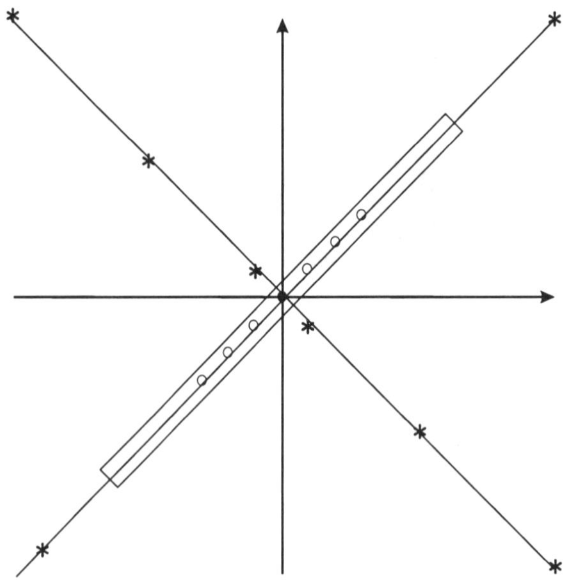

Figure 8.23. The bivariate boxplot for the artificial data

on the line $y = x$ are also such candidates. In addition, the value $\mathbf{r}_{\text{med}} = 1$ confirms the high qualitative robustness of the median correlation coefficient.

9

Applications

In this chapter, we present two kinds of applications. First, the above-developed methods may be used in some areas of applied statistics apparently providing stability of data processing algorithms, and sometimes giving a new point of view at inherent problems of the area of knowledge considered—this kind of applications may be called theoretical. Here we touch on such respectable sciences as the statistical theory of reliability and detection of signals.

Second, the applications in the common sense of word, dealing with some practical problems, desirably of a general interest. The problem of the dependence of sudden cardiac deaths on the meteorological and solar activity is surely of this kind.

9.1. On robust estimation in the statistical theory of reliability

Main results in robust statistics refer to the normal and its neighborhood models of data distributions. The aim of this section is to apply robust minimax approach to one of the traditional for the statistical reliability theory models of data distributions.

9.1.1. Robust minimax estimator of a scale parameter of the exponential distribution

The exponential distribution is the simplest model for the description of distributions of time before failure in the reliability theory(Barlow and Proschan, 1965; Gnedenko *et al.*, 1969; Barlow and Proschan, 1975; Gertsbakh, 1998). The efficient maximum likelihood estimator of the failure intensity parameter λ is the inverse value of the sample mean of the time to failure data: $\widehat{\lambda} = 1/\overline{T}$, $\overline{T} = \sum t_i$. The linear structure of the sample mean type estimators results in their great instability to the occurrence of rare outliers in the data. From the statistical point of view, this instability causes a sharp loss of the estimators' efficiency in the case of small deviations from the accepted stochastic model of data distribution, which in its turn may lead to significant errors in designed reliability characteristics.

Below we apply the Huber minimax approach to robust estimation of the scale parameter of the exponential distribution in the model of gross errors (Shevlyakov, 1995; Shevlyakov, 2000).

Problem setting. Consider a sample $t_1, \dots, t_n$ of random variables from the model of ε-contaminated exponential distributions

$$\mathcal{F} = \left\{ f\colon f(t,T) = \frac{1-\varepsilon}{T} \exp\left(-\frac{t}{T}\right) + \varepsilon h(t),\ 0 \le \varepsilon < 1 \right\}, \tag{9.1.1}$$

where ε is the contamination parameter characterizing the level of uncertainty of the accepted exponential model; $h(t)$ is an arbitrary distribution density; T is the scale parameter to be estimated.

Following (Huber, 1964), we consider M-estimators of the scale parameter T

$$\sum_{i=1}^{n} \chi\left(\frac{t_i}{\widehat{T}_n}\right) = 0, \tag{9.1.2}$$

where χ is the score function.

Assume that regularity conditions (F1) and (F2) are imposed on the class $\mathcal{F}$ of distribution densities, and conditions (χ1)–(χ4), on the class of score functions χ (Section 4.3).

The least informative density in the class of ε-contaminated exponential distributions. The least informative density $f^*(t)$ is the solution of the variational problem

$$f^* = \arg\min_{f\in\mathcal{F}} I(f), \qquad I(f) = \int_0^\infty t^2 \left(\frac{f'(t)}{f(t)}\right)^2 f(t)\,dt - 1 \tag{9.1.3}$$

with the side conditions of normalization and approximate exponentiality

$$\int_0^\infty f(t)\,dt = 1, \tag{9.1.4}$$

$$f(t) \ge (1-\varepsilon)e^{-t}, \tag{9.1.5}$$

where, without loss of generality, we set $T = 1$.

Restriction (9.1.1) on the class $\mathcal{F}$ is written in the inequality form (9.1.5), and it apparently includes the condition of non-negativeness.

THEOREM 9.1.1. *The solution of the variational problem* (9.1.3) *with side conditions* (9.1.4) *and* (9.1.5) *is of the form*

- *for* $0 \le \varepsilon < \varepsilon_0 = (1+e^2)^{-1}$,

$$f^*(t) = \begin{cases} (1-\varepsilon)e^{-t}, & 0 \le t < \Delta, \\ Ct^k, & t \ge \Delta, \end{cases} \tag{9.1.6}$$

where the parameters C, k, and Δ *are functions of* ε

$$C = (1-\varepsilon)e^{-\Delta}\Delta^{\Delta}, \quad k = -\Delta, \quad \frac{e^{-\Delta}}{\Delta - 1} = \frac{\varepsilon}{1-\varepsilon};$$

- *for* $\varepsilon_0 \le \varepsilon < 1$,

$$f^*(t) = \begin{cases} C_1 t^{k_1}, & 0 \le t < \Delta_1, \\ (1-\varepsilon)e^{-t}, & \Delta_1 \le t < \Delta_2, \\ C_2 t^{k_2}, & t \ge \Delta, \end{cases} \tag{9.1.7}$$

where the parameters C_1, Δ_1, k_1, C_2, Δ_2, *and* k_2 *are determined from the equations*

$$C_1 = (1-\varepsilon)e^{-1+\delta}(1-\delta)^{1-\delta}, \quad \Delta_1 = 1-\delta, \quad k_1 = -1+\delta,$$
$$C_2 = (1-\varepsilon)e^{-1-\delta}(1+\delta^{1+\delta}), \quad \Delta_2 = 1+\delta, \quad k_2 = -1-\delta,$$
$$\frac{e^{\delta}+e^{-\delta}}{e\delta} = \frac{1}{1-\varepsilon}. \tag{9.1.8}$$

In formulas (9.1.8), the auxiliary parameter δ, $0 \le \delta \le 1$, is introduced. The expressions for the Fisher information are of the following form, for the solutions (9.1.6) and (9.1.7), respectively:

$$I(f^*) = 1 - \varepsilon\Delta^2, \qquad I(f^*) = \frac{2\delta}{\tanh\delta} - \delta^2. \tag{9.1.9}$$

For small values of ε, the least informative density f^* corresponds to the exponential distribution in the zone $0 \le t < \Delta$; in the 'tail' area it is similar to the one-sided t-distribution. For large values of ε, a rather whimsical distribution minimizing the Fisher information appears—its density is ∞ at $t = 0$ (see also the Huber minimax solution in Section 4.4). The threshold values of the parameters are

$$\Delta_1 = 0, \quad \Delta_2 = 2, \quad \varepsilon = 0.119.$$

Some numerical results are given in Table 9.1.

The values of the distribution function

$$F^*(t) = \int_0^t f^*(t)\,dt,$$

evaluated at the points of 'glueing' of the extremals $C_1 t^{k_1}$ and $C_2 t^{k_2}$ with the constraint $(1-\varepsilon)e^{-t}$ are also given in Table 9.1.

Table 9.1. ε-contaminated exponential distributions minimizing the Fisher information for the scale parameter

ε	Δ_1	Δ_2	$F^*(\Delta_1)$	$F^*(\Delta_2)$	$1/I(f^*)$
0	0	∞	0	1	1
0.001	0	5.42	0	0.995	1.03
0.002	0	4.86	0	0.990	1.05
0.005	0	4.16	0	0.979	1.09
0.01	0	3.63	0	0.964	1.15
0.02	0	3.15	0	0.938	1.25
0.05	0	2.52	0	0.874	1.47
0.1	0	2.11	0	0.791	1.80
0.119	0	2	0	0.762	1.91
0.15	0.110	1.890	0.094	0.727	2.12
0.20	0.225	1.775	0.185	0.679	2.47
0.25	0.313	1.687	0.250	0.659	2.88
0.30	0.384	1.616	0.297	0.635	3.39
0.40	0.503	1.497	0.367	0.595	4.81
0.50	0.603	1.397	0.416	0.565	7.03
0.65	0.733	1.267	0.462	0.532	14.7
0.80	0.851	1.149	0.488	0.511	45.9
1	1	1	0.5	0.5	∞

The structure of the robust minimax estimator. The robust minimax estimator $\widehat{T}_n$, obtained from equation (9.1.2), is defined by the score function $\chi^*(t)$

$$\chi^*(t) = -t\,\frac{(f^*(t))'}{f^*(t)} - 1 = \begin{cases} \Delta_1 - 1, & 0 \le t < \Delta_1, \\ t - 1, & \Delta_1 \le t < \Delta_2, \\ \Delta_2 - 1, & t \ge \Delta_2. \end{cases} \tag{9.1.10}$$

Formula (9.1.10) holds for both solutions (9.1.6) and (9.1.7): solution (9.1.7) turns into solution (9.1.6) with $\Delta_1 = 0$.

Fig. 9.1 gives the shape of the score function χ^*.

We introduce

$$\mathscr{I}_1 = \{i : t_i/\widehat{T}_n < \Delta_1\}, \qquad \mathscr{I}_2 = \{i : t_i/\widehat{T}_n \ge \Delta_2\},$$
$$\mathscr{I} = \{i : \Delta_1 \le t_i/\widehat{T}_n < \Delta_2\}.$$

Then equation (9.1.2) becomes

$$\sum_{i\in\mathscr{I}_1}(\Delta_1 - 1) + \sum_{i\in\mathscr{I}}\left(\frac{t_i}{\widehat{T}_n} - 1\right) + \sum_{i\in\mathscr{I}_2}(\Delta_2 - 1) = 0. \tag{9.1.11}$$

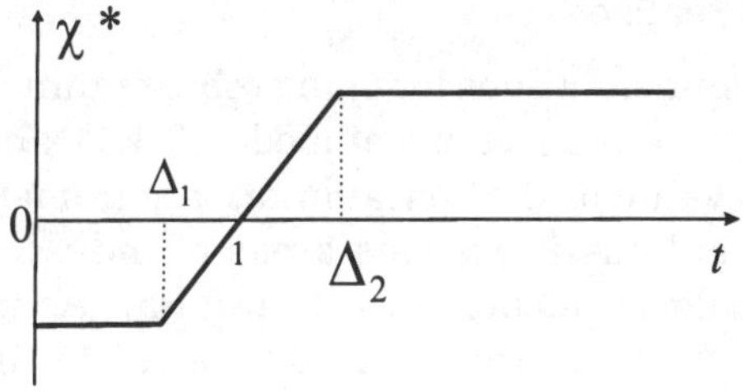

Figure 9.1. The score function of the robust minimax estimator for the scale parameter of the ε-contaminated exponential distribution

Denoting the numbers of observations belonging to the sets $\mathscr{I}_1$, $\mathscr{I}_2$, and $\mathscr{I}$ as n_1 n_2, and $n - n_1 - n_2$, from (9.1.11) we obtain

$$\widehat{T}_n = \frac{1}{n - n_1\Delta_1 - n_2\Delta_2} \sum_{i \in \mathscr{I}} t_i. \tag{9.1.12}$$

The structure of estimator (9.1.12) is similar to the structure of the trimmed mean with the correction term providing consistency

$$\widehat{T}_n(n_1, n_2) = \frac{1}{n - n_1\Delta_1 - n_2\Delta_2} \sum_{i=n_1+1}^{n-n_2} t_{(i)}, \tag{9.1.13}$$

where $t_{(i)}$ is the ith order statistic. If the numbers of the trimmed order statistics (left and right) are chosen as

$$n_1 = [F^*(\Delta_1)n], \qquad n_2 = [(1 - F^*(\Delta_2))n],$$

then the estimators $\widehat{T}_n$ derived from equation (9.1.11) and $\widehat{T}_n(n_1, n_2)$ are asymptotically equivalent. So, the simple estimator $\widehat{T}_n(n_1, n_2)$ (9.1.13) is recommended for the practical use.

Observe that in the limiting case where $\varepsilon \to 1$, the robust minimax estimator defined by the numerical solution of equation (9.1.11) (or by formula (9.1.13)) is the sample median $\widehat{T}_n = \operatorname{med} t_i$.

For $\varepsilon = 0$, the robust minimax estimator $\widehat{T}_n$ coincides with the sample mean $\overline{T}$, the efficient estimator for the scale parameter of the exponential distribution.

Summarizing the characteristics of the least informative density and score function, we can conclude that

- for small values of ε, the optimal algorithm provides the one-side sample trimming with subsequent averaging the remained observations;

- for large values of ε, the two-side trimming of the smallest and largest observations is realized.

The practical recommendations for using robust minimax estimator (9.1.13) are mainly defined by the restrictions of model (9.1.1), and within the context of this model, by the value of the contamination parameter ε. We recall that the results of investigations in various areas of industrial statistics show a good fit of ε-contamination models to actual data. As usual, the estimated and expected values of ε lie in the interval $(0.001, 0.1)$. If there is no a priori information about the value of ε, then one may set it equal to 0.1. In this case, according to formula (9.1.13) and to the results given in Table 9.1, the estimator is the one-sided trimmed mean at the level 21%.

REMARK 9.1.1. The rejection of those 21% of the largest time to failure values and the averaging of the remained gives perhaps not very optimistic but the guaranteed and reliable value of the mean time to failure characteristic.

9.1.2. On robust minimax estimators of the scale parameter for customary distributions of the statistical theory of reliability

The exponential distribution is an important but particular case of such models as the Weibull and gamma distributions (Ionescu and Limnios, 1999; Ushakov, 1994). In the problem of scale estimation, the parameters of location and form of these distributions are nuisance parameters, so in this setting we assume that their values are given. Now we apply the above approach to these general cases.

The structure of robust minimax estimators of the scale parameter in ε-contaminated models. We consider again the model of gross errors

$$\mathscr{F} = \{f\colon f(t) \geq (1-\varepsilon)g(t), 0 \leq \varepsilon < 1\}, \tag{9.1.14}$$

where $g(t)$ is a given distribution density satisfying the required regularity conditions (F1) and (F2); the scale parameter T is set equal to 1. To realize the Huber approach in its simplest version, we need an additional restriction on the tails of the density $g(t)$: they should be shorter than for t-type distributions, in other words, shorter than the extremals t^k of the variational problem of minimizing the Fisher information for the scale. Anyway, the Weibull and gamma distribution tails satisfy these restrictions.

In the general case of model (9.1.14), the solution of variational problem (9.1.3) can be rather complicated, but in the case of ε small enough (actually, this case precisely is of interest for applications), it is of the form

$$f^*(t) = \begin{cases} (1-\varepsilon)g(t), & 0 \leq t < \Delta, \\ Ct^k, & t \geq \Delta, \end{cases} \tag{9.1.15}$$

where the parameters C, k, and Δ are determined from the normalization conditions

$$\int_0^\infty f^*(t)\,dt = 1,$$

whereas the conditions of transversality provide the smoothness of glueing of the free extremals Ct^k with the constraint $(1-\varepsilon)g(t)$:

$$(1-\varepsilon)g(\Delta) = C\Delta^k, \qquad (1-\varepsilon)g'(t) = kC\Delta^{k-1}.$$

The concrete formulas for evaluating the parameters sought for can be more or less easily written out both for the ε-contaminated Weibull, gamma, or lognormal distributions, but the value of the upper bound ε_0, which guarantees the validity of solution (9.1.15) in the domain $\varepsilon \le \varepsilon_0$, depends on the concrete shape of the underlying distribution, and it seems not simple to get it in the general form.

However, assuming that solution (9.1.15) holds with small ε, we describe the structure of the minimax estimator for scale in this case.

The optimal score function is of the form

$$\chi^*(t) = -t\,\frac{(f^*(t))'}{f^*(t)} - 1 = \begin{cases} -t\frac{g'(t)}{g(t)} - 1, & 0 < t \le \Delta, \\ -k-1, & t > \Delta. \end{cases} \tag{9.1.16}$$

We introduce

$$\mathcal{I}_* = \{i : t_i/\widehat{T}_n > \Delta\}, \qquad \mathcal{I} = \{i : 0 < t_i/\widehat{T}_n \le \Delta\}.$$

Then equation (9.1.2) takes the form

$$\sum_{i\in\mathcal{I}}\left(-\frac{t_i}{\widehat{T}_n}\frac{g'(t_i/\widehat{T}_n)}{g(t_i/\widehat{T}_n)} - 1\right) + \sum_{i\in\mathcal{I}_*}(-k-1) = 0. \tag{9.1.17}$$

Denoting the numbers of observations belonging to the sets $\mathcal{I}_*$ and $\mathcal{I}$ by n_* and $n - n_*$, we find from (9.1.17) that the robust minimax estimator with score function (9.1.16) has the structure of the trimmed maximum likelihood estimator where n_* out of largest observations t_i are rejected, the rest of the sample being processed by the maximum likelihood method.

In the limiting case with $\varepsilon \to 0$, this estimator is the maximum likelihood estimator of scale for the $g(t)$.

9.2. Robust detection of signals based on optimization criteria

In this section we apply the results on the robust minimax estimation of location to the problems of detection of known signals. This problem itself can be also re-formulated in a non-traditional for nonparametric statistics way, allowing to choose the decision on detection by comparing the measures of closeness of the signals to the observed data. Much attention is paid to the application of the L_1-norm criterion to the detection of known signals both in discrete and continuous cases.

9.2.1. Preliminaries

Consider the problem of coherent binary detection of a known signal in the discrete case

$$x_i = \theta s_i + e_i, \qquad i = 1, 2, \dots, n, \tag{9.2.1}$$

and in the continuous case

$$x(t) = \theta s(t) + e(t), \qquad 0 \le t \le T. \tag{9.2.2}$$

The values of s_i are assumed to be known, and the independent errors e_i are from a common distribution F with density f.

The problem of detection is formulated as follows: given $\{x_i\}$ or $\{x(t)\}$, it is necessary to decide whether the signal $s(t)$ is observed or not, which value does the parameter θ take, $\theta = 0$ or $\theta = 1$. In this setting, the problem of detection is equivalent to the problem to test the hypothesis

$$H_0\colon \theta = 0 \quad \text{against} \quad H_1\colon \theta = 1. \tag{9.2.3}$$

Given prior information on error distributions, the classical theory of hypotheses testing yields various optimal (in the Bayesian, minimax, Neyman–Pearson senses) algorithms to solve problem (9.2.3) (Lehmann, 1959; Kendall and Stuart, 1963). In this case, all optimal algorithms are based on the likelihood ratio statistic (LHR): for solution, one must evaluate the value of this statistic and compare it with a certain bound. The differences between the above-mentioned approaches result only in the values of that bound.

Let us look back at problem (9.2.1); the LHR statistic is of the form

$$T_n(\mathbf{x}) = \prod_{i=1}^{n} \frac{f(x_i - s_i)}{f(x_i)}, \qquad \mathbf{x} = (x_1, x_2, \dots, x_n). \tag{9.2.4}$$

Observe that in this case it is necessary to know the true density f.

As a rule, in the practical problems of radio-location, acoustics, and communication, the error (or signal) distributions are only partially known. For instance, it may be known that the underlying distribution is close to normal, or/and there is some information on its behavior in the central zone, and nothing is known about the distribution tails, etc. In similar cases, distributions may be distorted by impulsive noises, which is typical for acoustics (Milne,

1962), sound location (Olshevskie, 1967), radio-location and communication (Mertz, 1961; Engel, 1965; Hall, 1966; Bello and Esposito, 1969).

The classic methods based on the LHR statistic behave poorly in the above situations (see some examples in (Huynh and Lecours, 1975; Krasnenker, 1980). There exists a great body of researches (e.g. (Krishnaiah and Sen, 1984)) on nonparametric and rank procedures of detection, which provide protection from impulsive noise and gross errors in the data, but, usually, at the sacrifice of considerable lack of efficiency as compared with the parametric LHR statistic tests in the case of absence of outliers. Therefore, designing efficient and sufficiently simple nonparametric tests for detection of signals still remains an important problem.

Huber minimax approach. There are some other alternatives to the parametric approach. In his fundamental work (Huber, 1965), Huber noticed that the LHR statistic is not robust, since some observations x_i making the term $f(x_i - s_i)/f(x_i)$ close to zero or very large destroy the LHR statistic. This happens, for example, when there are heavy-tailed deviations from normality.

Consider the problem of detection of a known constant signal in an additive noise

$$\begin{aligned} H_0\colon &\quad x_i = e_i, \\ H_1\colon &\quad x_i = \theta + e_i, \quad i = 1, 2, \dots, n, \end{aligned} \tag{9.2.5}$$

where $\theta > 0$ is a known signal.

Huber suggests to use the robust trimmed version of the LHR statistic

$$T_n(\mathbf{x}) = \prod_{i=1}^{n} \pi(x_i),$$

$$\pi(x) = \begin{cases} c', & x < c', \\ f_1(x)/f_0(x), & c' \le x \le c'', \\ c'', & x > c'', \end{cases} \tag{9.2.6}$$

where $f_0(x)$ and $f_1(x)$ are the distribution densities of observations under the hypotheses H_0 and H_1, respectively; $0 \le c' < c'' < \infty$ are some parameters. Observe that the lack of information about error distributions makes it necessary to consider the composite hypotheses $H_j, j = 0, 1$, that is, to deal with the problems of nonparametric nature. Usually, it is assumed that the hypotheses H_j are formed by the deviations of the true distribution densities $f_j(x)$ from the assumed (estimated or model) ones $\widehat{f}_j(x)$. Huber considers the case where the composite hypotheses H_j are set in the form of ε-contaminated distributions

$$H_j = \{f_j(x)\colon f_j(x) \ge (1 - \varepsilon_j)\widehat{f}_j(x)\}, \quad 0 \le \varepsilon_j \le 1, \quad j = 0, 1,$$

and constructs the minimax decision rule minimizing the risk in the least favorable case. This rule is of form (9.2.6).

Observe that the parameters c' and c'' can be uniquely determined only for sufficiently small ε_j, and, vice versa, with the fixed values of ε_j, they are correctly defined for sufficiently large $|s|$.

The minimax approach is also used in the following problem of testing the composite hypotheses (Kuznetsov, 1976)

$$H_j = \{f_j(x)\colon \underline{f}_j(x) \le f_j(x) \le \overline{f}_j(x)\}, \qquad j = 0, 1,$$

where $\underline{f}_j(x)$ and $\overline{f}_j(x)$ are the fixed functions determining the bounds for the true density $f_j(x)$. For instance, these bounds can be obtained with the use of the confidence limits for the Parzen density estimators. The minimax decision rule minimizing the Bayes risk in the least favorable case is reduced to the algorithms of the type 'supremum of the density by supremum', 'infimum by infimum', etc. Such algorithms can be applied to detection of any approximately known signal.

Asymptotically optimal robust detection. The detection of weak signals is of a particular interest, since the robust Huber test does not work in the zone of weak signals, where the classes of distribution densities for the hypothesis and alternative are overlapped.

Consider the problem to test the hypotheses

$$\begin{aligned} H_0&: \quad x_i = e_i, \\ H_1&: \quad x_i = \theta s_i + e_i, \quad i = 1, 2, \dots, n, \end{aligned} \tag{9.2.7}$$

for an arbitrary $\theta > 0$. Let f be a distribution density from the class $\mathcal{F}$, and d be some decision rule from the class $\mathcal{D}$ of randomized decision rules. The power function P_D for $d \in \mathcal{D}$ is then defined as

$$P_D = \beta_d(\theta \mid f) = \mathsf{E}_\theta\{d(\mathbf{x}) \mid f\}, \qquad f \in \mathcal{F}, \tag{9.2.8}$$

and the probability of false alarm is $P_F = \alpha = \beta_d(0 \mid f)$.

Now we set the following problem of detection: find a decision rule that guarantees the asymptotic level of quality of detection independently of the chosen f from the class $\mathcal{F}$, i.e., find the solution of the following maximin problem (El-Sawy and Vandelinde, 1977)

$$\max_{d \in \mathcal{D}} \min_{f \in \mathcal{F}} \beta_d(\theta \mid f) \tag{9.2.9}$$

with the side condition

$$\beta_d(0 \mid f) \le \alpha \quad \text{for all} \quad f \in \mathcal{F}. \tag{9.2.10}$$

Choose the value $\sqrt{n}\theta_n$ as the statistic for the decision rule d_ρ, where θ_n is an M-estimator of location of the form

$$\theta_n = \arg\min_\theta \sum_{i=1}^n \rho(x_i - \theta s_i). \qquad (9.2.11)$$

The function ρ should satisfy the conditions

(D1) the convex and symmetric function of contrast $\rho(u)$ strictly increases with positive u;

(D2) the score function $\psi(u) = \rho'(u)$ is continuous for all u;

(D3) $\mathsf{E}_F\psi^2 < \infty$ for all $f \in \mathscr{F}$;

(D4) $\dfrac{\partial \mathsf{E}_F\psi(x-\theta)}{\partial\theta}$ exists and is nonzero in some neighborhood of θ.

It is proved in (El-Sawy and Vandelinde, 1977) that if f^* minimizes the Fisher information over the class $\mathscr{F}$

$$f^* = \arg\min_{f\in\mathscr{F}} \int_{-\infty}^{\infty} \left(\frac{f'(x)}{f(x)}\right)^2 f(x)\,dx, \qquad (9.2.12)$$

the function $\rho^* = -\ln f^*$ satisfies conditions (D1)–(D4), and the inequality

$$A(f, \psi^*) \le A(f^*, \psi^*) \qquad (9.2.13)$$

holds for all $f \in \mathscr{F}$, where

$$A(f, \psi) \propto \frac{\mathsf{E}_F\psi^2}{(\partial \mathsf{E}_F\psi(x-\theta)/\partial\theta|_{\theta=0})^2},$$

then the following relations hold for each v greater than some bound γ:

$$\beta_d(0 \mid f) \le \beta_{d_{\rho^*}}(0 \mid f^*) \quad \text{for all} \quad f \in \mathscr{F},$$
$$\inf_{f\in\mathscr{F}} \beta_{d_{\rho^*}}(v \mid f) = \beta_{d_{\rho^*}}(v \mid f^*) = \sup_{d\in\mathscr{D}} \beta_d(v \mid f^*).$$

From these relations it follows that the pair (d_{ρ^*}, f^*) is the saddle point of the function $\beta_d(\theta \mid f)$, and also it is the solution of problem (9.2.9). The bound γ is defined by the false alarm probability α.

REMARK 9.2.1. Since the term $A(f, \psi)$ is proportional to the asymptotic variance $V(f, \psi)$ for M-estimators of location, this assertion allows for direct application of all results on minimax estimation of the location parameter to the problem of detection of known signals.

9.2.2. Asymptotic robust minimax detection on the basis of the optimization criteria for M-estimators.

Formulation of the algorithm for binary detection. Consider again the problems of testing hypotheses (9.2.1), (9.2.2), and (9.2.3). We suggest the following nonparametric algorithm of detection for these problems:

$$d(\mathbf{x}) = \begin{cases} 1, & \sum_{i=1}^n \rho(x_i) - \sum_{i=1}^n \rho(x_i - s_i) > 0, \\ \frac{1}{2}, & \sum_{i=1}^n \rho(x_i) - \sum_{i=1}^n \rho(x_i - s_i) = 0, \\ 0, & \sum_{i=1}^n \rho(x_i) - \sum_{i=1}^n \rho(x_i - s_i) < 0, \end{cases} \tag{9.2.14}$$

in the discrete case, and

$$d(\mathbf{x}) = \begin{cases} 1, & \int_0^T \rho[x(t)]\,dt - \int_0^T \rho[x(t) - s(t)]\,dt > 0, \\ \frac{1}{2}, & \int_0^T \rho[x(t)]\,dt - \int_0^T \rho[x(t) - s(t)]\,dt = 0, \\ 0, & \int_0^T \rho[x(t)]\,dt - \int_0^T \rho[x(t) - s(t)]\,dt < 0, \end{cases} \tag{9.2.15}$$

in the continuous case, where $\rho(u)$ is the contrast function (Shevlyakov, 1976; Chelpanov and Shevlyakov, 1983).

Randomized decision rules (9.2.14) and (9.2.15) are of clear structure: the choice of the hypotheses H_0 or H_1 depends on the distance of the signals $\{0\}$ and $\{s_i\}$ from the observed data $\{\mathbf{x}\}$ measured by the value of the optimization criterion.

Formulation of the algorithm for multi-alternative detection. Consider the problem of multi-alternative detection

$$\begin{aligned} H_0&: \quad x_i = s_0 + e_i, \\ H_1&: \quad x_i = s_1 + e_i, \\ &\ \ \vdots \\ H_k&: \quad x_i = s_k + e_i, \quad i = 1, \dots, n, \end{aligned} \tag{9.2.16}$$

where the signals $s_0, s_1, \dots, s_k$ are known.

In this case, the decision is made in favor of the signal s_j (the hypothesis H_j) that minimizes the distance from the observed data

$$s_j = \arg \min_{\theta = s_0, \dots, s_k} \sum_{i=1}^n \rho(x_i - \theta). \tag{9.2.17}$$

One may think on the transition from the problem of multi-alternative detection (9.2.16) to the problem of estimation of the location parameter θ if to consider the continuum set of hypotheses in (9.2.16)

$$\theta_n = \arg\min_\theta \sum_{i=1}^n \rho(x_i - \theta).$$

Thus the proposed non-traditional forms (9.2.14), (9.2.15), and (9.2.17) of the algorithm for detection have the obvious connection with the problem of estimation of the location parameter.

There is another motivation for the introduced decision rules. In statistics, there exist two general approaches to estimation of parameters: point and interval estimation. Evidently, the above procedure of hypotheses testing may be referred to as 'point hypothesis testing'. Henceforth we call such a procedure *ρ-test*.

Asymptotic normality of ρ-test statistics. Consider the problem of detection of a known constant signal $\theta > 0$, or the problem of testing the simple hypothesis H_0 against the simple alternative H_1

$$\begin{aligned} H_0: &\quad x_i = e_i, \\ H_1: &\quad x_i = \theta + e_i, \quad i = 1, 2, \dots, n. \end{aligned} \tag{9.2.18}$$

Given the error distribution density f, the quality of detection by ρ-tests is completely determined by the contrast function ρ. It is easy to see that the choice $\rho(u) = -\log f(u)$ leads to the optimal LHR test statistic minimizing the Bayesian risk with equal costs and a priori probabilities of hypotheses

$$\begin{aligned} \sum_{i=1}^n \rho(x_i) < \sum_{i=1}^n \rho(x_i - \theta) &\iff \sum_{i=1}^n \log f(x_i) > \sum_{i=1}^n \log f(x_i - \theta) \\ &\iff \prod_{i=1}^n \frac{f(x_i - \theta)}{f(x_i)} < 1. \end{aligned} \tag{9.2.19}$$

In this setting, under distributions symmetric about zero, the power function P_D and the false alarm probability P_F are mutually complementary so that $P_D = 1 - P_F$.

The following particular cases are of interest:

- $\rho(u) = u^2$ defines the LS or *L_2-norm test*, which is optimal under normal errors with the sample mean as the test statistic

$$\sum_{i=1}^n x_i^2 < \sum_{i=1}^n (x_i - \theta)^2 \iff T_n(\mathbf{x}) = \frac{1}{n}\sum_{i=1}^n x_i < \frac{\theta}{2};$$

- $\rho(u) = |u|$ yields the LAV or *L_1-norm test*, which is optimal under the double-exponential or Laplace error distribution with the Huber-type test statistic

$$\sum_{i=1}^n |x_i| < \sum_{i=1}^n |x_i - \theta| \iff T_n(\mathbf{x}) = \frac{1}{n}\sum_{i=1}^n \psi_H(x_i; 0, \theta) < \frac{\theta}{2},$$

where $\psi(x; a, b) = \max(a, \min(x, b))$;

- the Chebyshev norm $\rho(u) = \max |u|$ defines the L_∞-*norm test*, which is optimal under the uniform error distribution.

Now we establish the asymptotic normality of the ρ-test statistic.

Let a distribution density $f \in \mathcal{F}$ and a score function $\psi \in \Psi$ satisfy the regularity conditions ($\mathcal{F}$1), ($\mathcal{F}$2), (Ψ1)–(Ψ4) of Section 1.2. Consider the function

$$q(\theta_n) = n^{-1}\sum_{i=1}^{n} \rho(\theta_n + e_i) - n^{-1}\sum_{i=1}^{n} \rho(e_i).$$

By the Taylor expansion, we obtain

$$q(\theta_n) = q(0) + q'(0)\theta_n + \frac{q''(\xi\theta_n)}{2}\theta_n^2, \qquad 0 < \xi < 1.$$

Furthermore,

$$q(\theta_n) = \theta_n n^{-1}\sum_{i=1}^{n} \psi(e_i) - (\theta_n^2/2)n^{-1}\sum_{i=1}^{n} \psi'(e_i + \xi\theta_n).$$

Consider the infinitesimal alternative

$$\theta_n \to 0 \quad \text{as} \quad n \to \infty: \quad \theta_n = \theta/\sqrt{n}.$$

Set $T_n = n^{-1}\sum_{i=1}^{n} \psi(e_i)$ and $T_n' = n^{-1}\sum_{i=1}^{n} \psi'(e_i + \xi\theta_n)$. Then we can write for the power function

$$P_D = \mathsf{P}\left(q(\theta_n) > 0\right) = \mathsf{P}\left(T_n > -\frac{\theta_n}{2}T_n'\right).$$

Since $n^{1/2}T_n$ is asymptotically normal with mean 0 and variance $\mathsf{E}_F\psi^2$ and T_n' tends in probability to the positive constant $c = \mathsf{E}_F\psi' > 0$, we obtain

$$P_D = \Phi\left(\frac{\theta}{2}\,\frac{\mathsf{E}_F\psi'}{\sqrt{\mathsf{E}_F\psi^2}}\right). \tag{9.2.20}$$

From (9.2.20) it follows that the maximin problem

$$\max_{\psi\in\Psi}\min_{f\in\mathcal{F}} P_D(f,\psi) \tag{9.2.21}$$

is equivalent to the minimax problem

$$\min_{f\in\mathcal{F}}\max_{\psi\in\Psi} V(f,\psi), \tag{9.2.22}$$

where $V(f,\psi) = \mathsf{E}_F\psi^2/(\mathsf{E}_F\psi')^2$ is the asymptotic variance for M-estimators of a location parameter.

Thus, all results on minimax estimation of location are also true in this case, i.e., they provide the guaranteed level of the power function P_D

$$P_D(f, \psi^*) \geq P_D(f^*, \psi^*) \quad \text{for all} \quad f \in \mathscr{F},$$

and/or of the false alarm probability P_F

$$P_F(f, \psi^*) \leq P_D(f^*, \psi^*) \quad \text{for all} \quad f \in \mathscr{F},$$

since $P_D = 1 - P_F$.

9.2.3. The L_1-norm test

In this section we present some results on the properties of the L_1-test both in small samples and asymptotics. Certainly, it is easy to foresee these results: the behavior of the L_1-norm test as compared to the L_2-norm procedure must be similar to the comparative behavior of the sample median and sample mean due to the structures of the corresponding test statistics. Moreover, we recall that, since the L_1-norm procedure is optimal for the class $\mathscr{F}_1$ of nondegenerate distributions and for the ε-contaminated models as $\varepsilon \to 1$, the L_1-test inherits these minimax properties.

Discrete case: detection of a known signal under gross errors. Consider the problem of detection of a known signal $\theta > 0$ (9.2.18)

$$\begin{aligned} H_0&: \quad x_i = e_i, \\ H_1&: \quad x_i = \theta + e_i, \quad i = 1, 2, \dots, n. \end{aligned}$$

in the gross error model

$$f_e(x) = (1 - \varepsilon)\mathscr{N}(x; 0, 1) + \varepsilon\mathscr{N}(x; 0, k), \quad 0 \leq \varepsilon < 1, \quad k \geq 1 \qquad (9.2.23)$$

for various values of the contamination parameters ε and k in small samples ($n = 5, 7, 9$), large samples ($n = 100$), and in asymptotics as $n \to \infty$.

Comparative studies of the power functions for the L_1- and L_2-tests in small samples show that $P_D(L_1) > P_D(L_2)$ (recall that the power function completely characterize the quality of detection in this setting) from $k \approx 3$ ($\varepsilon = 0.1, 0.2$), and, only in the normal case, the L_1-test is inferior to the optimal L_2-test (see Fig. 9.2).

In asymptotics, we use the Pitman *asymptotic relative efficiency* (ARE) (Pitman, 1948) as the measure to compare the tests. For ARE(L_1, L_2),

$$\text{ARE}(L_1, L_2) = 4f_e^2(0)\sigma_e^2, \qquad (9.2.24)$$

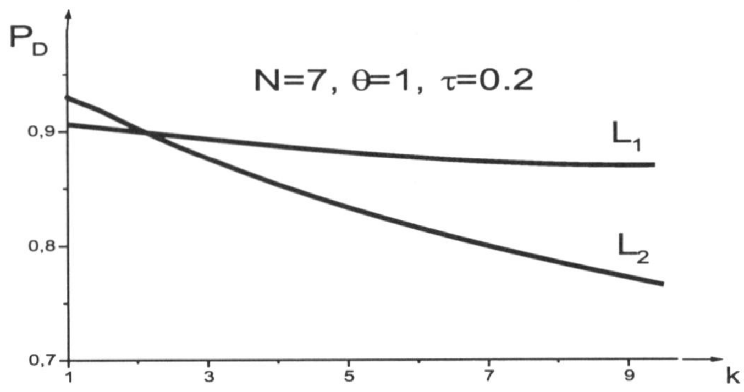

Figure 9.2. The power functions of the L_1- and L_2-tests under gross errors

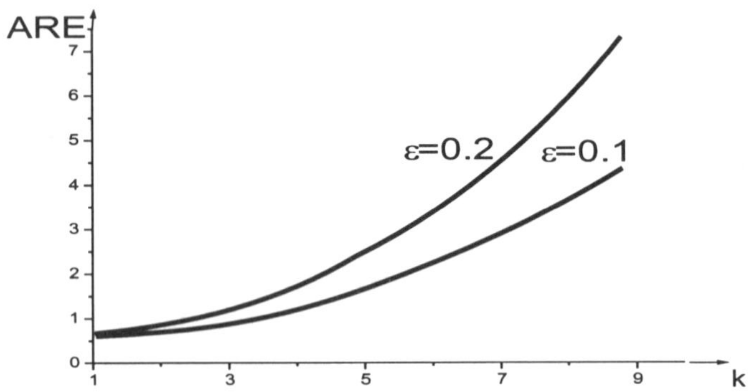

Figure 9.3. Asymptotic relative efficiency of the L_1- and L_2-tests under gross errors

and for model (9.2.23)

$$\text{ARE}(L_1, L_2) = \frac{2}{\pi}\left(1 - \varepsilon + \frac{\varepsilon}{k}\right)^{-2}(1 - \varepsilon + \varepsilon k^2). \tag{9.2.25}$$

Fig. 9.3 presents the ARE-curves for two values of ε. From (9.2.25) it follows that the L_1-test is inferior to the optimal L_2-test under normal errors (ARE $= 2/\pi \approx 0.637$), and it is essentially superior to the L_2-test under gross errors.

Observe that formula (9.2.24) coincides with the asymptotic relative efficiency of the sample median and sample mean, which are the optimal L_1- and L_2-estimators of the location parameter—here the connection between the problems of detection and estimation is evident. It is also known (Kendall and

Stuart, 1963; Hájek and Šidák, 1967) that formula (9.2.24) gives the value of the ARE of the *sign test* to the *Student t-test* based on the sample mean. Hence it follows that the L_1-test is equivalent in the sense of ARE to the sign test for the problem (9.2.18).

Continuous case: detection of a known signal under the Gaussian white noise. Consider the problem of detection of a known signal $\theta(t)$

$$\begin{aligned} H_0: &\quad x(t) = e(t), \\ H_1: &\quad x(t) = \theta(t) + e(t), \quad 0 \le t \le T. \end{aligned} \tag{9.2.26}$$

Now we find the power of the L_1-test under the Gaussian white noise.

Let the noise $e(t)$ be the *white noise* with zero mean and $\mathsf{E}e(t)e(t+\tau) = N_0\delta(\tau)$, where N_0 is its *power spectrum*.

It is well known (van Trees, 1971) that the Gaussian white noise can be derived by the limit transition from the following discrete scheme. Let $[0, T]$ be the interval of processing, $\Delta t = T/n$, $\{e_k\}$ be a sequence of independent normal random variables with zero mean and variance $N_0/\delta t$, $k = 1, \dots, n$. Then the random process $\{e_k\}$ converges to the Gaussian white noise.

For problem (9.2.26), the power of the L_2-test is

$$P_D(L_2) = \Phi(d/2), \tag{9.2.27}$$

where $d = \sqrt{E/N_0}$, $E = \int_0^T \theta^2(t)\,dt$ is the signal energy.

For the power of the L_1-test,

$$P_D(L_1) = \Phi(d/\sqrt{2\pi}). \tag{9.2.28}$$

From (9.2.27) and (9.2.28) it follows that the L_1-test is inferior to the L_2-test under the Gaussian white noise, but all the qualitative peculiarities of the optimal detection by the L_2-test are preserved for the L_1-test as well. For instance, the power depends only on the *signal-noise ratio* d and does not depend on the signal shape.

Fig. 9.4 demonstrates the dependence of the false alarm probability $P_F = 1 - P_D$ on the signal-noise ratio d for the L_1- and L_2-tests.

Continuous case: detection of a constant signal under impulsive noise. Consider now the impulsive noise case. Let $\theta(t) = \theta > 0$ be a constant signal, and the noise $e(t)$ be the impulses of magnitude $h > \theta$ and total duration T^+.

In this case, for the L_1-test

$$P_D = 1, \qquad P_F = \mathsf{P}(T^+ > T/2). \tag{9.2.29}$$

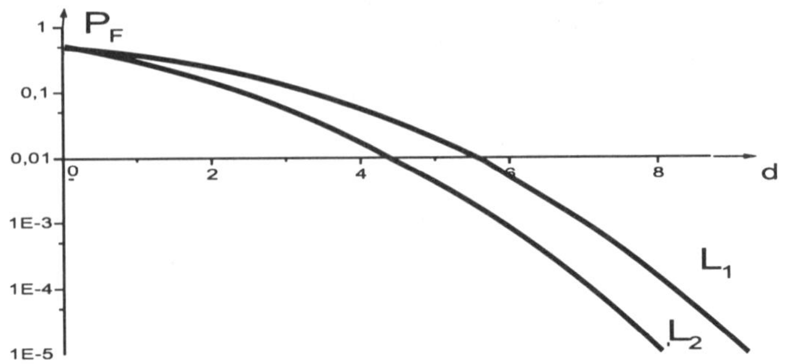

Figure 9.4. The false alarm probability for the L_1- and L_2-tests under the Gaussian white noise

For the impulsive noise of the opposite sign where $-h < -\theta$, these probabilities can be rewritten as

$$P_D = \mathsf{P}(T^- < T/2), \qquad P_F = 0, \tag{9.2.30}$$

where T^- is the total duration of negative impulses.

From (9.2.29) and (9.2.30) it is seen that the characteristics of the quality of detection do not depend on the magnitude of impulsive noise but are determined by the value of its total duration. Comparing the quality of detection by the L_2-test, for example, in the case of positive impulses, we obtain

$$P_D = 1, \qquad P_F = \mathsf{P}\left(T^+ > \frac{\theta}{2h}T\right).$$

Here with increasing h, the false alarm probability tends to 1. Given $0 < h < \theta$, the false alarm probabilities for the L_1- and L_2-tests coincide

$$P_F(L_1) = P_F(L_2) = \mathsf{P}\left(T^+ > \frac{\theta}{2h}T\right),$$

and they are equal to zero for $h < \theta/2$.

The marked independence of $P_F(L_1)$ of the magnitude of impulsive noise qualitatively repeats the peculiarities of the best L_1-norm approximations (see Chapter 6) and predetermines the robustness properties of detection by the L_1-test.

Setting a specific shape of the distributions for the times of impulse duration and pauses, it is possible to estimate numerically the quality of detection. For arbitrary distribution laws, the expression for the distribution of the total duration of impulses has a complicated form (Gnedenko *et al.*, 1969) and thus

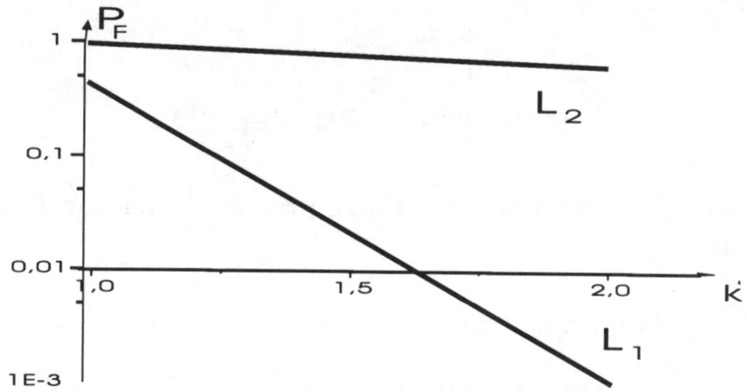

Figure 9.5. The false alarm probabilities for the L_1- and L_2-tests under impulsive noise

it is practically useless. However, given a large time of processing, one can use the normal approximation to this distribution function. In the case where the duration of impulses and pauses are distributed by the exponential law with parameters λ and μ, and the noise $e(t)$ is the Markov process with two states $\{0\}$ and $\{h\}$, the expectation and variance of T^+ are (see Section 6.1):

$$\mathsf{E}T^+ = \frac{T_1}{T_1 + T_2} T, \qquad \operatorname{Var} T^+ = \frac{2T_1^2 T_2^2}{(T_1 + T_2)^3} T,$$

where T_1, T_2 are the mean duration of impulses and pauses: $T_1 = 1/\lambda$ and $T_2 = 1/\mu$. Setting $k = T_2/T_1$ and $n = T/\sqrt{T_1 T_2}$, for the false alarm probability we obtain

$$P_F(L_1) = 1 - \Phi((k-1)(k+1)^{1/2} n^{1/2}/k^{3/4}). \tag{9.2.31}$$

Observe that

$$\begin{aligned} P_F(L_1) &= 1/2, && k = 1; \\ \lim_{n\to\infty} P_F(L_1) &= 0, && k > 1; \\ \lim_{n\to\infty} P_F(L_1) &= 1, && k < 1. \end{aligned}$$

Hence it follows that the L_1-test is consistent only for $k > 1$.

Fig. 9.5 demonstrates dependence of $P_F(L_1)$ and $P_F(L_2)$ on the value of k for $n = 100$ and $h = 1.5\theta$. The superiority of the L_1-test is obvious here.

Continuous case: detection of a periodic signal under impulsive noise. We introduce

$$\mathcal{T}_\theta^+ = \{t \mid \theta(t) \geq 0\}, \quad \mathcal{T}_\theta^- = \{t \mid \theta(t) < 0\}, \quad \mathcal{T}_e^+ = \{t \mid e(t) \geq 0\}.$$

Assume that the signal $\theta(t)$ and the impulsive noise $e(t)$ satisfy the conditions

$$\int_0^T \theta(t)\,dt = 0, \qquad e(t) > \max_{t\in \mathcal{T}_e} |\theta(t)|. \tag{9.2.32}$$

Then the power function and false alarm probability for the L_1-test can be represented as

$$P_F(L_1) = (1-p)\mathsf{P}(\mathcal{T}_e^+ \supset \mathcal{T}_\theta^+), \quad P_D(L_1) = 1 - p\mathsf{P}(\mathcal{T}_e^+ \supset \mathcal{T}_\theta^-), \tag{9.2.33}$$

where p is a priori probability of the hypothesis H_0.

The robustness of detection by the L_1-test is manifested in the independence of the power and false alarm probability on the magnitude of noise.

Relations (9.2.33) describe the quality of detection of any signal satisfying the first condition in (9.2.32). If a one-sided impulsive noise does not satisfy the second condition in (9.2.32), then expressions (9.2.33) can be used as the upper and lower bounds for the false alarm probability and power function of the test. It is almost obvious that reducing the noise magnitude makes the quality of detection only better.

Now we apply these results to the problem of detection of the sine-signal $\Theta(t) = A\sin\omega t$ (the interval $T = 2\pi k/\omega$ and the integer number k of periods) under the impulsive noise with exponential distributions of duration and pauses. Under these conditions, it is possible to obtain rather simple relations for the characteristics $P_F(L_1)$ and $P_D(L_1)$ sought for in the following particular cases:

$\lambda = \mu$:

$$P_F = \frac{1}{4}\exp\left(-\frac{\lambda(2k-1)\pi}{\omega}\right),$$
$$P_D = 1 - \frac{1}{4}\left[\left(\cosh\lambda\frac{\pi}{\omega}\right)^k + \left(\sinh\lambda\frac{\pi}{\omega}\right)^k\right]\exp\left(-\frac{\lambda 2k\pi}{\omega}\right);$$

$\mu \ll \lambda$:

$$P_F = \frac{1}{4}\exp\left(-\frac{\lambda(2k-1)\pi}{\omega}\right), \qquad P_D = 1 - \frac{1}{4}\exp\left(-\frac{\lambda 2k\pi}{\omega}\right);$$

$\lambda \ll \mu$:

$$P_F = \frac{1}{4}\exp\left(-\frac{\lambda(2k-1)\pi}{\omega}\right), \qquad P_D = 1 - \frac{1}{2}\exp\left(-\frac{\lambda\pi k}{\omega}\right).$$

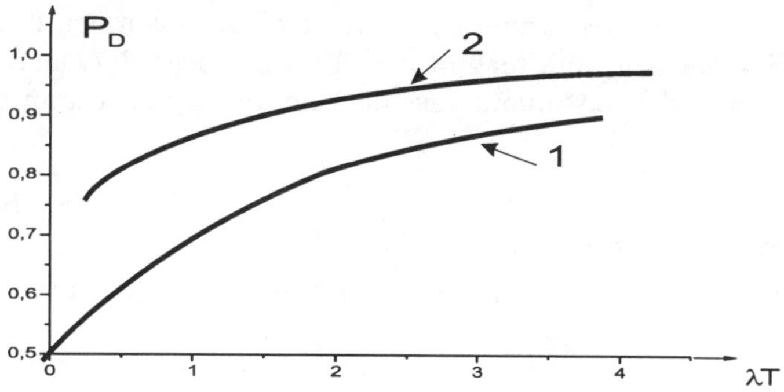

Figure 9.6. The lower (1) and upper (2) boundaries for the power of the L_1-test under impulsive noise with exponential distributions of duration and pauses

From the abovesaid it follows that, for any μ, the power P_D is bounded below and above, namely between the limits

$$1 - \frac{1}{2}\exp\left(-\frac{\lambda T}{2}\right) \le P_D \le 1 - \frac{1}{4}\exp(-\lambda T), \tag{9.2.34}$$

and $P_D \to 1$ as $\lambda T \to \infty$, as well as $P_F \to 0$. Also we have a better detection with increasing frequency ω of the signal under the fixed remained parameters.

Fig. 9.6 illustrates the abovesaid.

Now we give a simple qualitative analysis of detection of a periodic signal by the L_2-test under impulsive noise.

Let us choose $\theta(t) = \theta_0 \operatorname{sgn}(\sin \omega t)$ on the interval of processing $[0, T = 2\pi/\omega]$. For the impulsive noise of magnitude h and total duration T^+, we set $T_1^+ + T_2^+ = T^+$, where T_1^+ and T_2^+ are the parts of T^+ corresponding to the intervals $[0, T/2]$ and $[T/2, T]$ respectively. Then for the false alarm probability we obtain

$$P_F(L_2) = \mathsf{P}(T_1^+ - T_2^+ > \theta_0/(2h)),$$

and, for sufficiently large values of h, assuming closeness of the distributions for T_1^+ and T_2^+, we obtain

$$P_F = \mathsf{P}(T_1^+ - T_2^+ > 0) \approx 1/2.$$

Here the power is also close to 1/2. Thus we observe an extremely poor detection by the L_2-test like in the case of a constant signal. Furthermore, non-robustness of the L_2-test is obvious under an arbitrary number of periods.

Summarizing the above, we conclude that the L_1-test is highly robust under a rare impulsive noise, and it is slightly inferior to the optimal L_2-test under the Gaussian white noise. The obtained results allow us to expect a good performance of the L_1-test under the mixture of Gaussian white noise with a rare impulsive noise of high magnitude. This has been demonstrated in the discrete case, and the continuous case does not differ much from the discrete one here.

9.2.4. The L_p-norm tests

The Log-test. Now we revert to the problem of detection of a constant signal $\theta > 0$

$$\begin{aligned} H_0: &\quad x(t) = e(t), \\ H_1: &\quad x(t) = \theta + e(t), \quad 0 \le t \le T, \end{aligned} \tag{9.2.35}$$

where the impulsive noise $e(t)$ with impulses of magnitude $h > \theta$ and total duration T^+ is of the form

$$e(t) = \begin{cases} h, & t \in E_h, \\ 0, & t \in [0, T] \setminus E_h, \end{cases} \tag{9.2.36}$$

such that $\mu(E_h) = T^+$ ($\mu(x)$ is the ordinary Lebesgue measure on the real line).

Using the L_1-test, we obtain

$$P_D = 1, \qquad P_F = \mathsf{P}(T^+ > T/2),$$

i.e., the detection is true if the noise occupies less than half of the interval of processing. It is possible to improve the quality and robustness of detection of a constant signal under impulsive noise if to consider the nonparametric L_p-test ($p \ge 0$)

$$d(\mathbf{x}) = \begin{cases} 1, & \int_0^T |x(t)|^p\,dt - \int_0^T |x(t) - \theta|^p\,dt > 0, \\ \frac{1}{2}, & \int_0^T |x(t)|^p\,dt - \int_0^T |x(t) - \theta|^p\,dt = 0, \\ 0, & \int_0^T |x(t)|^p\,dt - \int_0^T |x(t) - \theta|^p\,dt < 0. \end{cases} \tag{9.2.37}$$

Assume that the hypothesis H_0 is true. Then the L_p-test takes the form

$$T^+ < \frac{\theta^p}{h^p - (h - \theta)^p + \theta^p} T. \tag{9.2.38}$$

Under the condition $h > \theta > 0$, the maximum of the expression

$$\max_p \frac{\theta^p}{h^p - (h - \theta)^p + \theta^p} = 1$$

is attained at $p = 0$. Hence test (9.2.37) can be written as $T^+ < T$, i.e., the decision is always true except $T^+ = T$. For equal a priori probabilities of the hypotheses H_0 and H_1, the characteristics of detection are

$$P_F = \tfrac{1}{2}\mathsf{P}(T^+ = T), \qquad P_D = 1 - \tfrac{1}{2}\mathsf{P}(T^+ = T).$$

Now we transform the test statistic for the case $p = 0$. The L_p-norm takes the following form at $p = 0$

$$\lim_{p\to 0}\left(\sum_{i=1}^{n} |x_i|^p\right)^{1/p} = \prod_{i=1}^{n} |x_i|.$$

Hence for the problem (9.2.35), we can use the logarithmic form of the test statistic or the Log-*test*

$$d(\mathbf{x}) = \begin{cases} 1, & \int_0^T \log|x(t)|\,dt - \int_0^T \log|x(t) - \theta|\,dt > 0, \\ \frac{1}{2}, & \int_0^T \log|x(t)|\,dt - \int_0^T \log|x(t) - \theta|\,dt = 0, \\ 0, & \int_0^T \log|x(t)|\,dt - \int_0^T \log|x(t) - \theta|\,dt < 0. \end{cases} \tag{9.2.39}$$

It follows from the above that the quality the Log-test under the impulsive noise is much better than that of the L_1-test: the Log-test fails only when the impulses occupy the whole interval $[0, T]$.

Now let the noise be described as

$$e(t) = \begin{cases} h, & t \in E_h, \\ h_1, & t \in [0, T] \setminus E_h, \end{cases} \tag{9.2.40}$$

where $h \gg \theta$ and $0 < h_1 \ll \theta$. It is easy to see that the quality of detection by the Log-test in this case is close to the quality of detection in model (9.2.36). Observe that model (9.2.40) is more realistic than (9.2.36). Thus extremely high robustness of the Log-test is explained by its structure and not by the degenerate character of the model.

Consider now the detection of an arbitrary signal $\theta(t)$ under impulsive noise (9.2.36) by the Log-test. It is easy to see that if the condition

$$\left|\int_0^T \log|\theta(t) \pm h|\,dt\right| < \infty \tag{9.2.41}$$

is satisfied, then the characteristics of the detection quality lie within the boundaries

$$P_D(\mathrm{Log}) \geq 1 - \mathsf{P}(T^+ = T), \qquad P_F(\mathrm{Log}) \leq \mathsf{P}(T^+ = T). \tag{9.2.42}$$

This result holds also in the case where $e(t)$ is an arbitrary impulsive noise satisfying the condition

$$\left| \int_0^T \log |\theta(t) + e(t)|\, dt \right| < \infty.$$

The Log-test manifests higher robustness with respect to impulsive noise than the L_1-test. However, the use of the L_1-test under the Gaussian noise yields a small lack in efficiency of detection as compared to the optimal L_2-test. It is interesting to know how the Log-test behaves in this case.

Consider the discrete version of the Log-test for detection of a constant signal θ under the Gaussian noise

$$d(\mathbf{x}) = \begin{cases} 1, & \sum_{i=1}^n \log |x_i| - \sum_{i=1}^n \log |x_i - \theta| > 0, \\ \frac{1}{2}, & \sum_{i=1}^n \log |x_i| - \sum_{i=1}^n \log |x_i - \theta| = 0, \\ 0, & \sum_{i=1}^n \log |x_i| - \sum_{i=1}^n \log |x_i - \theta| < 0. \end{cases} \tag{9.2.43}$$

The quality of detection is examined in small ($n = 5, 7, 9$) and large samples ($n = 100$) in ε-contaminated normal models for the fixed level of the signal θ. In the first case, Monte Carlo modeling is used; in the second, we apply the normal approximation to the distribution of the Log-test statistic

$$T(\mathbf{x}) = \frac{1}{n} \sum_{i=1}^n \log \left| \frac{x_i - \theta}{x_i} \right|. \tag{9.2.44}$$

Fig. 9.7 and 9.8 present the power function curves for the Log- and L_2-tests; hence the considerable loss of efficiency of the Log-test as compared to the L_2-test is seen, for example, $P_D(L_2) = 0.933$, $P_D(\mathrm{Log}) = 0.655$ for $\theta = 0.3$ and $n = 100$. Nevertheless, the Log-test is consistent because of the asymptotic normality of its test statistic.

The results of modeling demonstrate high robustness of the *Log*-test under contamination, in particular, it begins to dominate over the L_2-test only with sufficiently large values of the contamination parameters k and ε, namely from $k \approx 7$ with $\varepsilon = 0.2$. Recall that the L_1-test dominates over the L_2-test from $k \approx 3$ under the same conditions.

We conclude that the use of the Log-test is preferable in a clearly expressed non-normality of the impulsive noise.

The L_∞-test. Now we consider another particular case of the L_p-test as $p \to \infty$, and the L_p-norm becomes the Chebyshev uniform norm

$$\lim_{p\to\infty} \left(\sum_{i=1}^n |x_i|^p \right)^{1/p} = \max_i |x_i|.$$

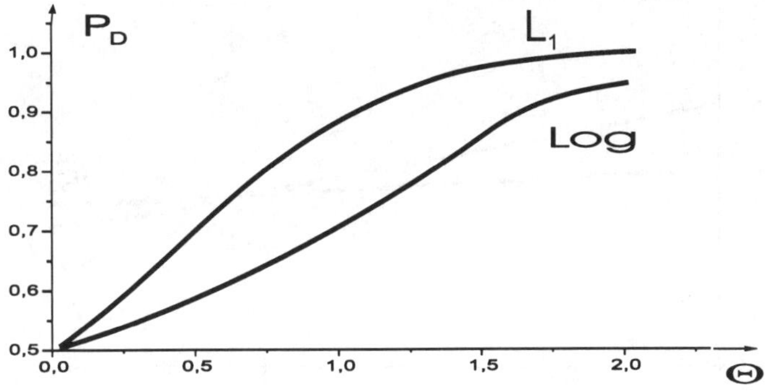

Figure 9.7. The power of the Log- and L_2-tests in Gaussian noise for $\theta = 0.3$, $n = 100$

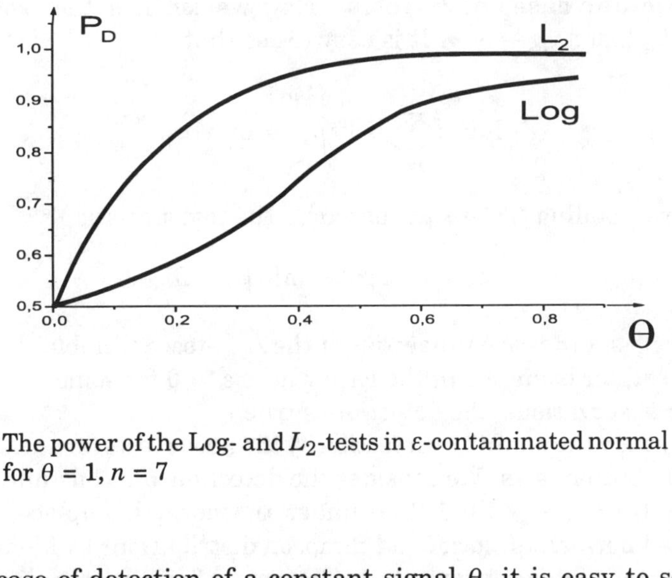

Figure 9.8. The power of the Log- and L_2-tests in ε-contaminated normal model for $\theta = 1$, $n = 7$

In the case of detection of a constant signal θ, it is easy to see that the L_∞-test statistic is half of the sum of extremal order statistics optimal for the uniform distribution

$$P_F(L_\infty) = \mathsf{P}\left(\frac{e_{(1)} + e_{(n)}}{2} > \frac{\theta}{2}\right), \qquad P_D(L_\infty) = \mathsf{P}\left(\frac{e_{(1)} + e_{(n)}}{2} < \frac{\theta}{2}\right).$$

Obviously, it follows from the above that if the underlying uniform distribution is defined on the interval $(-a, a)$ with $\theta > 2a$, then $P_F = 0$ and $P_D = 1$.

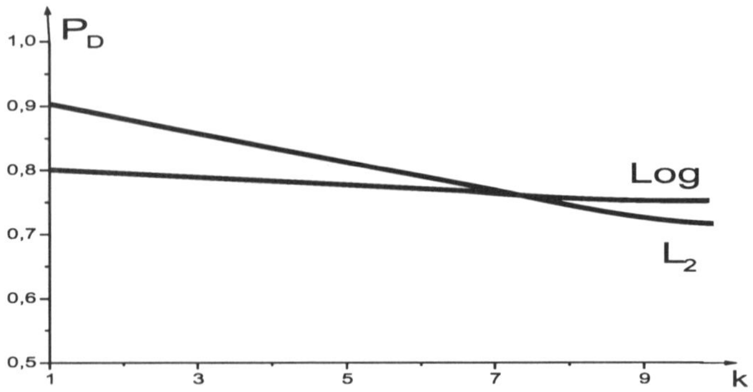

Figure 9.9. The power of the L_p-tests under the normal (upper) and Laplace (lower) distributions

Some particular cases of L_p-tests. Now we consider the exotic extreme case of the L_p-test as $p \to -\infty$. It is easy to see that

$$\lim_{p\to-\infty} \left(\sum_{i=1}^{n} |x_i|^p \right)^{1/p} = \min_i |x_i|,$$

and the corresponding $L_{-\infty}$-test is based on the test statistic

$$T(\mathbf{x}) = \min_i |x_i| - \min_i |x_i - \theta|.$$

Intuition suggests that the properties of the $L_{-\infty}$-test resemble the properties of the Log-test, for example, in the case where $e_i = 0$ for some i and $\theta \pm e_i \neq 0$ for all other observations, the detection is true.

Now we study the quality of detection by the above tests for a wide spectrum of noise distribution laws. We consider the detection of a constant signal θ by the L_p-tests ($p = -\infty, -2, 0, 1, 2, \infty$) under the normal, Laplace, Cauchy, ε-contaminated normal, uniform, and Simpson distributions by Monte Carlo for $n = 9$. The value of θ is set equal to 1. The power function curves are given in Figures 9.9–9.11.

Summarizing these graphs, we arrive at the following:

- As expected, the L_1-test is the best under the Laplace distribution, the L_2-test, under the normal distribution, and the L_∞-test, under the uniform distribution.

- The L_∞-test is the best for finite distributions like the Simpson and uniform ones; the quality of detection by other tests in this case decreases as $p \to -\infty$, and it is the least for $p = -\infty$.

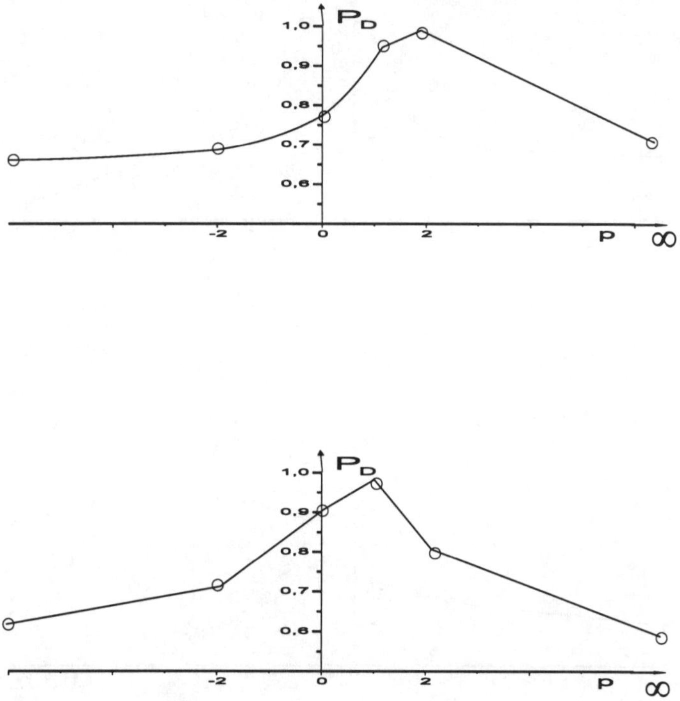

Figure 9.10. The power of the L_p-tests under the Cauchy (upper) and uniform (lower) distributions

- Under heavy-tailed distributions such as the Cauchy, Laplace, and contaminated normal ones, the L_p-tests with $p < 1$ possess greater power than those with $p > 2$.

- The L_1-test is also the best under the Cauchy distribution; the Log-test is a little inferior to it in this case.

In general, the use of the L_p-tests with $p < 0$ seems senseless, these tests behave poorly in the majority of cases. As for the remained group of the L_p-tests with $p \geq 0$, we may recommend the Log-test for a specific impulsive noise, and the L_∞-test for finite distributions, but the latter test fails if there occurs an outlier in the data. The L_1-test performs uniformly well under nearly all the distributions due to its minimax properties, so we may recommend it as a simple robust test for detection of known signals.

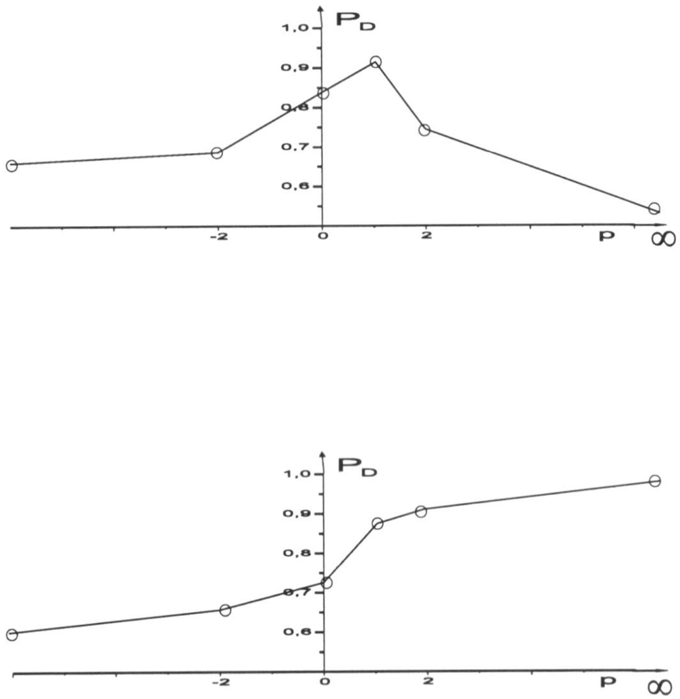

Figure 9.11. The power of the L_p-tests under the ε-contaminated normal ($\varepsilon = 0.1$, $k = 5$) and Simpson distributions

9.3. Statistical analysis of sudden cardiac death risk factors

In this section we expose the meteorological and solar sudden cardiac death risk factors by classical and robust statistical methods. The methodological aspects of their use are discussed.

9.3.1. Preliminaries

The exposure of sudden cardiac death (SCD) risk factors (RF) is a rather old problem. There are many researches treating this problem and similar questions mainly studying the influence of cardio-pathological factors (blood pressure, atherosclerosis), psychological factors (stresses), social factors (smoking, alcoholism), etc. (Mindlin and Kosagovskaya, 1986).

Here we consider only the RFs connected with meteorological and solar factors. This direction appeared due to the pioneering works (Tchijevsky, 1928; Tchijevsky, 1930; Tchijevsky, 1934; Tchijevsky, 1936).

The actuality of studying of this problem depends on the following:

- the mortality rate by coronary heart diseases is about 50% of the general mortality;
- the rate of SCD is about the 70% of the coronary heart mortality.

Another problem concerns the choice of adequate statistical methods for data processing. The high level of meteorological and solar data distribution uncertainty forces us to use nonparametric and robust procedures. The comparative study of classical and robust solutions of the above problem may be of interest both for physicians and statisticians.

9.3.2. Sudden cardiac death data and the problem formulation

There are different medical definitions of the sudden cardiac death (Mindlin and Kosagovskaya, 1986), and in this study, we accept as sudden the deaths occurring within 6 hours after appearing of onset symptoms.

The data under processing are the daily measured SCDs and meteo-solar factors in Arkhangelsk for 1983–85, totally 1096 days:

N is the daily number of the SCD, ($0 \le N \le 5$);

T is the average temperature (°C);

ΔT is the daily increment of temperature;

ΔP is the daily increment of pressure (mbar);

v is the average wind speed (m/s);

AK is the terrestrial magnetic activity index ($H \cdot 10^{-5}$);

W is the Wolf number;

S is the area of sunspots (in the solid angle units = $2\pi \cdot 10^{-6}$ steradian);

AS is the integral solar activity index (0—low, 1—moderate, 2—high, 3—very high);

PS is the number of sun flares.

The level of mortality N varies from $N = 0$ (the class $N = 0$ when there are no sudden deaths, totally 682 days) up to $N = 5$. The class $N \neq 0$ when sudden deaths occur includes 414 days: one SD a day $N = 1$—289 cases, two SDs a day $N = 1$—92, more than two SDs a day $N > 2$—33.

The exploratory data analysis shows that

- the data distributions vary much;

Table 9.2. Factor means and their standard errors for the class $N = 0$

Factors	T	ΔT	**ΔP**	v	AK	W	S	**AS**	PS
$\overline{x}$	1.09	2.69	**4.89**	2.85	25.3	64.2	447	**0.67**	4.89
$s_{\overline{x}}$	.46	.09	**.16**	.05	1.17	1.96	22	**.03**	.24
med	1.40	2.50	**4.10**	2.80	16.0	55.0	200	**0**	4.00
s_{med}	.53	.11	**.15**	.06	.52	1.89	18	-	.20

- they are mostly asymmetric and have heavy tails;
- there are outliers in the data.

The factors are measured in different scales (interval, ordinal, and mixed), the measurement procedures for some factors (AK, AS, W) are poorly provided from the metrological point of view (Ol, 1971). These data characteristics give definite warning against the use of classical statistical methods based on the least squares procedures and, in addition, give favor for the use of nonparametric and robust statistical procedures to provide the stability of inference.

9.3.3. Sudden cardiac death risk factors

Three types of statistical data characteristics are evaluated while exposing the SCDRF: the characteristics of location, scale, and correlation. Parallel with classical estimators, the robust median-type ones are used. Recall that the sample median possesses not only high qualitative (its breakdown point equals 1/2) and quantitative (the B- and V-robustness) robust properties, but what is more important here, the sample median is the unique estimator to be used with the ordinal scale: it is equivariant under monotonic data transformations (see Chapter 2) and therefore it is stable under comparisons in ordinal scales (Pfanzagl, 1969; Orlov, 1976).

Discrimination between the classes $N = 0$ and $N > 0$. The sudden death risk factors are exposed by comparing the values of sample means $\overline{x}$, the sample medians, and their standard errors for the classes $N = 0$ and $N > 0$ (Chirejkin and Shevlyakov, 1990; Chirejkin and Shevlyakov, 1993). The results are given in Tables 9.2 and 9.3.

The classes $N = 0$ and $N > 0$ obviously differ in the risk factors **ΔP**, the pressure increment, and **AS**, the integral index of solar activity, both for classical and robust estimators.

Table 9.3. Factor means and their standard errors for the class $N > 0$

Factors	T	ΔT	**ΔP**	v	AK	W	S	**AS**	PS
$\bar{x}$	.26	2.98	**11.80**	2.89	23.09	60.9	416	**1.17**	4.55
$s_{\bar{x}}$	.62	.07	**.39**	.07	1.26	2.57	27	**.02**	.32
med	2.70	2.50	**10.30**	2.75	16.0	44.0	140	**1**	1.50
s_{med}	.55	.15	**.32**	.07	.50	2.10	20	-	.21

Correlation and factor analysis. The sample correlation matrix **R** and the robust correlation matrix $\mathbf{R}_{\text{med}}$ with elements $\mathbf{r}_{\text{med}}$ evaluated for the classes $N = 0$ and $N > 0$ are qualitatively similar but there are definite quantitative differences.

All factors of solar activity W, S, AS, and PS are strongly correlated with each other in both classes, and the values of correlation coefficients (classical and robust) are greater than 0.6, for example, $\mathbf{r}(W,S) = 0.742$, $\mathbf{r}_{\text{med}}(W,S) = 0.936$ and $\mathbf{r}(W,PS) = 0.700$, $\mathbf{r}_{\text{med}}(W,PS) = 0.914$. The greater values of the robust estimator $\mathbf{r}_{\text{med}}$ can be explained by presence of outliers in the data: the real correlation is stronger, and it is estimated by the robust median correlation coefficient.

The correlation matrices **R** and $\mathbf{R}_{\text{med}}$ are used in the principal component procedure of factor analysis, which gives the following principal components:

SOLAR formed by all factors of solar activity W, S, AS, PS, AK;

TEMPERATURE formed by T and ΔT;

PRESSURE formed by ΔP, v, and ΔT.

These three components explain 70% and 80% of the data variance by classical and robust methods respectively.

On some details of discrimination. The results of discrimination between the classes $N = 0$ and $N > 0$ by the values of the integral index of solar activity AS (one of the main risk factors) are shown in Table 9.4.

Table 9.4 demonstrates that the low solar activity (AS=0) practically induces absence of the SCDs, but there is an essential specification: with AS > 0, occurrence of the SCDs is more plausible during the periods of the lower solar activity.

The last row of Table 9.4 for 1985 (the year of the 'calm' Sun) indicates that if the AS-index exceeds 0 then the SCDs occur inevitably. The analysis of the periods of low and high solar activity shows that the most dangerous factors with respect to the SCD are the leaps of the AS-index against the background

Table 9.4. Discrimination between the classes $N = 0$ and $N > 0$ by the integral index of solar activity AS

The numbers of days with sudden deaths

1983 N=0 AS=0	33	1983 N>0 AS=0	1	$\overline{AS} = 1.11$
1984 N=0 AS=0	82	1984 N>0 AS=0	0	$\overline{AS} = 0.98$
1985 N=0 AS=0	219	1985 N>0 AS=0	0	$\overline{AS} = 0.41$
1983 N=0 AS>0	202	1983 N>0 AS>0	127	$\overline{AS} = 1.11$
1984 N=0 AS>0	111	1984 N>0 AS>0	151	$\overline{AS} = 0.98$
1985 N=0 AS>0	5	1985 N>0 AS>0	134	$\overline{AS} = 0.41$

of the low solar activity (from 0 to 1). During the period of the high solar activity, such leaps are not essential.

9.3.4. Conclusions

(1) The daily rate of SCDs is significantly influenced by the solar activity factors, especially by the AS-index and by the daily increment of pressure ΔP. Influence of the terrestrial magnetic activity (the AK-index) has not manifest itself.

(2) Robust median-type estimators prove to be more efficient than classical estimators in the factor analysis procedure.

(3) The sample median should be used for estimation of location while data processing in order to provide:

   (a) the stability of statistical inference;

   (b) the possibility to compare the results of various studies.

Recall that the latter property can be provided only due to the equivariancy of the sample median under monotonic transformations of the data.

REMARK 9.3.1. The problem of statistical analysis of the SDCRF is regarded here as the problem of multivariate statistical analysis, though the observed data are the multivariate time series, so the above results should be considered only as preliminary. Various mysterious aspects of prediction for sudden deaths lie aside this study.

In addition, we may put forward the conjecture that the meteorological risk factors seem to be of lunar origin, not solar.

Bibliography

Abdelmalek, N. N. (1980). A FORTRAN subroutine for the L_1 solution of overdetermined systems of linear equations. *ACM Trans. Math. Software* **6**, 228–230.

Abramowitz, M., and Stegun, I. (1972). *Handbook of Mathematical Functions.* Dover, New York.

Adatia, A. (1988). Robust estimators of the 2-parameter gamma distribution. *IEEE Trans. Rel.* **37**, 234–238.

Adichie, J. N. (1967). Estimation of regression coefficients based on rank tests. *Ann. Math. Statist.* **38**, 894–904.

Afifi, A. A., and Azen, S. P. (1979). *Statistical Analysis. A Computer Oriented Approach.* Academic Press, New York.

Aivazyan, S. A., Bukhshtaber, V. M., Enyukov, I. S., and Meshalkin, L. D. (1989). *Applied Statistics. Classification and Reduction of Dimensionality. Reference Edition.* Finansy i Statistika, Moscow (in Russian).

Akhiezer, N. I. (1958). *Theory of Approximation.* Frederick Ungar, New York.

Anderson, T. W. (1971). *The Statistical Analysis of Time Series.* Wiley, New York.

Andrews, D. F., Bickel, P. J., Hampel, F. R., Huber, P. J., Rogers, W. H., and Tukey, J. W. (1972). *Robust Estimates of Location.* Princeton Univ. Press, Princeton.

Armstrong, R. D., Frome, E. L., and Kung, D. S. (1979). A revised simplex algorithm for the absolute deviation curve fitting problem. *Commun. Stat.* **B8**, 175.

Arthanari, T. S., and Dodge, Y. (1981). *Mathematical Programming in Statistics.* Wiley, New York.

Astrom, J. K. J., and Eykhoff, P. (1971). System identification — a survey. *Automatica* **7**, 123–162.

Atkinson, A. C. (1985). *Plots, Transformations, and Regression.* Clarendon Press, Oxford.

Atkinson, A. C. (1986). Masking unmasked. *Biometrika* **73**, 533–541.

Atkinson, A. C. (1994). Fast very robust methods for the detection of multiple outliers. *J. Amer. Statist. Assoc.* **89**, 1329–1339.

Atkinson, A. C., and Mulira, H.-M. (1993). The stalactite plot for the detection of multiple outliers. *Statist. Comput.* **3**, 27–35.

Atkinson, A., and Riani, M. (2000). *Robust Diagnostics Regression Analysis.* Springer, New York.

Azencott, R. (1977a). Estimation d'un paramètre de translation à partir de tests de rang. *Austérisque.* #43–44, 41–64.

Azencott, R. (1977b). Robustesse des R-estimateurs. *Austérisque.* #43–44, 189–202.

Baker, G. A. (1975). *Essentials of Padé Approximants.* Academic Press, New York.

Barlow, R. E. (1998). *Engineering Reliability.* ASA–SIAM, Philadelphia.

Barlow, R. E., and Proschan, F. (1965). *Mathematical Theory of Reliability.* Wiley, New York.

Barlow, R. E., and Proschan, F. (1975). *Statistical Theory of Reliability and Life Testing: Probability Models.* Holt, New York.

Barnett, V., and Lewis, T. (1978). *Outliers in Statistical Data.* Wiley, New York.

Barrodale, I. (1968). L_1-approximation and the analysis of data. *Appl. Statist.* **17**, 51–57.

Barrodale, I., and Roberts, F. D. K. (1974). Algorithm 478: Solution of an overdetermined system of equations in the L_1-norm. *Commun. ACM* **17**, 319–320.

Barrodale, I., and Roberts, F. D. K. (1978). An efficient algorithm for discrete L_1 linear approximation with linear constraints. *SIAM J. Numer. Anal.* **15**, 603.

Bateman, H., and Erdélyi, A. (1953). *Higher Transcendental Functions*, **2.** McGraw–Hill, New York.

Bebbington, A. C. (1978). A method of bivariate trimming for estimation of the correlation coefficient. *Appl. Statist.* **27**, 221–226.

Bello, P. A., and Esposito, R. (1969). A new method for calculating probabilities of errors due to impulsive noise. *IEEE Trans. Comm. Technology* **17**, 368–379.

Beran, R. (1974). Asymptotically efficient adaptive rank estimates in location models. *Ann. Stat.* **2**, 63–74.

Bernoulli, D. (1777). Dijudicatio maxime probabilis plurium observationum discrepantium atque verisimillima inductio inde formanda. *Acta Acad. Sci. Petropolit.* **1**, 3–33.

Bernstein, S. N. (1911). Sur l'approximation des fonctions continues par des polynomes. *C. R. Acad. Sci.* **152.**

Bernstein, S. N. (1926). *Leçons sur les Propriétés Extrémales et la Meilleure Approximation des Fonctions Analytiques d'une Variable Reélle.* Gaurthier–Villars, Paris.

Bhattacharya, P. K. (1967). Efficient estimation of a shift parameter fom grouped data. *Ann. Math. Stat.* **38**, 1770-1787.

Bickel, P. J. (1973). On some analogues to linear combination of order statistics in the linear model. *Ann. Statist.* **1**, 597–616.

Bickel, P. J. (1976). Another look at robustness: a review of reviews and some new developments. *Scand. J. Statist. Theory and Appl.* **3**, #4, 145–168.

Bickel, P. J., and Lehmann, E. L. (1973). Measures of location and scale. In: *Proc. Prague Symp. Asymptotic Statist.* **I**, Prague, Charles Univ., pp. 25–36.

Bickel, P. J., and Lehmann, E. L. (1975). Descriptive statistics for nonparametric models. *Ann. Statist.* **3**, 1045–1069.

Birnbaum, A., and Laska, E. (1967). Optimal robustness: a general method with application to linear estimators of location. *J. Amer. Statist. Assoc.* **62**, 1230–1240.

Blomqvist, N. (1950). On a measure of dependence between two random variables. *Ann. Math. Statist.* **21**, 593–600.

Bloomfield, P., and Steiger, W. L. (1983). *Least Absolute Deviations: Theory, Applications, and Algorithms.* Birkhäuser, Boston.

Bokk, H. O. (1990). The extension of robust methods in the case of a multivariate parameter. In: *Nonparametric and Robust Statistical Methods in Cybernetics and Informatics: Proc. VII All-Union Seminar*, **1.** Tomsk University, Tomsk, pp. 111–114 (in Russian).

Boscovich, R.J. (1757). De litteraria expeditione per pontificiam ditionem, et synopsis amplioris operis, ac habenturplura eius ex exemplaria etiam sensorum impressa. *Bononiensi Scientiarum et Artium Instituto Atque Academia Commentarii* **4**, 353–396.

Bunyakovsky, W. (1859). *Mémoires de l'Académie des sciences de St-Pétersbourg, 7 série.*

Box, G. E. P. (1953). Non-normality and test on variances. *Biometrika* **40**, 318–335.

Box, G. E. P., and Jenkins, G. M. (1970). *Time Series Analysis: Forecasting and Control.* Holden Day, San Francisco.

Brewer, K. R. W. (1986). *Likelihood Based Estimation of Quantiles and Density Estimation.* Unpublished manuscript.

Bustos, O. H., and Yohai, V. J. (1986). Robust estimates for ARMA models. *J. Amer. Statist. Assoc.* **81**, 155–168.

Campbell, N. A. (1980). Robust procedures in multivariate analysis I: Robust covariance estimation. *Appl. Statist.* **29**, 231–237.

Campbell, N. A. (1982). Robust procedures in multivariate analysis II: Robust canonical variate analysis. *Appl. Statist.* **31**, 1–8.

Cauchy, A.-L. (1992) *Cours d'analyse de l'Ecole Royale Polytechnique. Premiere partie: Analyse algebrique.* Editrice CLUEB, clxvii, Bologna.

Chambers, J. M., Cleveland, W. S., Kleiner, B., and Tukey, P. A. (1983). *Graphical Methods for Data Analysis.* Wadsworth, Belmont.

Chapman D. G., and Robbins, H. E. (1951). Minimum variance estimation without regularity assumptions. *Ann. Math. Statist.* **22**, 581–586.

Cheng, C. (1995). The Bernstein polynomial estimator of a smooth quantile function. *Statist. Probab. Lett.* **24**, 321-330.

Chelpanov, I. B., and Shevlyakov, G. L. (1983). Robust recognition algorithms based on approximation criteria. *Autom. Remote Control* **44**, 777–780.

Chernoff, H., Gastwirth, J. L., and Johns, M. V. (1976). Asymptotic distribution of linear combinations of functions of order statistics with applications to estimation. *Ann. Math. Statist.* **38**, 52–72.

Chirejkin, L. V., and Shevlyakov, G. L. (1990). Risk factors of sudden cardiac deaths in Arkhangelsk. In: *Ishemic Disease.* Leningrad Institute of Cardiology, pp. 15–25 (in Russian).

Chirejkin, L. V., and Shevlyakov, G. L. (1993). Classical and robust statistical analysis of the risk factors of sudden deaths by cardiovascular diseases. *Autom. Remote Control* **55**, 130–137.

Collins, J., and Wiens, D. (1985). Minimax variance M-estimators in ε-contamination models. *Ann. Statist.* **13**, 1078–1096.

Costa, V., and Deshayes, J. (1977). Comparison des R-L-M-estimateurs. *Austérisque.* #43–44, 209–238.

Cramér, H. (1946). *Mathematical Methods of Statistics.* Princeton Univ. Press, Princeton.

Crow, E. L., and Siddiqui, M. M. (1967). Robust estimation of location. *J. Amer. Statist. Assoc.* **62**, 353–389.

Cypkin, Ja. Z. [Tsypkin Ya. Z.] (1976). Optimization in conditions of uncertainty. *Sov. Phys. Dokl.* **21**, 317–319.

Daniel, C. (1920). Observations weighted according to order. *Amer. J. Math.* **42**, 222–236.

Davies, P. L. (1987). Asymptotic behavior of S-estimators of multivariate location parameters and dispersion matrices. *Ann. Statist.* **15**, 1269–1292.

Davies, P. L., and Gather, U. (1993). The identification of multiple outliers. (with discussion). *J. Amer. Statist. Assoc.* **88**, 782–801.

Deniau, C., Oppenheim, G., and Viano, C. (1977a). M-estimateurs. *Austérisque.* #43–44, 31–40.

Deniau, C., Oppenheim, G., and Viano, C. (1977b). Robustesse: π-robustesse et minimax-robustesse. *Austérisque.* #43–44, 51–166.

Deniau, C., Oppenheim, G., and Viano, C. (1977c). Robustesse des M-estimateurs. *Austérisque.* #43–44, 167–187.

Deniau, C., Oppenheim, G., and Viano, C. (1977d). Courbes d'influence et sensibilité. *Austérisque.* #43-44, 239–252.

Devlin, S. J., Gnanadesikan, R., and Kettenring, J. R. (1975). Robust estimation and outlier detection with correlation coefficients. *Biometrika* **62**, 531–545.

Devlin, S. J., Gnanadesikan, R., and Kettenring, J. R. (1981). Robust estimation of dispersion matrices and principal components. *J. Amer. Statist. Assoc.* **76**, 354–362.

Diday, E. (1972). *Nouvelles méthodes et nouveaux concepts en classification automatique et reconnaissance des formes.* These de doctorat d'état, Univ. Paris IX.

Dixon, W. J. (1950). Analysis of extreme values. *Ann. Math. Stat.* **21**, 488-506.

Dixon, W. J. (1960). Simplified estimation from censored normal samples. *Ann. Math. Stat.* **31**, 385–391.

Dodge, Y. (1987). An introduction to L_1-norm based statistical data analysis. *Comput. Statist. and Data Anal.* **5**, 239–253.

Dodge, Y., and Jurečkovà, J. (2000). *Adaptive Regression.* Springer, New York.

Donoho, D. L. (1982). *Breakdown Properties of Multivariate Location Estimators.* Ph.D. qualifying paper. Harvard University, Dept. of Statistics.

Donoho, D. L., and Huber, P. J. (1983). The notion of breakdown point. In: *A Festschrift for Erich L. Lehmann (Bickel, P. J., Doksum, K. A., Hodges, J. L., Eds.).* Wadsworth, Belmont, pp. 157–184.

Draper, N. R., and Smith, H. (1966). *Applied Regression Analysis.* Wiley, New York.

Dutter, R. (1977). Numerical solution of robust regression problems: Computational aspects, a comparison. *J. Statist. Comput. Simul.* **5**, 207–238.

Eddington, A. S. (1914). *Stellar Movements and the Structure of the Universe.* Macmillan, London.

Ekblom, H., and Henriksson, S. (1969). L_p-criteria for the estimation of location parameters. *SIAM J. Appl. Math.* **17**, 1130–1141.

El-Sawy, A. H., and Vandelinde, D. V. (1977). Robust detection of known signals. *IEEE Trans. Inform. Theory* **23**, 722–727.

Engel, J. S. (1965). Digital transmission in the presense of impulsive noise. *Bell Syst. Tech. J.* **44**, 1699–1743.

Ershov, A. A. (1979). Stable methods of estimating parameters. *Autom. Remote Control* **39**, 1152–1181;

Eykhoff, P. (1974). *System Identification. Parameter and State Estimation.* Wiley, New York.

Falk, M. (1985). Asymptotic normality of the kernel quantile estimator. *Ann. Statist.* **13**, 428–433.

Fedorov, E. D. (1994). Least absolute values estimation: computational aspects. *IEEE Trans. Autom. Control* **39**, 626–630.

Fernholz, L. T. (1983). *Von Mises Calculus for Statistical Functionals.* Springer, New York.

Forsythe, A. B. (1968). Robust estimation of straight line regression coefficients by minimizing *p*th power deviations. *Technometrics* **14**, 159-166.

Fox, A. J. (1972). Outliers in time series. *J. R. Statist. Soc.* **B34**, 350–363.

Gallagher, M. A., and Moore, A. H. (1990). Robust minimum distance estimation using the 3-parameter Weibull distribution. *IEEE Trans. Rel.* **39**, 575–580.

Gehrig, W., and Hellwig, K. (1982). Eine charakterisieung der gewichteten L_r-distanz. In: *Or-Spectrum*. Springer, Berlin, pp. 233–237.

Gelfand, I., and Fomin, S. (1963). *Calculus of Variations*. Prentice–Hall, Englewood Cliffs, NJ.

Gentle, J. E. (1977). Least absolute value estimation: An introduction. *Commun. Statist.* **B6**, 313-328.

Gentle, J. E., Kennedy, W. J., and Sposito, V. A. (1977). On least absolute values estimation. *Commun. Statist. (Theory and Methods)* **6**, 839–845.

Gentle, J. E., Narula, S. C., and Sposito, V. A. (1988). Algorithms for unconstrained L_1 linear regression. *Comput. Statist. Data Anal.* **6**, 335–339 (1988).

Gentleman, W. M. (1965). *Robust Estimation of Multivariate Location by Minimizing pth Power Deviations*. Ph.D. Thesis. Princeton University, Princeton, NJ.

Gertsbakh, I. B. (1998). *Statistical Reliability Theory*. Marcel Dekker, New York.

Gilewicz, J. (1978). *Approximants de Padé*. Springer, Berlin.

Gnanadesikan, R. (1977). *Methods for Statistical Data Analysis of Multivariate Observations*. Wiley, New York.

Gnanadesikan, R., and Kettenring, J. R. (1972). Robust estimates, residuals, and outlier detection with multiresponse data. *Biometrics* **28**, 81–124.

Gnedenko, B. V., Belyaev, Yu. K., and Solovyev, A. D. (1969). *Mathematical Methods in Reliability Theory*. Academic Press, New York.

Goldberg, K. M., and Iglewicz, B. (1992). Bivariate extensions of the boxplot. *Technometrics* **34**, 307–320.

Green, P. J. (1984). Weighted least squares procedures: A survey. *J. Roy. Statist. Soc.* **46**, 149–170.

Grubbs, F. E. (1950). Sample criteria for testing outlying observations. *Ann. Math. Stat.* **21**, 27–58.

Guilbo, E. P., and Chelpanov, I. B. (1975). *Signal Processing on the Basis of Ordinal Choice*. Soviet Radio, Moscow (in Russian).

Guilbo, E. P., and Shevlyakov, G. L. (1977). Efficient criteria for detection of outliers. In: *Proc. VII All-Union Conf. Inform. Redundancy in Autom. Syst.* Leningrad, pp. 151–154 (in Russian).

Guilbo, E. P., Chelpanov, I. B., and Shevlyakov, G. L. (1979). Robust approximation of functions in case of uncertainty. *Autom. Remote Control* **40**, 522–529.

Hadi, A. S. (1992). Identifying multiple outliers in multivariate data. *J. Royal Statist. Soc.* **B54**, 761–771.

Hájek, J., and Šidák, Z. (1967). *Theory of Rank Tests*. Academic Press, New York.

Hall, H. M. (1966). *A New Model for "Impulsive" Phenomena: Application to Atmospheric-Noise Communication Channels*. Stanford Elec. Lab. Techn. Rept 3412-8 and 7050-7. Stanford Univ., Stanford, CA.

Hampel, F. R. (1968). *Contributions to the Theory of Robust Estimation.* Ph.D. Thesis, University of California, Berkeley.

Hampel, F. R. (1971). A general qualitative definition of robustness. *Ann. Math. Statist.* **42**, 1887–1896.

Hampel, F. R. (1973). Robust estimation: A condensed partial survey. *Z. Wahrsch. verw. Geb.* **27**, 87–104.

Hampel, F. R. (1974). The influence curve and its role in robust estimation. *J. Amer. Statist. Assoc.* **69**, 383–393.

Hampel, F. R., Rousseeuw, P. J., and Ronchetti, E. (1981). The change-of-variance curve and optimal redescending M-estimators. *J. Amer. Statist. Assoc.* **76**, 643–648.

Hampel, F. R., Ronchetti, E., Rousseeuw, P. J., and Stahel, W. A. (1986). *Robust Statistics. The Approach Based on Influence Functions.* Wiley, New York.

Hawkins, D. M. (1980). *The Identification of Outliers.* Chapman & Hall, London.

Hawkins, D. M. (1993a). A feasible solution algorithm for the minimum volume ellipsoid estimator. *Comput. Statist.* **9**, 95–107.

Hawkins, D. M. (1993b). A feasible solution algorithm for the minimum covariance determinant estimator in multivariate data. *Comput. Statist. Data Anal.* **17**, 197–210.

Hodges, Z. L., and Lehmann, E. L. (1963). Estimates of location based on rank tests. *Ann. Math. Statist.* **34**, 598–611.

Holland, P. W., and Welsch, R. E. (1977). Robust regression using iteratively reweighted least squares. *Commun. Statist. (Theory and Methods)* **6**, 813–828.

Hogg, R. V. (1972). More light on the kurtosis and related statistics. *J. Amer. Statist. Assoc.* **67**, 422–424.

Hogg, R. V. (1974). Adaptive robust procedures: a partial review and some suggestions for future applications and theory. *J. Amer. Statist. Assoc.* **69**, 909–923.

Huber, P. J. (1964). Robust estimation of a location parameter. *Ann. Math. Statist.* **35**, 73–101.

Huber, P. J. (1965). A robust version of the probability ratio test. *Ann. Math. Statist.* **36**, 1753–1758.

Huber, P. J. (1967). The behaviour of maximum likelihood estimates under nonstandart conditions. In: *Proc. 5th Berkeley Symp. on Math. Statist. Prob.* **1.** Berkeley Univ. California Press, pp. 221–223.

Huber, P. J. (1972). Robust statistics: A review. *Ann. Math. Statist.* **43**, 1041–1067.

Huber, P. J. (1973). Robust regression: Asymptotics, conjectures, and Monte Carlo. *Ann. Math. Statist.* **1**, 799–821.

Huber, P. J. (1974). Fisher information and spline interpolation. *Ann. Statist.* **2**, 1029–1033.

Huber, P. J. (1979). Robust smoothing. In: *Robustness in Statistics (Launer, R. L., Wilkinson G. N., Eds.)* Academic Press, New York, pp. 33–47.

Huber, P. J. (1981). *Robust Statistics.* Wiley, New York.

Huber, P. J. (1985). Projection pursuit (with discussion). *Ann. Statist.* **13**, 435–525.

Huynh, T. H., and Lecours, M. (1975). Digital system performance on impulsive noise environment: a survey. In: *Proc. 13th Annual Allerton Conf. on Circuit and Systems Theory*, pp. 792–794.

Ibragimov, I. A., and Khasminskii, R. Z. (1981). *Statistical Estimation. Asymptotic Theory.* Springer, New York.

Ionescu, D., and Limnios, N., Eds. (1999). *Statistical and Probabilistic Models in Reliability.* Birkhäuser, Boston.

Jaeckel, L. A. (1971a). Robust estimates of location: Symmetry and asymmetric contamination. *Ann. Math. Statist.* **42**, 1020–1034.

Jaeckel, L. A. (1971b). Some flexible estimates of location. *Ann. Math. Statist.* **42**, 1540–1552.

Jaeckel, L. A. (1972). Estimating regression coefficients by minimizing the dispersion of residuals. *Ann. Math. Statist.* **43**, 1449–1458.

Jeffreys, H. (1932). An alternative to the rejection of outliers. *Proc. Royal Soc. London* **A137**, 78–87.

Jung, J. (1955). On linear estimates defined by a continuous weight function. *Ark. Math.* **3**, 199–209.

Jurečkovà, J. (1971). Nonparametric estimate of regression coefficients. *Ann. Math. Statist.* **42**, 1328–1338.

Jurečkovà, J. (1977). Asymptotic relations of M-estimates and R-estimates in linear regression model. *Ann. Statist.* **5**, 364–372.

Jurečkovà, J. (1984). M-L-R-estimators. In: *Handbook of Statistics (Krishnaiah, P. R., and Sen, P. K, Eds.)* **4.** Elsevier, Amsterdam, pp. 463–485.

Kagan, A. M., Linnik, Yu. V., and Rao, S. R. (1973). *Characterization Problems in Mathematical Statistics.* Wiley, New York.

Kaigh, W. D., and Cheng, C. (1991). Subsampling quantile estimators and uniformity criteria. *Comm. Statist.* **A20**, 539–560.

Kashyap, P. L., and Rao, A. R. (1976). *Dynamic Stochastic Models from Empirical Data.* Academic Press, New York.

Katkovnik, V. Ya. (1979). Linear and nonlinear methods of nonparametric regression analysis. *Sov. Autom. Control* **12**, 25–34.

Katkovnik, V. Ya. (1985). *Nonparametric Identification and Data Smoothing. The Method of Local Approximation.* Nauka, Moscow (in Russian).

Kendall, M. G., and Stuart, A. (1962). *The Advanced Theory of Statistics. Distribution Theory.* **1.** Griffin, London.

Kendall, M. G., and Stuart, A. (1963). *The Advanced Theory of Statistics. Inference and Relationship.* **2.** Griffin, London.

Kendall, M. G., and Stuart, A. (1968). *The Advanced Theory of Statistics. Design and Analysis, and Time Series.* **3.** Griffin, London.

Kharin, Yu. S. (1996a) *Robustness in Statistical Pattern Recognition.* Kluwer, Dordrecht.

Kharin, Yu. S. (1996b). Robustness in discriminant analysis. In: *Robust Statistics, Data Analysis, and Computer Intensive Methods (Rieder, H., Ed.).* Springer, New York, pp. 225–233.

Kleiner, B., Martin, R. D., and Thomson, D. J. (1979). Robust estimation of power spectra. *J. R. Statist. Soc.* **B41**, 313–351.

Klir, G. J., and Folger, T. (1990). *Fuzzy Sets, Uncertainty, and Information.* Prentice–Hall, Englewood Cliffs, NJ.

Koenker, R., and Bassett, G. J. (1978). Regression quantiles. *Econometrica* **46**, 33–50.

Kolmogorov, A. N. (1931). On the method of median in the theory of errors. *Math. Sbornik* **38**, #3/4, 47–50.

Krasnenker, V. M. (1980). Robust methods for detection of signals (a survey). *Avtom. Telemekh.* #5, 65–88 (in Russian).

Kreinovich, V. Ya. (1986). A general approach to analysis of uncertainty in measurements. In: *Proc. 3rd USSR National Symposium on Theoretical Metrology*. Mendeleyev Metrology Institute (VNIIM), Leningrad, pp. 187-188 (in Russian).

Krishnaiah, P. R., and Sen, P. K., Eds. (1984). *Handbook of Statistics. Vol.. 4: Nonparametric Methods.* Elsevier, Amsterdam.

Kuznetsov, V. P. (1976). Robust detection of an approximately known signal. *Probl. Inform. Transmission* **12**, #2, 47–61 (in Russian).

Kuznetsov, V. P. (1991). *Interval Statistical Models.* Radio i Svyaz, Moscow (in Russian).

Lebart, L., Morineau, A., and Warwick, K. M. (1984). *Multivariate Descriptive Statistical Analysis: Correspondence Analysis and Related Techniques.* Wiley, New York.

Lee, R. C. K. (1964). *Optimal Estimation, Identification and Control.* MIT Press, Cambridge, MA.

Lehmann, E. L. (1959). *Testing Statistical Hypothesis.* Wiley, New York.

Ljung, L. (1987). *System Identification—Theory for User.* Prentice–Hall, Englewood Cliffs, NJ.

Ljung, L. (1995). *System Identification. Toolbox for Use with* MATLAB. MathWorks, Natick, MA.

Lopuhaä, H. P. (1989). On the relation between S-estimators and M-estimators of multivariate location and covariance. *Ann. Statist.* **17**, 1662–1683.

Lopuhaä, H. P. (1992). Highly efficient estimators of multivariate location with high breakdown point. *Ann. Statist.* **20**, 398–413.

Lopuhaä, H. P., and Rousseeuw, P. J. (1991). Breakdown points of affine-equivariant estimators of multivariate location and covariance matrices. *Ann. Statist.* **19**, 229–248.

Lorentz, G. G. (1986). *Bernstein Polynomials.* Chelsea, New York.

Luneva, N. V. (1983). On a problem of robust estimation from correlated observations. *Avtom. Telemekh.* #2, 167–170.

Makshanov, A. V., Smirnov, A. V., and Shashkin, A. K. (1991). *Robust Methods of Data Processing in Radio-Technical Systems.* St-Petersburg Univ. Press, St-Petersburg (in Russian).

Mallows, C. L. (1975). *On Some Topics in Robustness.* Bell Lab Technical Memorandum. Bell Telephone Labs, Murray Hill, NJ.

Maronna, R. A. (1976). Robust m-estimators of multivariate location and scatter. *Ann. Statist.* **4**, 51–67.

Martin, R. D. (1979). Robust estimation for time series. In: *Robustness in Statistics (Launer, R. L., and Wilkinson, G. N., Eds.).* Academic Press, New York, pp. 147–176.

Martin, R. D. (1980). Robust estimation of autoregressive models. In: *Directions in Time Series (Brillinger, D. R., and Tiao, G. C., Eds.).* Inst. Math. Statist. Publ., Hayward, CA, pp. 228–254.

Martin, R. D. (1981). Robust methods for time series. In: *Applied Time Series Analysis (Findley, D. F., Ed.)* **2**, pp. 683-760.

Martin, R. D., and Yohai, V. J. (1984). Robustness in time series and estimating ARMA models. In: *Handbook of Statistics, Vol.* **5** *(Hannan, E. J., Krishnaiah, P. R., and Rao, M. M, Eds.).* Elsevier, Amsterdam, pp. 119–155.

Maronna, R. A. (1975). Robust M-estimators of multivariate location and scatter. *Ann. Statist.* **4**, 51–67.

Mendeleyev, D. I. (1895). Course of work on the renewal of prototypes or standard measures of lengths and weights *Vremennik Glavnoi Palaty Mer i Vesov* **2**, 157–185 (in Russian).

Mertz, P. (1961). Model of impulsive noise for data transmission. *IRE Trans. Inform. Theory.* **9**, 130–137.

Meshalkin, L. D. (1971). Some mathematical methods for the study of non-communicable diseases. In: *Proc. 6th Int. Meeting of Uses of Epidemiol. in Planning Health Services* **1.** Primosten, Yugoslavia, pp. 250–256.

Milne, A. R. (1962). Sound propagation and ambient noise under ice. In: *Underwater Acoustics (Albers, V. M., Ed.)* **2.** Plenum, New York, pp. 103–138.

Mindlin Ya. S., and Kosagovskaya I. I. (1986). *Sudden Death by Coronary Diseases as a Social-Hygienic Problem.* VNIIMI, Moscow (in Russian).

Mosteller, F., and Tukey, J. W. (1977). *Data Analysis and Regression.* Addison–Wesley, New York.

Mudrov, V. I., and Kushko, V. L. (1983). *Methods of Data Processing: the Quasi-Likelihood Estimators.* Radio and Svyaz, Moscow (in Russian).

Nemirovskii, A. S. (1981). Recursive estimation of parameters of linear plants. *Autom. Remote Control* **42**, 472–480.

Nemirovskii, A. S. (1985). Nonparametric estimation of smooth regression functions. *Sov. J. Comput. Syst. Sci.* **23**, #6, 1–11.

Nemirovskii, A. S., Polyak, B. T., and Tsybakov, A. B. (1983). Estimators of maximum likelihood type for nonparametric regression. *Sov. Math. Dokl.* **28**, 788–792.

Newcomb, S. (1886). A generalized theory of the combination of observations so as to obtain the best result. *Amer. J. Math.* **8**, 343–366.

Ol, A. I. (1971). The manifestation of solar activity in the magnitosphere and ionosphere of the Earth. In: *The Influence of Solar Activity on Atmosphere and Biosphere of the Earth.* Nauka, Moscow, pp. 104-118 (in Russian).

Olshevskie, V. V. (1967). *Characteristics of Sea Reverberation.* Consultants Bureau, New York.

Onishchenko, V. F., and Tsybakov, A. B. (1987). Asymptotic normality of M-estimates. *Autom. Remote Control* **48**, 926–935.

Orlov, V. I. (1976). *Stability in Social and Economic Models.* Nauka, Moscow (in Russian).

Papadimitriou, C. H., and Steiglitz, K. *Combinatorial Optimization: Algorithms and Complexity.* Prentice–Hall, Englewood Cliffs, NJ.

Parzen, E. (1979a). Nonparametric statistical data modeling (with comments). *J. Amer. Statist. Assoc.* **74**, 105-131.

Parzen, E. (1979b). A density-quantile function perspective on robust estimation. In: *Robustness in Statistics (Launer, R. L., and Wilkinson, G. N., Eds.).* Academic Press, New York, pp. 237–258.

Pashkevich, M. E., and Shevlyakov, G. L. (1993). On a median analogue of the Fourier transform. In: *Modern Problems of Data Analysis and Simulation.* Minsk University, pp. 80–85 (in Russian).

Pasman, V. R., and Shevlyakov, G. L. (1987). Robust methods of estimation of correlation coefficients. *Autom. Remote Control* **48**, 332–340.

Perez, M. J., and Palacin, F. (1987). Estimating the quantile function by Bernstein polynomials. *Comput. Statist. Data Anal.* **5**, 391–397.

Pfanzagl, J. (1969). On measurability and consistency of minimum contrast estimates. *Metrika* **14**, 248–278.

Pitman, E. J. G. (1948). *Non-Parametric Statistical Inference.* Univ. North Carolina, Institute of Statistics.

Polyak, B. T., and Tsypkin, Ya. Z. (1978) Robust identification. In: *Identif. Syst. Parameter Estim., Part 1, Proc. 4th IFAC Symp.* Tbilisi, 1976, pp. 203–224.

Polyak, B. T., and Tsypkin, Ya. Z. (1979). Adaptive estimation algorithms (convergence, optimality, stability). *Autom. Remote Control* **40**, 378–389.

Polyak, B. T., and Tsypkin, Ya. Z. (1980). Robust identification. *Automatica* **16**, #10, 53–65.

Polyak, B. T., and Tsypkin, Ya. Z. (1983). Optimal methods of estimating autoregression coefficients in the case of incomplete information. *Eng. Cybern.* **21**, 100–109.

Prokhorov, Yu. V. (1956). Convergence of random processes and limit theorems in probability theory. *Theory Probab. Appl.* **1**, 157–214.

Randles, R. H., and Hogg, R. V. (1973). Adaptive distribution-free tests. *Commun. Statist.* **2**, 337–356.

Randles, R. H., Ramberg, T. S., and Hogg, R. V. (1973). An adaptive procedure for selecting the population with largest location parameter. *Technometrics* **15**, 337–356.

Rao, R. C. (1965). *Linear Statistical Inference and its Applications.* Wiley, New York.

Rao, R. C. (1989). *Statistics and Truth.* Council of Scientific and Industrial Research, New Delhi.

Rey, W. J. J. (1978). *Robust Statistical Methods.* Springer, Berlin.

Rey, W. J. J. (1983). *Introduction to Robust and Quasi-Robust Statistical Methods.* Springer, Berlin.

Rice, J. R. (1964). *The Approximation of Functions. Linear Theory.* **1.** Addison–Wesley, New York.

Rice, J. R. (1965). The characterization of nonlinear L_1-approximations. *Archiv. Rat. Mech. Anal.* **17**, 61–66.

Rieder, H. (1994). *Robust Asymptotic Statistics.* Springer, New York.

Rocke, D. M., and Woodruff, D. L. (1993). Computation of robust estimates of multivariate location and shape. *Statistica Neerlandica* **47**, 27–42.

Rocke, D. M., and Woodruff, D. L. (1994). Computable robust estimation of multivariate location and shape in high dimension using compound estimators. *J. Amer. Statist. Assoc.* **89**, 888–896.

Rocke, D. M., and Woodruff, D. L. (1996). Identification of outliers in multivariate data. *J. Amer. Statist. Assoc.* **91**, 1047–1061.

Rousseeuw, P. J. (1981). A new infinitisemal approach to robust estimation. *Z. Wahrsch. verw. Geb.* **56**, 127–132.

Rousseeuw, P. J. (1984). Least median of squares regression. *J. Amer. Statist. Assoc.* **79**, 871–880.

Rousseeuw, P. J. (1985). Multivariate estimation with high breakdown point. In: *Mathematical Statistics and Applications (Grossman, W., Pflug, G., Vincze, I., and Wertz, W., Eds.)* Reidel, Dodrecht, pp. 283–297.

Rousseeuw, P. J., and Croux, C. (1993). Alternatives to the median absolute deviation. *J. Amer. Statist. Assoc.* **88**, 1273–1283.

Rousseeuw, P. J., and Leroy, A. M. (1987). *Robust Regression and Outlier Detection.* Wiley, New York.

Rousseeuw, P. J., and van Zomeren, B. C. (1990). Unmasking multivariate outliers and leverage points. *J. Amer. Statist. Assoc.* **85**, 633–639.

Rousseeuw, P. J., and Yohai, V. (1984). Robust regression by means of S-estimators. In: *Robust and Nonlinear Time Series Analysis (Franke, J., Härdle, W., and Martin, R. D., Eds.).* Springer, New York, pp. 256–272.

Rychlik, T. (1987). An asymptotically most bias-stable estimator of location parameter. *Statistics* **18**, 563–571.

Sacks, J. (1975). An asymptotically efficient sequence of estimators of a location parameter. *Ann. Statist.* **3**, 285–298.

Sacks, J., and Ylvisaker, D. (1972). A note on Huber's robust estimation of a location parameter. *Ann. Math. Statist.* **43**, 1068–1075.

Sage, A. P., and Melsa, J. L. (1971). *System Identification.* Academic Press, New York.

Schum, D. A. (1994). *Evidential Foundations of Probabilistic Reasoning.* Wiley, New York.

Seki, T., and Yokagama, S. (1996). Robust parameter-estimation using the bootstrap method for the 2-parameter Weibull. *IEEE Trans. Rel.* **45**, 34–41.

Sen, P. K. (1968). Estimates of the regression coefficient based on Kendall's tau. *J. Amer. Statist. Assoc.* **63**, 1379–1389.

Sheather, S. J., and Marron, J. S. (1990). Kernel quantile estimators. *J. Amer. Statist. Assoc.* **85**, 410–416.

Shevlyakov, G. L. (1976). *Robust Estimation and Detection of Signals Based on the Method of Least Absolute Deviations.* Ph.D. Thesis, Leningrad Polytechnical Institute (in Russian).

Shevlyakov, G. L. (1982a). On nonparametric methods of detection of signals based on approximation criteria. *Proc. Leningrad Polytechnic Inst.* #388, 50–54 (in Russian).

Shevlyakov, G. L. (1982b). Robust methods of estimation of the parameters of transition processes in electric machines. *Proc. Leningrad Polytechnic Inst.* #388, 67–71 (in Russian).

Shevlyakov, G. L. (1988). Robust properties of the median correlation coefficient. In: *Math. Methods of Optimal Control and Data Processing.* RRTI, Ryazan, pp. 109–112 (in Russian).

Shevlyakov, G. L. (1991). *Methods and Algorithms of Data Analysis under the Conditions of Uncertainty (With Applications in Cardiology).* Dr.Sci. Thesis, St. Petersburg State Technical University (in Russian).

Shevlyakov, G. L. (1992). Breakdown points of L_1-regression. *Theory Probab. Appl.* **37**, 140–141 (in Russian).

Shevlyakov, G. L. (1995). Robust minimax adaptive approach to regression problems in interval computations. In: *Abstr. APIC'95, El Paso*, pp. 185–187.

Shevlyakov, G. L. (1996). Stability of L_1-approximations and robustness of least modules estimates. *J. Math. Sci.* **81**, 2293–2999.

Shevlyakov, G. L. (1997a). On robust estimation of a correlation coefficient. *J. Math. Sci.* **83**, 434–438.

Shevlyakov, G. L. (1997b). Robust estimation of a scale parameter of the exponential distribution in models of faults. *Autom. Remote Control* **58**, 273–277.

Shevlyakov, G. L. (2000). Robust minimax estimation of scale in time to failure distribution models. In: *Abstr. 2nd Int. Conf. on Math. Methods in Reliability. Bordeaux, France, July 4–7, 2000,* **2**. Université Victor Segalen Bordeaux 2, pp. 956–959.

Shevlyakov, G. L., and Jae Won Lee (1997). Robust estimators of a correlation coefficient: Asymptotics and Monte Carlo, *Korean J. Math. Sci.* **4**, 205–212.

Shevlyakov, G. L., and Khvatova, T. Yu. (1998a). On robust estimation of correlation matrices. In: *Proc. Intern. Conf. "Computer Data Analysis and Modeling"* **2**. Minsk, pp. 101–106.

Shevlyakov, G. L., and Khvatova, T. Yu. (1998b). On robust estimation of a correlation coefficient and correlation matrix. In: *MODA 5—Advances in Model-Oriented Data Analysis. (Atkinson, A. C.,* et al., *Eds.).* Physica, Heidelberg, pp. 153–162.

Shevlyakov, G. L., and Khvatova, T. Yu. (1998c). Detection of multivariate outliers using bivariate boxplots. In: *Resumés des exposés 9-éme Colloque Russe–Français "Analyse des données et statistique appliquée", 24 août–1 septembre 1998.* Saratov, pp. 17–18.

Shulenin, V. P. (1993). *Introduction to Robust Statistics.* Tomsk Univ. Press, Tomsk (in Russian).

Shurygin, A. M. (1994a). New approach to optimization of stable estimation. In: *Proc. I US / Japan Conf. on Frontiers of Statistical Modeling.* Kluwer, Dordrecht, pp. 315–340.

Shurygin, A. M. (1994b). Variational optimization of estimator's stability. *Avtom. Telemekh.* #11, 73–86 (in Russian).

Shurygin, A. M. (1995). Dimensions of multivariate statistics. *Avtom. Telemekh.* #8, 103–123 (in Russian).

Shurygin, A. M. (1996). Regression: the choice of a model and stable estimation. *Avtom. Telemekh.* #6, 104–115 (in Russian).

Shurygin, A. M. (2000). *Applied Stochastics: Robustness, Estimation, Forecasting.* Finansy i Statistika, Moscow (in Russian).

Siegel, A. F. (1982). Robust regression using repeated medians. *Biometrika* **69**, 242–244.

Smolyak, S. A., and Titarenko, B. P. (1980). *Stable Methods of Estimation.* Statistika, Moscow (in Russian).

Spearman, C. (1904). The proof and measurement of association between two things. *Amer. J. Psychol.* **15**, 88–93.

Stahel, W. A. (1981). *Robuste Schätzungen: Infinitesimale Optimalität und Schätzungen von Kovarianzmatrizen.* Ph.D. Thesis. ETH, Zurich.

Stigler, S. M. (1969). Linear functions of order statistics. *Ann. Math. Statist.* **40**, 770–788.

Stigler, S. M. (1973). Simon Newcomb, Percy Daniell and the history of robust estimation 1885–1920. *J. Amer. Statist. Assoc.* **68**, 872–879.

Stigler, S. M. (1974). Linear function of order statistics with smooth weight functions. *Ann. Statist.* **2**, 676–693.

Stigler, S. M. (1977). Do robust estimators work with real data? *Ann. Statist.* **5**, 1055–1078.

Stigler, S. M. (1981). Gauss and the invention of least squares. *Ann. Statist.* **9**, 465–474.

Stone, C. J. (1975). Adaptive maximum likelihood estimators of a location parameter. *Ann. Stat.* **3**, 267–284.

Tchijevsky, A. L. (1928). Sun-spot and history. In: *Bull. New York Acad. Sci.* New York.

Tchijevsky, A. L. (1930). Les périodes solaires et la mortalité. In: *La Côte d'Azur Médicale* **11**, Toulon.

Tchijevsky, A. L. (1934). Action de l'activité périodique solaire sur la mortalité générale. In: *Traité de Climatologie Biologique et Médicale* **11**. Paris.

Tchijevsky, A. L. (1936). *Les Épidémies et les Perturbations Électromagnétiques du Milieu Extérieur.* Hippocrate, Paris.

Theil, H. (1950). A rank-invariant method of linear and polynomial regression analysis. *Ned. Akad. Wetensch. Proc.* **A53**, 386–392, 521–525, 1397–1412.

Tsybakov, A. B. (1982). Robust estimates of function values. *Probl. Peredachi Inf.* **18**, #3, 39–52.

Tsybakov, A. B. (1983). Convergence of nonparametric robust algorithms of reconstruction of functions. *Autom. Remote Control* **44**, 1582–1591.

Tsypkin, Ya. Z. (1984). *Foundations of Information Theory of Identification.* Nauka, Moscow (in Russian).

Tsypkin, Ya. Z., and Poznyak, A. S. (1981). Optimal and robust optimization algorithms in the presence of correlated noise. *Sov. Phys. Dokl.* **26**, 570–571.

Tukey, J. W. (1960). A survey of sampling from contaminated distributions. In: *Contributions to Probability and Statistics. (Olkin, I., Ed.).* Stanford Univ. Press, Stanford, pp. 448–485.

Tukey, J. W. (1962). The future of data analysis. *Ann. Math. Statist.* **33**, 1–67.

Tukey, J. W. (1975). Usable resistant/robust techniques of analysis. In: *Proc. 1st ERDA Symp.* Los Alamos, New Mexico, pp. 1–31.

Tukey, J. W. (1977). *Exploratory Data Analysis.* Addison–Wesley, Reading, MA.

Tukey, J. W. (1979). Robust techniques for the user. In: *Robustness in Statistics. (Launer, R. L., and Wilkinson, G. N., Eds.).* Academic Press, New York, pp. 103–106.

Tyler, D. E. (1983). Robustness and efficiency properties of scatter matrices. *Biometrika* **70**, 411–420.

Tyler, D. E. (1991). Some issues in the robust estimation of multivariate location and scatter. In: *Directions in Robust Statistics and Diagnostics (Stahel, W., and Weisberg, S., Eds.)* Springer, New York.

Ushakov, I. A., Ed. (1994). *Handbook of Reliability Engineering.* Wiley, New York.

Ushakov, I. A. (2000). Reliability: past, present, future. In: *Recent Advances in Reliability Theory. (Ionescu, D., and Limnios, N., Eds.).* Birkhäuser, Boston, pp.. 3–21.

Van Eeden, C. (1970). Efficiency-robust estimation of location. *Ann. Math. Statist.* **41**, 172–181.

Van Trees, H. L. (1971). *Detection, Estimation and Modulation Theory.* Wiley, New York.

Velleman, P.F., and Hoaglin, D.C. (1981). *Applications, Basics, and Computing of Exploratory Data Analysis.* Duxbury Press, Boston.

Vilchevski, N. O., and Shevlyakov, G. L. (1984). Robust estimation of the shift parameter under a bounded noise variance. *Autom. Remote Control* **45**, 1048–1053.

Vilchevski, N. O., and Shevlyakov, G. L. (1985a). Application of the Bernstein polynomials in adaptive estimation of distributions. In: *Proc. V All-Union Conf. on Nonparametric and Robust Methods in Cybern.* Tomsk University, Tomsk, pp. 44–47 (in Russian).

Vilchevski, N. O., and Shevlyakov, G. L. (1985b). Application of Bernstein polynomials in adaptive estimation. VINITI 5258–85, p. 6 (in Russian).

Vilchevski, N. O., and Shevlyakov, G. L. (1986). Robust adaptive approach to identification of regression models. In: *IFAC Proc.* Pergamon Press, pp.. 81–86.

Vilchevski, N. O., and Shevlyakov, G. L. (1987). Axiomatic approach to the choice of an estimation criterion. In: *Proc. VI All-Union Conf. Nonparametric and Robust Methods in Cybern.* Tomsk University, Tomsk, pp. 93–97 (in Russian).

Vilchevski, N. O., and Shevlyakov, G. L. (1990a). Adaptive robust algorithms of identification. In: *Proc. VII All-Union Conf. on Nonparametric and Robust Methods in Cybern.* Tomsk University, Tomsk, pp. 130–136 (in Russian).

Vilchevski, N. O., and Shevlyakov, G. L. (1990b). Minimisation of Fisher information in some classes of continuous and discrete distributions. *Notes in Math. Statist.* **54**, 240–243 (in Russian).

Vilchevski, N. O., and Shevlyakov, G. L. (1994). Robust minimax estimation of the location parameter with a bounded variance. In: *Stability Problems for Stochastic Models (Zolotarev, V. M.,* et al., *Eds.).* Moscow/Utrecht, TVP/VSP, pp.. 279–288.

Vilchevski, N. O., and Shevlyakov, G. L. (1995a). On the choice of an optimization criterion under uncertainty in interval computations. In: *Abstr. APIC'95, El Paso*, pp. 187–188.

Vilchevski, N. O., and Shevlyakov, G. L. (1995b). Robust minimax adaptive M-estimators of regression parameters. In: *MODA 4—Advances in Model-Oriented Data Analysis (Kitsos, C. P., and Muller, W. G., Eds.)*. Physica, Heidelberg, pp. 235–239.

Vilchevski, N. O., and Shevlyakov, G. L. (1997). On Rao–Cramér inequality, Fisher information and lattice least favourable distributions. *Theory Probab. Appl.* **42**, 387–388.

Vilchevski, N. O., and Shevlyakov, G. L. (1998). On the characterization of Fisher information and stability of the least favorable lattice distributions. *J. Math. Sci.* **92**, 4104-4111.

Vilchevski, N. O., and Shevlyakov, G. L. (2000). On the Bernstein polynomial estimators of distributions and quantile functions. *J. Math. Sci.* (to appear).

Walley, P. (1990). *Statistical Reasoning with Imprecise Probabilities.* Chapman & Hall, London.

Walter, E., and Pronzato, L. (1997). *Identification of Parametric Models from Experimental Data.* Springer, Berlin.

Weibull, W. (1939). A statistical theory of the strength of materials. *Ing. Vetenskaps Akad. Handl.* #151.

Weichselberger, K. (1995). Axiomatic foundations of the theory of interval probability. In: *Proc. 2nd Gauss Symp. (Mammitzsch, V., and Schneeweiss, H., Eds.)*. de Gruyter, Berlin.

Weiss, L., and Wolfowitz, J. (1970). Asymptotically efficient nonparametric estimators of location and scale parameters. *Z. Wahrsch. verw. Geb.* **16**, 134–150.

Weiszfeld, E. (1937). Sur le point par leque la somme des distansis de *n* points donnes est minimum. *Tôhoku Mathematics J.* **37**, 355–386.

Whittle, P. (1962). Gaussian estimation in stationary time series. *Bull. Int. Statist. Inst.* **39**, 105–129.

Wolfowitz, J. (1974). Asymptotically efficient nonparametric estimators of location and scale parameters. II. *Z. Wahrsch. verw. Geb.* **30**, 117–128.

Yang, S. S. (1985). A smooth nonparametric estimator of a quantile function. *J. Amer. Statist. Assoc.* **80**, 1004–1011.

Yohai, V. J., and Maronna, R. A. (1977). Asymptotic behavior of least-squares estimates for autoregressive processes with infinite variances. *Ann. Statist.* **5**, 554–560.

Zadeh, L. A. (1965). Fuzzy sets. *Information and Control* **8**, 338–353.

Zadeh, L. A. (1975). The concept of a linnguistic variable and its application to approximate reasoning. *Information Science* **8**, 199–249.

Zani, S., Riani, M., and Corbellini, A. (1998). Robust bivariate boxplots and multiple outlier detection. *Comp. Statist. Data Anal.* **28**, 257–270.

Zieliński, R. (1977). Robustness: a quantitative approach. *Bull. Acad. Polon. Sci. Ser. Math., Astr. Phys.* **25**, 1281–1286.

Zieliński, R. (1981). Robust statistical procedures: a general approach. Preprint 254. Institute of Mathematics, Polish Academy of Sciences, Warsaw.

Zieliński, R. (1987). Robustness of sample mean and sample median under restrictions on outliers. *Applicationae Mathematicae* **19**, 239–240.

Zolotarev, V. M. (1977). General problems of the stability of mathematical models. In: *Proc. 41st Session of the ISI.* New Delhi, pp. 382–401.

Zolotarev, V. M. (1997). *Modern Theory of Summation of Random Variables.* VSP, Utrecht.

Zolotukhin, I. V. (1988). Asymptotic properties of M-estimators for regression parameters and their applications to the optimal experimental design. In: *Asymptotic Methods in Probability Theory and Mathematical Statistics.* Mathematical Institute of the Uzbek SSR Academy of Sciences, Tashkent, pp. 77–86 (in Russian).

Index